*Also by Thistle Y. Harris*

WILD FLOWERS OF AUSTRALIA
AUSTRALIAN PLANTS FOR THE GARDEN
NATURE PROBLEMS
EASTERN AUSTRALIAN WILD FLOWERS
PLANT LIFE IN AUSTRALIA
NATURECRAFT IN AUSTRALIA (Ed.)

THISTLE Y. HARRIS

# Alpine Plants
# OF AUSTRALIA

## including subalpine and montane plants

LINE DRAWINGS BY LINDA TURNER

ANGUS AND ROBERTSON

*First published in 1970 by*

ANGUS AND ROBERTSON (PUBLISHERS) PTY LTD

221 George Street, Sydney
54 Bartholomew Close, London
107 Elizabeth Street, Melbourne
89 Anson Road, Singapore



National Library of Australia
Card number & ISBN 0 207 12071 4

*Registered in Australia for transmission by post as a book*
PRINTED IN AUSTRALIA BY WATSON FERGUSON & CO., BRISBANE

# ACKNOWLEDGMENTS

In preparing this book for publication I have been generously assisted by many people, including Government officials and personal friends.

For the examination and study of material in New South Wales I wish to thank particularly the one-time Kosciusko State Park Trust and the Park Superintendent, Mr Neville Gare, and Dr D. F. McMichael, Director of the National Parks and Wildlife Service. For help with field work in Tasmania I owe thanks to the Scenery Preservation Board, especially to its then Superintendent, Mr J. B. Thwaites.

For supplementary material that I was unable to examine in the field, and for material for illustrations, I am indebted to a number of personal friends including Mr Alan M. Gray, Mr and Mrs Herbert J. King, Mrs Kay Geeves, Mrs Betty Gee, Miss Kathleen Elger, and Miss Essie Huxley. Mrs G. M. Holley and Mrs Susan Heins did valuable work with preliminary field sketches.

Checking of plant names and identification is an exacting and time-absorbing task, and I am particularly indebted to the experts who assisted in this work: these include Mr N. K. Mair, Superintendent of the Royal Botanic Gardens and Director of the National Herbarium, Sydney, and his staff including Mr L. A. S. Johnson, Miss Nerida Ford, Dr Barbara Briggs, and Dr Mary Tindale; also Mr H. J. Willis of the Melbourne Herbarium, Dr Hj. Eichler of the State Herbarium of South Australia, Dr R. C. Carolin of the Botany Department, University of Sydney, and Dr Winifred Curtis and Miss Heather Guilline of the University of Tasmania.

Thanks are due to Mr Norton Harvey of Launceston, Tasmania, for permission to use the photograph of *Geum renifolium*.

Dr A. B. Costin, Chief of the Division of Plant Industry, Commonwealth Scientific and Industrial Research Organization, Canberra, made useful comments on the manuscript and supplied the foreword. I am indebted to him for this.

Finally my thanks go to Mrs H. E. Brown and Miss Eva Symes who prepared a fair typewritten copy from almost indecipherable notes.

THISTLE Y. HARRIS

# FOREWORD

Through her several books on the wildlife of Australia, Thistle Harris is already well known as one of the foremost exponents of this country's unique flora. Few botanists have her gift of being able to translate by word and illustration exact botanical information to the public at large. In this, her latest book, *Alpine Plants of Australia*, Miss Harris has again been able to do just this.

*Alpine Plants of Australia* is in two main parts, Part I describing the general environment of the mountain areas of mainland Australia and Tasmania, the features of the plant life, and the plant communities or vegetation in which the individual species of plants described in Part II occur. Such environmental information, often lacking in other botanical works, provides an essential backdrop against which the plants themselves become alive to the reader. Part II, the main section of the book, describes each of the main plant species, its habitat and distribution.

In addition to treating the strictly alpine species of mainland Australia and Tasmania—that is, those found above the treeline—species of subalpine, montane and talbeland zones are also included. The traveller throughout south-eastern Australia, as well as the traveller to the high country itself, will therefore find Miss Harris's latest book an invaluable companion.

The 1970s are the decade in which man is becoming deeply concerned with the quality and conservation of his environment. The first steps toward effective action, whether it be in the control of air and water pollution, or in the protection of wildlife, are concern and then interest and understanding. *Alpine Plants of Australia* will do much to promote the interest and knowledge of the Australian public in its unique alpine flora.

*Canberra,*
*September, 1970*

A. B. COSTIN
*Chief, Division of Plant Industry*
*C.S.I.R.O.*

# CONTENTS

# PHOTOGRAPHS

BLACK AND WHITE (*between pages 34 and 35*)



COLOUR (*between pages 98 and 99*)

# INTRODUCTION

AUSTRALIA is renowned for its rich and varied flora: the subtropical rain-forests, the open forests of the east coast with eucalypts, banksias and angophoras, the magnificent sandplain flora of the west, and the flamboyant though ephemeral plant population of the dry inland characterize for many Australia's floral heritage.

Let us not forget, however, the plants that flourish on the high moors. Here, in those parts of the mainland where the altitude exceeds 5,000 feet and in the more southern parts of Tasmania rarely more than 4,000 feet, on the very roof of Australia, are to be found species of wild flowers unknown elsewhere, exquisite in their delicate beauty yet sturdy enough to withstand the snow-laden blizzards of winter and the hot drying winds of summer. In a good season they cover the moors in a blaze of colour, often brought to fulfilment by a few days of summer temperatures after a prolonged period of snow and sleet. From neat compact bushes, from twisted branches scrambling over rocks, from the sides of rippling streams along the line of the melting snowdrifts and from the heart of the peat bogs a riot of colourful blossoms appear as by a miracle, while the air is laden with the delicate perfume which so many of them possess. For a few weeks all is gaiety until the early snows nip the tender shoots and all but the hardiest and most persistent plants settle down to await the thaw of the following spring.

The area above the treeline, at 6,000 feet or over, to which this true alpine flora is confined occupies some 2,000 square miles of the mainland, about 1,000 of which are in the south-east of New South Wales, 140 in the Australian Capital Territory, and 860 in the Victorian Alps. In Tasmania many small areas scattered over the island carry a similar flora at elevations between 3,000 and 5,000 feet.

On the mainland the areas between 5,000 and 6,000 feet support a subalpine vegetation in which many of the true alpine plants are present in considerable numbers and some new species appear. Below this level and down to about 2,000 feet is the region known as the montane: here fewer of the true alpine species persist, with many new species—that is, species which do not occur at the higher elevations. Below the montane region are the tablelands, a region where a few of the more adaptable alpine plants persist.

In Tasmania the subalpine vegetation occupies elevations between 2,000 and 3,000 feet and the montane vegetation elevations roughly between 1,000 and 2,000 feet, with some few species persisting to the coast. The high country flora, developing in an atmosphere of extreme isolation caused by the heavy winter snowfalls, presents a range of species differing from those in any other part of the continent. Where such isolation of plants does not occur there is a much lesser tendency towards specialization.

# I

# General Survey of Alpine Habitats

# GENERAL TOPOGRAPHY

AUSTRALIA has no truly high mountains. The ranges consist of raised, eroded peneplains with a few residuals, and the highest elevation, that of Mount Kosciusko in the Southern Alps of New South Wales, is only a little over 7,300 feet. Above 6,000 feet on the Kosciusko plateau and above 5,750 feet on Mount Bogong in Victoria trees cease to exist. In some of the more sheltered places, such as on Mount Jagungal in the Perisher Ranges of New South Wales, a few straggling snow gums, *Eucalyptus niphophila*, survive at 6,500 feet where they are protected by the main range to the west. Some isolated mountain peaks are fairly steep—notably Mount Kosciusko, just mentioned, and Mount Bogong, Mount Loch, Mount Hotham, Mount Feathertop and Mount Buller in Victoria. The Victorian mountains are relatively dry on the lower slopes, with few bogs and fens and great stretches of grassland and heath. They are rich in alpine herbfields but have little fjaeldmark (or fell-field of low-growing tufted plants), which is more extensive on the Kosciusko plateau.

In general it may be said that the high mountain country is rich in herbfields extending over several diverse geographical zones with a wide range of habitats independent of the rock material from which the soils are derived. These herbfields are much richer in vegetation than the European alpine country.

The climax of the alpine zone on the mainland is the tall alpine herbfield, indicated by large stands of snow daisy, *Celmisia longifolia*, amidst the most common of the snow grasses, *Poa caespitosa*. This region is traversed by streams emptying into typical sphagnum bogs and opening out into fjaeldmark in extremely exposed areas.

In Tasmania the high tops of Cradle Mountain, Mount Field and other peaks, none reaching a greater altitude than 5,000 feet, bear the same type of vegetation as occurs at altitudes of 6,000 feet or more on the mainland and, as we have seen, the true alpine flora is restricted to these elevations (6,000 feet and over on the mainland, 3,000 feet and over in the more southern parts of Tasmania). In most of this elevated area plants persist to the greatest heights and only a few isolated patches are completely devoid of vegetation, such as on Ben Lomond in Tasmania, where rounded weathered boulders of dolerite occur—a zone known as potato field.

In the area above the snowline where the true alpine flora occurs, the ground is covered with snow continuously for at least one month each year but usually for considerably longer periods. While such areas are occupied exclusively by these specialized alpine plants, quite sizeable pockets of the same species are to be found also at considerably lower elevations where the cold air and biting winds produce the same conditions as occur normally at much greater elevations. Such little pockets of typical alpine plants are quite surprising to come upon amidst a characteristic subalpine or even montane vegetation.

# SOILS AND TOPOGRAPHY

The rocks of the Australian Alps consist mainly of granites, gneisses, and porphyries, and in some localities slates, basalt and limestone. The predominating granites weather to form a light acid soil. On extensive plateau areas the soils can become comparatively dry in summer, but interspersed with them are numerous fens and sphagnum bogs constantly fed by the streams formed from the melting snow.

Although there is considerable variation in rock types in these areas, and consequently variation in soil, the vegetation is not markedly affected by this.

In elevated areas of Tasmania the predominant rock type is dolerite forming sills overlying sandstone. These sills are broken to talus slopes of considerable extent, providing well-drained soils interspersed with semi-acid swamps of flat, loamy sand collecting drainage from the tops. Glacial valleys occur, and glacial lakes are a feature of the landscape.

On the mainland the peneplain presents exposed faces with gentle slopes showing evidence of glaciation, the old surface being deeply dissected by rivers and terminal and lateral moraines, while glacial lakes such as Albina, Blue and Cootapatamba in the southern alps are common. Later glaciation caused cirques in some regions over 6,000 feet in height, leaving the rivers and their tributaries to wander through senile valleys.

# CLIMATE ABOVE THE SNOWLINE

The vegetation of the high moors is subject to great extremes of climate. The lower atmospheric pressure at high altitudes accelerates evaporation, and plants have to adapt themselves to excessive loss of water during the summer months. Winds are high in both winter and summer, the prevailing winds on the Australian Alps being north and south-west, and there are no effective natural windbreaks on this vast expanse of plateau. As well there is the high precipitation, 70 to 120 inches in the alpine and 30 to 80 inches in the subalpine regions, with a range of 130 to 180 rain days per year in the former and 120 to 140 in the latter. This gives many days on which the evaporation rate will be low. On other days the air is commonly clear and cloudless; this, together with its rarefied nature, additional ultra-violet light occurring at high elevations, reflection from the snow and the considerable expanse of shiny-surfaced rock in some places, gives a great intensity of light. In summer the long, and in winter the very restricted, hours of daylight added to these atmospheric conditions give a set of factors for plant development very different from those prevailing in most other communities. Heavy dews and frosts also occur at different times.

Temperature variations are great, particularly in the highest regions where summer temperatures may range from below 32 degrees to over 80 degrees Fahrenheit. The mean winter temperature in alpine parts is below 25 degrees Fahrenheit and may remain below 32 degrees for four to six months. In subalpine regions it varies between 30 and 60 degrees in winter. The midsummer range is respectively 50 to 60 and 60 to 75 degrees Fahrenheit. One of the factors that affect plants greatly is the summer fluctuation of temperature: in one day in summer—indeed in a few hours—temperatures may fall by as much as sixty degrees. The effect of these low temperatures on the soils is tempered somewhat by the insulating effect of the blanket of snow. However, a succession of severe autumn frosts followed by heavy and persistent snow, a not unusual occurrence in these parts, may freeze the soil until late spring.

# PLANT ADAPTATIONS

Given such unusual conditions it is not surprising that the alpine areas should produce a flora very different from that found elsewhere. One interesting aspect is the broad similarity in form of so many of the plants.

Because of the geographical position and the long winter there is virtually only one season of development, late spring and early summer. The optimum flowering period varies with the time of the late snowfalls. In addition, local falls of snow will determine the time of development of many plants, and species in an isolated patch of territory protected from a late heavy fall may flower up to a month earlier than those in a similar but unprotected area in close proximity to it. Thus one cannot predict the exact time when a particular area will exhibit its best flowering. When the season does commence, however, the rate of development is phenomenal, and plants open their flower buds with a rapidity only equalled by desert plants in a good season.

In general the plants of the high country have to adapt themselves to low water availability on account of the high acid content of the soils, which reduces intake; plants of the depressed areas have to contend with the intense cold of the soils and those of the exposed plateaux with dry and shallow soils. Above the soil surface the factors that seriously affect plant growth are high and frequent winds and rapid temperature changes. The plants of the high moors exhibit characteristic adaptations such as nanism (dwarf formation), tussock formation and rock-climbing habit, while their leaves and sometimes the whole plant are frequently covered with a thick coat of wool or hair, and in addition the leaves are often ericoid (fine), divided, reduced to spines, or leathery.

The characteristic low form of growth of alpine plants and their rapidity of development make the high moors a summer wonderland with massed displays of vivid colour covering the slopes of glacial lakes and exposed plateaux, while the bogs and fens contribute their share of colour to the general scene. The compact form of so many of the plants, the symmetrical habit they so frequently adopt and the fact that they are often entirely covered with brilliant blossoms combine to make charming individual pictures that are a delight to the nature photographer. The exposed rocks, too, add their own touch to the general beauty of the terrain, providing a fine foil for the many plants which scramble over them or nestle in their crevices for protection against the harsh winds.

A feature of many of these plants that never fails to amaze the summer visitor is their delicacy of appearance combined with their ability to withstand the extreme exigencies of the climate without fading or tearing. The eyebrights and the buttercups, as well as the bluebells and veronicas, look as though the merest breath of wind or lightest touch would mar their delicate beauty, yet they stand unsullied through days of bitterly cold strong winds and fierce sun. And each year the most spectacular patches occur in a different place, flowering time for plants at this altitude being largely determined by the time of occurrence of the late snows.

These plants are indeed an extraordinary and magnificent heritage: relatively small in number and fleeting in their display, but something to be cherished and preserved for all time, for they are unique as well as beautiful and once lost they are gone for ever.

# DISTRIBUTION OF PLANTS

We have already spoken of the major alpine areas in Australia and their extent. In New South Wales the largest alpine areas occur on the Kosciusko plateau and are wholly within the Kosciusko National Park. Small areas of high mountain vegetation occur also at Barrington Tops and in the highest parts of the New England district, where snow may lie for some time. In southern and south-western parts of the Australian Capital Territory, extending into New South Wales and to the headwaters of the Murrumbidgee and Goodradigbee rivers there is high mountain vegetation at places such as Brindabella and the Mount Franklin district, linking up with the Kosciusko area. To the west, in Victoria, is the stretch of the Bogong High Plains and Mount Buffalo and south-west Mount Buller; south of Bogong High Plains, Mount Feathertop and Mount Hotham carry a fine array of high moor vegetation. Still farther south this type of vegetation is to be found on the Dargo High Plains and southwards through a number of small patches to the Snowy Plains and, most southerly, to the big plain near Mount Wellington.

In Tasmania high mountain vegetation occupies a greater percentage of the total area than it does on the mainland. In the north-west the Cradle Mountain reserve with individual peaks reaching more than 5,000 feet in height covers an area of 525 square miles. Seventy miles to the east is Ben Lomond, 5,160 feet, the second highest peak on the island. Mount Ossa, 5,305 feet, the highest peak in Tasmania, is situated near the centre of this reserve. In the central south is the readily accessible Mount Field and in the south-east the Hartz Mountains. In the central west is Frenchman's Cap reaching over 4,700 feet in height, and in the rugged south-west the fascinating peaks of the Anne, Arthur and Frankland ranges.

# ALPINE COMMUNITIES

There are several fairly clearly defined zones in the alpine tract and these zones are repeated in the subalpine and, to some extent, in the montane areas.

The greater part of the alpine tract is occupied by herbfields, sometimes referred to as grasslands, where snow grasses are dominant and the vegetation consists almost entirely of herbs and other small plants. The less exposed of these areas, where the herbs are taller and more conspicuous above the grasses, are called tall alpine herbfield; the more exposed areas, where the dwarf herbs allow the grasses to be more apparent, are known as short alpine herbfield.

Fens, on level or gently undulating slopes, and bogs, on irregular raised patches, occupy the wetter patches in alpine and subalpine habitats and are generously dispersed among the herbfields. They carry a particular and very rich flora.

On the most exposed areas, at the very top of the alpine tract, where the rugged rocks are often unclothed except with snow, the flora is sparse and highly specialized. This zone is the fjaeldmark or fell-field.

In the more protected parts, along watercourses in the moraines and other valleys, the vegetation is thicker and individual species often occur in considerable stands. This zone is called heathland.

The moist patches of the heathland, which are by no means uncommon, support grasses able to withstand perpetual water in the soil, together with a variety of smaller adaptable plants. This zone is known as sod-tussock grassland.

These are the main vegetation zones of the alpine habitats. Each carries its characteristic formation and its characteristic species; each has its own particular attraction as well as its own problems to combat.

Alpine communities proper occupy the completely treeless zone above 6,000 feet on the mainland and above 3,500 feet in Tasmania. These, of course, extend to lower altitudes in the less protected areas, while subalpine communities invade the higher elevations in extremely well-protected parts.

Although there is a broad general similarity of appearance among alpine plants, the great diversity of habitat within the area gives a great variety of species. There are, however, a considerable number of plant families represented by a single genus only or even by a single species, several with a few genera only, and some of the families are not common in other Australian habitats.

A wide range of species occupies the alpine tract on the mainland, particularly on the Kosciusko plateau. The number of species in Tasmania is considerably fewer, but a large percentage is endemic (restricted to Tasmania) and of considerable interest on this account. Many of the alpine plants are entirely restricted to the alpine tract but others have a considerable range. This is particularly marked in some Tasmanian plants, to which reference will be made later.

## Alpine Herbfields

The alpine herbfields, rich in humus, occupy steep rocky faces, crevices and ledges on the sides of high mountains such as Mount Bogong, Mount Hotham and Mount Buller in Victoria, the Kosciusko plateau in New South Wales, and the dolerite slopes of such mountains in Tasmania as Mount Rufus, Ben Lomond, Cradle Mountain and Mount Field East.

### *Tall Alpine Herbfield*

The wide, exposed plateau areas of the mainland are covered with snow grass of several species, *Poa caespitosa* being dominant. This is interrupted by the beautiful silver foliage and straight-stemmed white flowerheads of the snow daisy, *Celmisia longifolia*, which frequently takes complete control of considerable stretches, particularly in the Kosciusko area, and descends in lesser abundance to lower elevations.

Here the giant wallaby grass, *Danthonia frigida*, mingles with snow grasses and grasses of other kinds. The hoary-leafed small everlasting daisy or sunray, *Helipterum albicans* ssp. *alpinum*, makes a magnificent showing with its gleaming bronze buds and smooth white papery flowers with rich golden centres, and the starry little *Stellaria pungens* with pure white open flowers occurring through a range of elevations is sometimes quite spectacular.

In the wetter parts rushes and sedges occur, together with small ground orchids of many kinds, and the alpine violet, *Viola betonicifolia*, is not uncommon.

The magnificent white buttercup or windflower, *Ranunculus anemoneus*, follows the line of the melting snows. Its pink buds peep out from the accumulated debris among the sphagnum moss and along the edge of the snow to open out into fragile-looking, though in reality remarkably tough, many-petalled white flowers 2 or more centimetres in diameter. Here, too, occasional patches of white and purple eyebright, *Euphrasia* spp., and alpine celery, *Aciphylla glacialis*, appear. Bluebells, *Wahlenbergia*, dance in brilliant blue. Two species of the alpine lily or pineapple plant, *Astelia alpina* and *A. psychrocharis,* form a silver carpet over wide stretches; they are completely prostrate, with small heads of creamy white flowers and brilliant red fruits close to the ground, the latter being much relished by wallabies.

In rocky outcrops and crevices the lovely little white mountain daisy, *Brachycome nivalis*, and the two common alpine ferns, *Blechnum penna-marina* and *Polystichum prolifera*, find shelter.

In Tasmania much of this habitat, in the dolerite country, is covered with a rush-like plant, *Restio australis*, found abundantly at lower elevations also, and with snow grasses interrupted by patches of the varied and colourful heath, *Richea scoparia.* The Tasmanian herbfield is rich in other heaths also; there are several species of *Trochocarpa*, *Leucopogon* and *Monotoca*, most of them compact, prickly or fine-leafed plants with clusters of small white, cream or pink flowers and an abundance of attractive, brightly coloured "berries".

The rocket plant, *Bellendena montana*, is one of the taller members of the Tasmanian herbfield and also penetrates to considerably lower elevations. A symmetrical rounded bush with attractive foliage, it reaches a height of 30 to 60 centimetres, with long heads of creamy white or pink flowers followed by brilliant red fruits.

Occasional patches of snow daisies occur, too, and the lovely little Tasmanian alpine lily, *Milligania*, grows in clumps along the snow-melt streams.

### *Short Alpine Herbfield*

In those parts of the alpine herbfields where rocks do not afford shelter short alpine herbfield occurs. Here the vegetation is carpetlike and fed constantly by the waters of the melting snow. Small streams and wet gravelly patches intersect the whole area. Vast stretches of this formation cover the high moors. To my mind these stretches *are* the high moors, relentlessly exposed, torn by ferocious winds, offering no protection whatever to either plant or animal, on first viewing a desolate expanse. But a closer investigation reveals a wealth of colour and variety of form that makes this one of the most fascinating areas for study. Short alpine herbfield covers a large section of the Kosciusko plateau in New South Wales, of the Bogong High Plains, Mount Feathertop, Mount Buller and Mount Stirling in Victoria, and considerable areas of Tasmania between Waldheim and Lake St Clair, Ben Lomond, Mount Field and Mount Barrow as well as other high exposed parts of the island State.

Alongside streams on the mainland the small and dainty white daisy, *Brachycome stolonifera*, and the charming little buttercup-like plant with the almost unpronounceable name, *Dichoschiadium ranunculaceum*, make a pleasing show. Individual patches of smaller everlasting daisies such as *Helichrysum dealbatum* in Tasmania and *Helipterum albicans* ssp. *alpinum* on the mainland are common.

In Tasmania this habitat is the home of the cushion-bushes, of which there are five outstanding species of three genera representing three plant families. They resemble each other so closely in their peculiar habit of growth and, in some cases, in their foliage, that it is not easy to distinguish them except at flowering time.

They form dense, tufted hummocks, sometimes several feet in diameter, with closely packed leaves, and inhabit boggy places and stretches of running water, so raised above the water-level that it is possible to cross a stream dryshod over them. In some species the foliage is brilliant green, in others a soft sage green, and these often intermingle with each other to form a pleasing colour pattern. In flower several of them are dotted with white stars, and the whole appearance of a good patch in flower is not quickly forgotten. You will read more about them later. Taking advantage of the soft moist bed, sundews, especially *Drosera arcturi*, richeas and other small plants are not uncommon among these cushions.

The short alpine herbfield is also renowned for its mat-forming plants, and carpets of starry white or blush pink *Neopaxia australasica* on the mainland and *Montia fontana* in Tasmania sometimes cover considerable areas. The distribution and flowering of *Neopaxia* is determined by the snow patches, so it may occur at different places and times each year.

Another low-growing plant is the sky lily, *Herpolirion novae-zelandiae*, with large white or delicately blue green flowers, common both on the mainland and in Tasmania.

In swiftly running streams the alpine marsh marigold, *Caltha introloba*, opens its white or pinkish four-petalled flowers, and growing among it sometimes is the small-flowered bright-yellow river buttercup, *Ranunculus inundatus*.

Mountain gentian, *Gentianella diemensis*, comes late in the season and dots the landscape with fragile-looking blue-veined white flowers like tiny tulips. The alpine plantain, *Plantago gunnii*, with its rosette of soft leaves, lives in the moister parts.

Another mat plant of considerable interest is the intensely sweet-scented *Stackhousia pulvinaris* with tiny deep-cream or yellow flowers on a mat of bright-green leaves.

Several heaths, rather similar to one another, grow in this association too, all with tiny, tubular, often hairy flowers and brilliant "berries" of various colours. Among these are three of wide distribution, *Leucopogon hookeri*, *Lissanthe montana*, and *Pentachondra pumila*, with another species, *P. ericifolia*, confined to Tasmania, all with red berries. They scramble over rocks or hug the ground, and require careful examination to distinguish them from one another.

The aromatic and attractive alpine mintbush, *Prostanthera cuneata*, spreads itself over the rocks in elegant abandon; it occurs commonly at lower elevations also, often assuming a more erect habit of growth there.

### Fens

Fens are a characteristic of the alpine area. They occur on wet level or gently sloping stretches, sometimes along permanent waterways or where water lies for the greater part of the year. Here the accumulated peat makes a boggy, poorly aerated soil which supports a vegetation rich in sedges, tussocky grasses, reeds, water millfoil, and bullrushes. The mountain purslane, *Neopaxia australasica*, covers quite large tracts of fen country with its starry white flowers. This genus was until recently regarded by botanists as two distinct genera, *Claytonia* and *Montia*. Ground orchids, including *Diuris pedunculata*, *Eriochilus autumnalis* and *Spiranthes sinensis*, are not uncommon, and a completely submerged form of *N. australasica* also occurs.

The ewartias, sometimes called Australian edelweiss, are mat plants with soft grey foliage, often found spread over rocks and bearing tiny white daisy flowers. There is one mainland species, *Ewartia nubigens*, and three Tasmanians, *E. meridithii, E. catipes* and *E. planchonii.*

The alpine pine, *Podocarpus lawrencii* (syn. *P. alpina*), sometimes frequents the fens but is found more commonly on the heathlands.

Some species of buttercups, such as *Ranunculus anemoneus* and the smaller bright-yellow *R. muelleri*, love the fens.

A Tasmanian member of this family, *Anemone crassifolia*, a dainty small plant with divided leaves and white open flowers, is a delight. In Tasmania, too, the pineapple plant, *Astelia alpina*, is common in the fens while several species of *Richea*, cushion-bushes and *Drosera* are common. *Drapetes tasmanica*, a small member of the Pimelea family, frequents the fens in the western mountains of Tasmania, less commonly on the mainland. *Coprosma pumila*, with tangled stems and rich red "berries", is found in this habitat both in Tasmania and on the mainland.

The alpine plantain, *Plantago muelleri*, forms patterns with its bright-green mats in mainland fens, while another species, *P. gunnii*, forms grey-green rosettes both here and in Tasmania.

Tasmania also has pernettyas in the fens. These are members of the Erica family, closely related to the epacrids. One, *Pernettya tasmanica*, is a prostrate, cushion-forming species with succulent red, yellow or white fruits. Another, *P. lanceolata*, is a small shrub with a deep rose-pink fruit.

Purple loosestrife, *Lythrum salicaria*, with its rich purple heads abounds in places, and *Isotoma* (*Lawrencia*) *fluviatilis* and pratias scramble over the wet surface. On the fen margins billy buttons (*Craspedia uniflora*) and the common yellow everlasting, *Helichrysum bracteatum*, make bright patches.

### Bogs

Bogs, particularly raised bogs, support a richer vegetation than that of the fens. They occur mainly on alpine and subalpine slopes which are constantly wet and irregular in their surfaces so that the bog is raised above the level of the prevailing water table. Sphagnum moss is a feature of these bogs. In Tasmania large areas of them are completely taken over by the fern *Gleichenia dicarpa*. Two species of *Boronia*, *B. citriodora* and *B. rhomboidea*, both strongly aromatic, are common in Tasmanian alpine bogs, as well as a number of other taller plants from other associations, such as the rocket plant, *Bellendena montana*, several species of *Orites* and many kinds of *Richea*. Both species of *Boronia* mentioned occur also on the heathlands. An endemic Tasmanian of the bogs of alpine quartzite country is the open-petalled *Geum renifolium*, a dainty little plant with white flowers belonging to the Rose family. Other Tasmanian denizens of the bog are the Tasmanian shrubby pine, *Diselma archeri*, and two plants common only at much lower elevations on the mainland, native dogrose (*Bauera rubioides*) and swamp heath (*Sprengelia incarnata*).

Both on the mainland and in Tasmania trigger plants, *Stylidium graminifolium*, invade the bogs as well as other associations; daisies of many kinds are common.

The smooth willow-herb, *Epilobium glabellum*, with its bright-pink flowers, makes a pleasing show in the wettest parts, and cotulas make bright-yellow patches in the pools.

On the mainland the yellow alpine bottlebrush, *Callistemon sieberi*, a late flowerer, makes a grand show in full bloom. Its Tasmanian counterpart is *C. viridiflorus*. Many species of *Epacris*, including *E. breviflora* and *E. serpyllifolia*, scent the air, and the only mainland species of *Richea*, *R. continentis*, often covers large areas of bog. A tea-tree, *Leptospermum arachnoideum*, is one of the commonest of the taller plants occurring here.

Ground orchids, including those of the fen, together with *Prasophyllum suttonii* and *Thelymitra venosa*, nestle among the grasses and sedges. A bright little buttercup, *Ranunculus pimpinellifolius*, has its home here, as have two species of *Drosera* or sundew.

Some of the plants of the drier areas, such as *Pimelea ligustrina, Baeckea gunniana* and species of *Euphrasia*, also invade the raised bog.

Bog patches are frequently made gay with the dancing purple flowers of the quaint little fairies' aprons or bladderworts (*Utricularia dichotoma*) or patches of the mountain gentian (*Gentianella diemensis*) with its masses of delicately veined white flowers.

This association is one of the richest of the alpine habitats.

### Fjaeldmark

In some of the most windswept parts of the high tops where the considerable falls of winter snow are removed rapidly by high and frequent winds, the flora is reduced to a few species able to withstand these extremely harsh conditions where alternate freezing and thawing of the snow cause drastic alterations to soil temperatures.

The community consists of mosses and lichens with interrupted patches of low-growing shrubs between boulders in the soil-filled depressions. This is the fjaeldmark, known also in Tasmania as fell-field.

The average height of the vegetation in this formation is usually 25 centimetres or less; many plants are completely prostrate or form rosettes on the ground. The few taller plants that survive here are sparsely branched with small, leathery leaves, often spiny or covered with short hairs and with considerably woody tissue.

Among the taller plants both on the mainland and in Tasmania are *Epacris petrophila* and *Coprosma pumila*, both of which occur also in more favoured situations. On the mainland the attractive little dwarf speedwell, *Veronica densifolia*, is found in rock pockets or scrambling over moist rock surfaces. The little alpine sunray, *Helipterum albicans* ssp. *alpinum*, and the edelweisses, *Ewartia*, penetrate this habitat also.

Among the rosette plants *Oreomyrrhis sessiliflora*, a member of the Carrot family, with its characteristic odour and a simple umbrella of small flowers, is a Tasmanian inhabitant. In Tasmania, too, the cushion-bushes or bolster plants fill the wet depressions of the fell-field with bright-green or grey cushions, and among them is often found *Drapetes tasmanica*.

**Alpine Heathland**

Heathland is widespread in both alpine and subalpine areas throughout Australia. In the true alpine areas this association occupies the rocky banks of watercourses, the shelter of rocky outcrops, glacial moraines and other protected parts. Even here, where snow may lie on the ground for one to eight months, conditions are severe and winds are strong.

The plants are well defined into two groups, rock-clinging plants and small erect shrubs.

One of the most interesting of the rock-clingers, occurring both on the mainland and in Tasmania, is the alpine pine, *Podocarpus lawrencii*, a slow-growing but long-lived species with tough, wiry stems and grey-green, fine rather fleshy leaves. Concealed between the rock or soil surface and the foliage are the cones, the pollen cones a delicate wine-red, the fruits, on short green bases, a rich orange-red, but both have to be uncovered for their beauty to be revealed. Another creeping pine, *Microcachrys tetragona*, occurs in Tasmania in the same environment.

Flowering plants of this rock-clasping habit are common here too. Among the most attractive is the alpine mintbush, *Prostanthera cuneata*, with small, light-green aromatic leaves and two-lipped white flowers blotched with purple and yellow in the throat, occurring also in the herbfields. When grown in less exposed situations this assumes the habit of a shapely, small, erect bush. Some boronias are also rock-clinging in this habitat, including the lemon boronia (*B. citriodora*) and *B. rhomboidea* in Tasmania and *B. algida* on the mainland. A related plant with starry white flowers and attractive pink buds, *Phebalium ovatifolium*, and another with yellow flowers, *P. ozothamnoides*, are also rock-clingers in this habitat. A more robust rock-scrambler from the mainland is *Orites lancifolia* of the Waratah family, while in the same association in Tasmania several other species of this genus occur. *Grevillea australis*, an interesting little rock-scrambler restricted to high areas on the mainland, is the only species of *Grevillea* in Tasmania. There are several rock-clinging rice flowers, including *Pimelea ligustrina*, the most widely distributed, found both on the mainland and in Tasmania; *P. alpina*, restricted to the mainland, with *P. pygmaea*, a mat former, and *P. sericea* restricted to Tasmania. Another common rock-clinger of this habitat is a member of the Tea-Tree family, *Baeckea gunniana*, with bright-red stems and starry white flowers. The native cherry, *Exocarpos humifusum*, with bright-red succulent fruits is another scrambler of this formation in Tasmania.

A plant of somewhat scrambling habit and of wide distribution both on the mainland and in Tasmania is *Acaena anserinifolia*, with heads of attractive reddish fruits. *A. montana* is often associated with it in Tasmania.

Among the small shrubs common in this association on the mainland are alpine wattle (*Acacia alpina*), several pea flowers, including *Bossiaea foliosa*, forming dense thickets in places, a few species of *Oxylobium*, and purple *Hovea longifolia* (sometimes rock-clinging). The little heath, *Lissanthe montana,* assumes shrub form; and the alpine pepper-bush, *Tasmannia vickeriana*, with bright-red stems and light-green foliage, is found.

Among the smaller plants are the beautifully blue native flax (*Linum marginale*), the trigger plant (*Stylidium*) and the willow herb (*Epilobium*).

In Tasmania the alpine heathland formation is sometimes described as dwarf alpine scrub and carries several of the same species as appear on the mainland in a similar formation. It often occurs in fan-shaped areas at the lower ends of valleys and flats on rising ground and sometimes in patches of grassland. The plants are frequently more robust and taller than those of the mainland heath because additional protection is often afforded by several species of dwarf pine, including *Diselma archeri*, surrounding them.

One of the most remarkable plants of this formation is the spreading guinea-flower, *Hibbertia procumbens*, a mat-forming plant with masses of brilliant yellow flowers. It flourishes from alpine elevations to the sea coast and appears to do equally well in all habitats. This feature of Tasmania's flora is, to me, one of the most interesting. While there are some species restricted to an altitude of 3,000 feet or over, a considerable number of others are capable of an almost total distribution throughout the island. The native dogrose, *Bauera rubioides*, is an example of this. On the mainland it is definitely restricted to coastal and coastal range habitats. In Tasmania, however, one meets it everywhere, most commonly the white-flowering form, and in a variety of shapes and sizes; on the coast a compact shrub, in alpine habitats a prostrate plant, in temperate forests and valleys a dense, almost impenetrable growth of up to 6 metres, a challenge to the bushwalker. Swamp heath, *Sprengelia incarnata*, also confined to coastal habitats on the mainland, has a somewhat similar distribution in Tasmania. *Epacris microphylla*, another plant reaching only to montane levels on the mainland, extends to the highest areas in Tasmania.

### Sod-tussock Grassland

Sod-tussock grassland occupies considerable stretches of level and undulating plateau land in the alpine tract. In the past it provided some of the richest grazing lands, and is still regarded by some local graziers as a potential source of feed for their stock in summer. Grazing here and on the heathlands has caused considerable damage to the vegetation, even without the frequent fires, both deliberate and accidental, which inevitably accompany such practices. Several species have no doubt disappeared from particular areas, and many were jeopardized. Summer grazing has not been permitted in the high areas on Kosciusko for some years, and many species which were becoming rare have made amazing recoveries.

Sod-tussock grassland fringes the raised bogs, and the distinction between the two zones is rarely clearly defined. The herbfields are interrupted by sod-tussock grassland at many points. It occurs in the Kosciusko and Bogong High Plains districts, in the Australian Capital Territory and on Barrington Tops, and is common in Tasmania. Persistent winter snow and a considerable range of climate are characteristic of the whole of the alpine area. These, together with cold air drainage and imperfect soil aeration, provide the conditions needed by sod-tussock grassland to produce its characteristic plant growth.

The zone is rich in grasses, with species of snow grass, wallaby grass and kangaroo grass interspersed with varieties of orchids found in other alpine zones. On the mainland at least two species of native geranium occur in sod-tussock grassland, with several species of buttercup (*Ranunculus*), and several kinds of native violets. *Kunzea muelleri*, common in other formations, too, sometimes occupies considerable stretches, while patches of the almost ubiquitous starry white *Stellaria pungens* often put on a good show. A small creeping pimelea with bright-pink flowers, *Pimelea alpina*, is hidden among the grasses or scrambles over banks and rocks.

On barer patches the completely prostrate *Scaevola hookeri* with bright-green semi-succulent leaves and dainty one-sided lilac flowers may make an atractive show. Another common prostrate plant is *Trachymene humilis*. Patches of bluebells, *Wahlenbergia*, occur frequently.

In some parts great stands of the trigger plant, *Stylidium graminifolium*, exclude almost all other species.

This zone is also the home of the willow herbs (*Epilobium*), and of purple matches, *Comesperma retusum*, among the somewhat taller species. Composites make up the greater part of the taller species, including several kinds of *Brachycome* and *Helichrysum*, two species of

*Senecio* (*S. lautus* and *S. pectinatus*), the native dandelion (*Microseris scapigera*), billy buttons (*Craspedia uniflora*), and the snow daisy (*Celmisia longfolia*).

Many of the same plants occur also in Tasmania's sod-tussock grassland. In addition the dogrose (*Bauera rubioides*) and the swamp heath (*Sprengelia incarnata*), occurring at lower elevations on the mainland, inhabit this grassland. The spectacular *Richea scoparia* and another species, *R. sprengelioides*, are common and so also are the endemic mountain rocket (*Bellendena montana*), several species of *Orites* and, among the prostrate plants, *Pentachondra pumila*, *Exocarpos humifusum* and *Pernettya tasmanica*.

# SUBALPINE COMMUNITIES

On the mainland the country occupying elevations between 5,000 and 6,000 feet and in Tasmania 2,500 and 3,500 feet may be regarded as subalpine habitat. Of course, in specialized areas where exposure is severe, characteristically alpine patches occur well below 5,000 feet on the mainland and below 2,500 feet in Tasmania, while in the more protected areas above 6,000 feet isolated patches of subalpine plants appear. Subalpine areas may be classified as heathland, grassland, woodland and raised bog; the temperate rainforests of Tasmania are also frequently found at the same elevations.

The subalpine communities cover a considerable area. The precipitation is lower than in the alpine areas proper; the climatic conditions in general are less severe than those of alpine areas but the plants that inhabit these subalpine parts are still exposed to a climate of considerable winter snows, high winds and fluctuating temperatures. Summer temperatures can be very high and also extremly variable, while in the winter temperatures can approach very close to those of alpine areas. The subalpine communities include plants of a wide tolerance. Many alpine dwellers descend to subalpine level, though usually less abundantly than in the higher parts, while many species that do not ascend to higher levels are encountered.

**Subalpine Heathland**

In the subalpine heathland of the mainland most of the plants found in similar alpine habitats occur, but their numbers tend to dwindle and they are replaced by other plant species not commonly occurring at higher altitudes. The formation consists of gently undulating slopes with rather poorly aerated soils.

Plants occurring at both alpine and subalpine elevations are generally taller at the lower altitudes, many species which appear as prostrate or scrambling forms above 6,000 feet assuming an erect habit here. *Baeckea gunniana* is a remarkable example of this; at 5,000 feet and under it is tall and slender, reaching a height of 2 metres, while in the alpine tract it is commonly completely prostrate.

Heathlands are covered with snow grasses and grasses of other kinds. Members of the Waratah family are quite abundant. *Hakea microcarpa*, a white-flowering, sweet-scented shrub 1 to 2 metres high with needle-like leaves, is very conspicuous in full flower. *Grevillea victoriae*, which is among the most variable of shrubs and is one of the most attractive of the grevilleas, appears here as a bushy plant 60 to 150 cm high with broadish leaves and large apricot-coloured flowers produced in great abundance. *G. floribunda*, with soft woolly foliage and cream and red flowers, favours the sheltered valleys; *G. lanigera*, a rather similar species with woolly foliage and cream and pink flowers, is found in similar situations though it will survive in less sheltered situations. Both these plants grow 30 to 90 cm in height and are rather bushy.

Among members of the Boronia family the phebaliums are predominant. *Phebalium phylicifolia* is an attractive plant 1.5 metres or more high with light-green foliage and an abundance of bright-yellow flowers which make a beautiful show in summer. *P. ozothamnoides*, which also occurs in the alpine heath, is a smaller shrub with small dainty white flowers.

A characteristic plant of this formation, occurring also in the alpine bogs, is the alpine bottlebrush, *Callistemon sieberi*, a member of the Myrtle family, a rather stiff shrub 1 to 2.5 metres high with small but dense rich yellow brushes which make a spectacular show in midsummer, particularly in the wetter areas. It occurs commonly in large stands to the exclusion of most other species. *Kunzea muelleri* appears either in its alpine prostrate form or as a semi-upright shrub.

Other members of the same family include two species of tea-tree which do not occur at higher elevations: *Leptospermum phylicoides*—until fairly recently known as *Kunzea peduncularis*—fringes the river beds with its fine foliage and dainty small white flowers; *L. lanigerum,* which has much larger flowers and coarser, woolly foliage, also likes the waterways but is more widespread.

A rather smooth form of the well-distributed alpine rice-flower, *Pimelea ligustrina*, is fairly common. A number of smaller species of *Pimelea* also occur. *Geranium antrorsum*, a small pink-flowering plant, is common in the grasslands and is often found in rock crevices.

The dainty *Parahebe perfoliata*, with attractive grey-green, opposite, stem-clasping leaves and heads of brilliantly blue flowers, is a feature of subalpine heathland. It is a clumpy plant, preferring moist pockets. A number of other veronicas, some of which also occur at alpine elevations, may be found in these heathlands. One of these is *Parahebe derwentiana*, a robust plant 1 to 1.5 metres high of compact habit, with spikes of white or pink flowers, and another is *V. nivea*, a smaller species with fine foliage and shorter heads of white or light-blue flowers. Both these species often occur in dense masses.

The whole heathland is covered with snow grass and other grasses, dotted with many kinds of everlasting daisies such as *Helichrysum bracteatum* and *H. scorpioides* both yellow-flowering species, and other composites, with scattered patches of the charming little gentian, *Gentianella diemensis*, and the intriguing trigger plant, *Stylidium graminifolium*, with its heads of bright-pink flowers.

Patches of the attractive but treacherous *Acaena anserinifolia* occur. This wicked little plant is surmounted by balls of red feathery-looking fruits which conceal the most persistent barbs with an extremely vigorous sock-clinging propensity.

In Tasmania these subalpine heathlands, occurring at 2,500 to 3,500 feet, are often called subalpine meadows.

A very lovely low-growing tea-tree, *Leptospermum humifusum*, with starry white flowers in great abundance, is a feature of the Tasmanian habitat.

In the very acid quartzite soils of the west and south-west of Tasmania vast stretches of button-grass, *Gymnoschoenus sphaerocephalus*, grow in dense tufts with stems reaching to a height of 1.52 metres in places. So vigorous is it that huge stretches of country are covered with it to the exclusion of practically everything else. In drier, less acid areas tufts of rushes and sedges are mixed with it. At higher elevations 3 small species of flannel flowers, *Actinotus suffocata*, *A. moorei* and *A. bellidioides,* occur.

Native irises, *Patersonia*, of several kinds and *Diplarrena moroea*, also a member of the Iris family, are very common in the Tasmanian subalpine heathland.

Fringing the lakes and swamps and on the slopes of Tasmanian subalpine heathland the deciduous beech, *Nothofagus gunnii*, is found; in autumn the dying leaves make a golden spectacle.

**Raised Bog**

The subalpine raised bogs of the mainland resemble rather strongly those of the alpine regions but contain some additional species. They occur on permanently wet sloping ground.

The delicate purple-flowered lily, *Arthropodium multiflorum*, sways gracefully in the ever-blowing breeze; the little sundew, *Drosera peltata*, is common.

Among the large plants a white-flowered needlebrush, *Hakea lissocarpa*, is quite spectacular at its best, and the alpine bottlebrush, *Callistemon sieberi*, common on the heathlands, extends to these bogs.

In Tasmania a fern, *Gleichenia dicarpa*, reaches from higher elevations and, in association with several kinds of rushes and sedges, occupies much of this territory to the exclusion of all else. Interspersed with it are mats of the alpine lily, *Astelia alpina* and, on the mainland, *A. psychrocharis*, also with their rich red fruits. Sphagnum moss is common in all the bogs.

In the running water, fairies' aprons or bladderworts, *Utricularia* spp., make a dainty picture with their bright-purple flowers.

### Subalpine Grassland

Subalpine grassland fairly closely resembles that of the alpine areas, with snow grass and wallaby grass growing luxuriantly on level and undulating land while several other smaller kinds of grasses and sedges occur. On sod-tussock grassland patches of the completely prostrate pea, *Pultenaea fasciculata*, make a carpet of rich gold; a lacy everlasting daisy-bush, *Helichrysum myrsinoides*, appears and a number of kinds of ground orchids not occurring at higher elevations are dotted among the grasses. In dry-tussock grassland the alpine form of the small *Geranium antrorsum* is not uncommon.

*Brachycome rigidula*, a small and dainty daisy with finely divided foliage and pink or purple small flowers replaces several larger flowering and more robust species. A smooth form of the large rice-flower, *Pimelea ligustrina*, occurs here also.

### Subalpine Woodland

While some of the subalpine area is at a sufficient elevation to be part of the treeless area, the lower slopes are wooded—indeed, considerable tracts of subalpine country are wooded. On the mainland they occur on the Kosciusko plateau, on the Victorian highlands and in the Australian Capital Territory. The subalpine formation is also fairly extensive in Tasmania. Strong winds prevail and their influence is shown in the gnarled and tortured stems of the trees and in their dwarfed and twisted forms. At the highest points at which these trees occur as isolated stalwarts unprotected by surrounding vegetation and almost buried in the snow which lies on the ground for one to four months of each year, the wind can rage unceasingly for days and it is almost unbelievable that a single tree can withstand such buffeting. The only species which can survive these extreme conditions is the true snow gum, *Eucalyptus niphophila*, on the mainland and its counterpart, *E. coccifera*, in Tasmania.

*E. niphophila* (Fig. 1) is a remarkable tree. Even in death it is beautiful. Standing forlorn at the highest elevations at which trees can grow at all, it is a stunted, twisted thing, often with

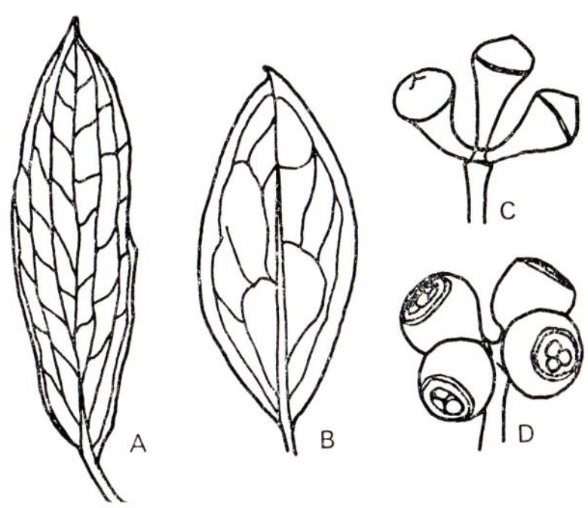


FIG. 1 *Eucalyptus niphophila*: A, mature leaf; B, sucker leaf; C, buds; D, fruits. (All actual size except A × ½.)

several stems springing from a single stock, mallee-like, in weirdly fantastic shapes, the stems with a smooth, satiny sheen and frequently mottled and coloured in rich reds, apricots, and lush greens. One has to see these trees in a wild wet blizzard to appreciate the exquisite beauty of both their form and their colouring.

It is undoubtedly a tree of moods, all of them alluring. To add to the fascination of the smooth trunks are the scribbles of the insect which writes its undecipherable message on these as it does on the bark of the coastal scribbly gum.

The tree is vulnerable. Sudden or unusual exposure may kill off a whole stand of them, just as did the disastrous fires of 1959 at Kosciusko—and there the skeletons stand and have stood for many years, grim reminders of relentless nature or the unimaginative action of the firebug, following in the steps of persistent grazing.

One can hardly err in identifying the species. In the first place it is either the only species in the area or is associated with two very different kinds. The leaves of the snow gum are thick and leathery, the veins conspicuous and almost parallel with the midvein. The sucker leaves are opposite, shortly stalked except in the very early stages, thick and leathery. The flowers are cream, rather small but produced in rich abundance. Many trees are prodigal with their flowers and there seems to be no regular flowering season; one can usually find a stray tree or two in good flower through the warmer months. The fruit is large, 7 to 10 millimetres in diameter, usually without a stalk, mealy or grey-green, the valves enclosed. The snow gum, like so many eucalypts, develops a lignotuber or thickened base to the stem, which makes it possible for the tree to recover from all but severe fires since the food stored there enables the tree to throw out suckers and so survive.

Descending to much lower altitudes and occurring in the lower parts of the subalpine woodland is a very similar species, *E. pauciflora* (Fig. 2), sometimes also called snow gum or cabbage gum or white sallee. This is found also

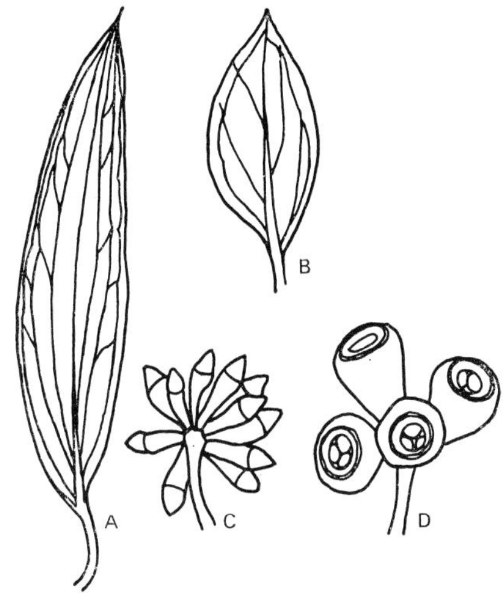


FIG. 2 *Eucalyptus pauciflora*: A, mature leaf (× ½); B, sucker leaf (× ½); C, buds (actual size); D, fruits (actual size).

in Tasmania from sea-level to 3,500 feet. Its general appearance resembles that of the true snow gum but it is a more robust, less twisted and picturesque tree, living as it does in more pampered situations. The sucker leaves are similar, though usually grey-green in this species and light-green in the true snow gum. The mature leaves are similar but tend to be longer. The flowers are also similar but the fruit, though similar otherwise, is shortly stalked and not grey-green. The chief distinguishing factor between the two species, however, is habitat.

Associated with the snow gums is the higher altitudes on the mainland is one other common species, the attractive and picturesque black

FIG. 3 *Eucalyptus stellulata*: A, sucker leaves; B, mature leaves; C, buds; D, fruits. (All actual size except B × ½.)

sallee, *E. stellulata* (Fig. 3). It is a smaller and more compact tree, with a fairly dense spreading crown of soft green leaves. It occurs as high as 5,500 feet, usually on damp flats or near watercourses, especially in fairly open places where it may form dense stands. At high elevations it is quite small and almost shrub-like, but on the lower slopes it is a shapely and handsome tree. One of its distinguishing features is its smooth, satiny, olive-green bark mottled with brown, always with some rough bark at the butt. The leaves are a soft light-green, thinner and less coarse than those of the snow gums; the veins are not longitudinal as they are in the latter. The flowers and fruits are small, the fruits are rounded and rather clustered, the valves enclosed.

A tree of lesser importance in this area because of its rather restricted habitat both on the mainland and in Tasmania is the whistling or spinning top or Dargo gum, *E. perriniana* (Fig. 4). In Tasmania it occurs at 1,000 to 2,000 feet but it is found also on the Kosciusko plateau at higher elevations at Munyang, Pretty Plains and one or two other places. It is an easy tree to identify, for its sucker leaves are most distinct and these it sometimes retains throughout its life; at other times it loses them entirely to be replaced by quite different mature leaves.

In exposed situations it may be a mallee with many stems, and at other times a small, crooked, single-stemmed tree. In protected habitats it is fairly tall. The trunk is smooth, white, often blotched with green and orange-brown and then very beautiful. The leaves, stems and branches are ashy-blue. In its early stages it is quaintly attractive, being completely clothed in soft grey sucker leaves in fused pairs round a stout slaty-grey stem. These fused leaves are often almost perfectly round and quite large, 6 to 12 centimetres across the complete circle. The mature leaves are also green but alternate and long and pointed, giving an altogether different appearance to the tree. The flowers are small and cream, without stalks, usually 3 together, often nestling at the base of the sucker leaves. The fruit is without a stalk, mealy-white, hemispherical, 5 to 6 millimetres in diameter, the valves enclosed or sometimes slightly protruding.

This species makes a fine garden or park tree and it is to be hoped that some stands will be perpetuated, for it is not a common species. The small stands at Tunnack and Strickland in Tasmania are well worth preserving because of the persistence of their sucker leaves. The specimens here are small and of no commercial value, but if a definite reserve is not declared it is conceivable that they will be removed to make way for crops or buildings.

The understorey of the subalpine woodland formation on the mainland abounds in pea flowers such as *Bossiaea foliosa*, *Daviesia ulicina* and *Oxylobium ellipticum*, often in dense stands. Representing the Boronia family are two kinds of *Phebalium* occurring in the heathlands of this elevation, *P. phylicifolium* and *P. ozothamnoides*, and bell-flowered *Correa lawrenciana* var. *rosea*. The latter is a medium-sized shrub in which the bells are commonly dusty-red although the species in other habitats usually has yellow or

FIG. 4 *Eucalyptus perriniana*: A, mature leaves (× ½); B, sucker leaves (× ½); C, fruits (actual size); D, buds (actual size).

green bells. *Tasmannia vickeriana*, an attractive shrub with light-green leaves and succulent-looking red stems, flourishes on the mainland also.

Many of the smaller plants of the alpine areas occur throughout the subalpine woodland, including several species of *Ranunculus, Viola, Veronica* and *Wahlenbergia*. *Acacia alpina*, a handsome little semi-prostrate wattle is common on the mainland.

In the wetter regions, rushes and sedges are abundant. Small forms of the purple eyebright, *Euphrasia*, show their dainty lipped flowers in profusion. A white-flowering mintbush, *Prostanthera microphylla*, occurs in patches among the grasses; the most widespread of all the buttercups, *Ranunculus lappaceus*, revels in this habitat, while the completely prostrate umbellifer, *Trachymene humilis*, favours this formation before all others.

The bulbine lily, *Bulbine bulbosa*, makes some of the wetter parts of the woodlands gay with its stalks of light-yellow flowers; the flax lily, *Dianella tasmanica*, with blue flowers and fruits favours the shelter of overhanging trees. *Grevillea australis* occurs here as a taller plant than it does at higher elevations; *G. victoriae* sometimes forms quite dense thickets. *Persoonia chamepeuce*, a prostrate geebung, does well in this formation.

Two species of groundsel, *Senecio lautus* and *S. pectinatus*, are common, as well as another yellow daisy with rather papery flower heads, *Podolepis acuminata*.

In Tasmania the indigenous snow gum, *Eucatyptus coccifera* (Fig. 5), with its associate *E. urnigera*, may be regarded as the counterpart of the *E. pauciflora—E. stellulata* association. This association is common at 3,000 to 4,000 feet but occurs most abundantly somewhat below 3,000 feet.

FIG. 5 *Eucalyptus coccifera*: A, mature leaf; B, buds; C, fruits; D, sucker leaves. (All actual size except D × ½.)

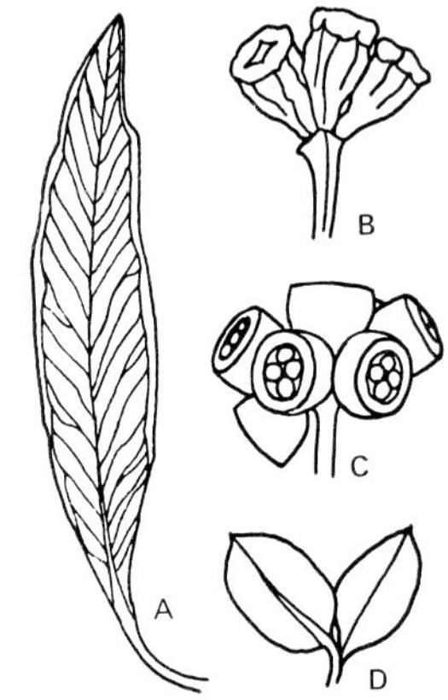


*E. coccifera* is usually rather stunted and slender, though sometimes shrubby, and, in protected areas, reaching a considerable height with a smooth bark which is often streaked with a rich red sometimes mingled with green on the main trunk and branches, while at other times it assumes a gleaming white surface which is often streaked with a satiny-grey. The sucker leaves are grey-green and opposite, the mature leaves greyish, alternate, 5 to 6 centimetres in length. The cream or white flowers 5 to 7 in a group, are sometimes borne at the base of the sucker leaves, often in great profusion. The fruit is hemisperical or broadly turban-shaped.

*E. urnigera* (Fig. 6), companion of *E. coccifera* in subalpine woodland, frequently occurs

FIG. 6 *Eucalyptus urnigera*: A, sucker leaves (× ½); B, mature leaves and fruits (× ½); C, buds (actual size).

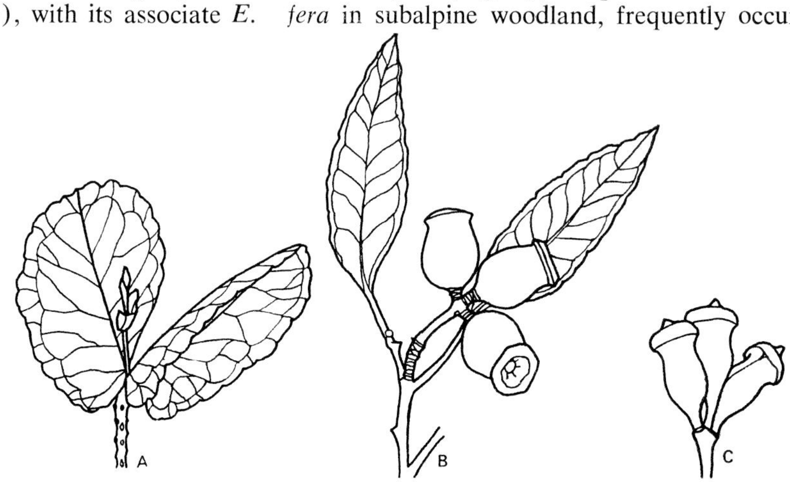

in quite extensive stands. In protected habitats it is fairly tall, reaching 15 metres at times, but exposed trees are stunted. The trunk is smooth, white blotched with green and orange-brown, often very beautiful. The sucker leaves are broad, rounded and stem-clasping, usually wavy in outline. The mature leaves are leathery, very variable in shape, alternate, 6 to 12 centimetres long. The flowers are small, in groups of three; the fruit is stalked, urn-shaped, grey-green, 7 to 10 millimetres in diameter, the valves deeply enclosed.

Another gum of this formation is varnished gum, *E. vernicosa* (Fig. 7). Restricted in range to Cradle Mountain and the mountains of the west and south-west, it is a very small tree, mallee-like with a smooth bark and opposite leaves even in the mature stage, shortly stalked, leathery, very glossy, crinkled on the margins. The buds are characteristically rough; the flowers are small, in groups of three. The fruit is hemispherical or bell-shaped, 5 to 6 millimetres long, borne singly, the valves level or slightly sunk below the surface of the fruit.

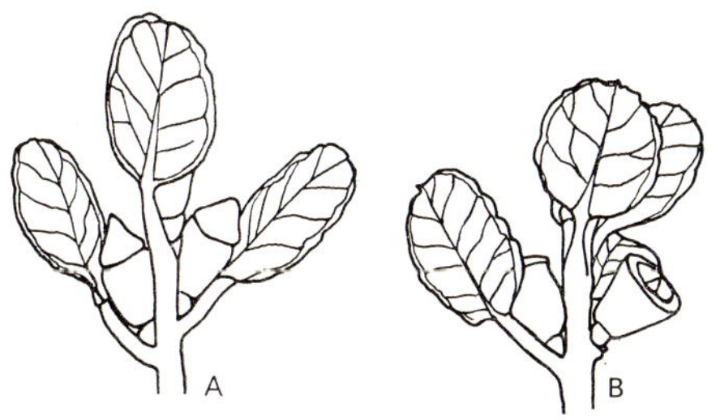


FIG. 7 *Eucalyptus vernicosa*: A, sucker leaves with buds; B, mature leaves with fruits. (Both actual size.)

At the lower level of this formation in Tasmania another eucalypt, yellow gum, *E. johnstonii* (Fig. 8) occurs in a tall tree up to 45 metres high on well-drained slopes. It has a smooth bark streaked with red and orange, often very handsome, and with rough bark at the butt. The mature leaves are broad and rough and glossy-green with wavy margins. The flowers are in threes. The fruit is hemispherical or bell-shaped, rough, 7 to 14 millimetres in diameter, the valves protruding.

Here, too, the deciduous beech, *Nothofagus gunnii*, presents an appearance quite unlike anything else in this forest. The lacy, delicate foliage with its rich red new growth in summer is remarkable; in autumn, when the leaves are gold prior to being shed, it spreads a glory round the lakes and watercourses, on the fringes of which it makes its home. At this level it reaches a height of some 6 metres. Often associated with this beech is the King William pine, *Athrotaxis selaginoides*, its erect and narrow form contrasting sharply with the flowing and spreading beech, just as its brilliant green foliage makes a foil for the dark-green and gold pattern made by the beech leaves.

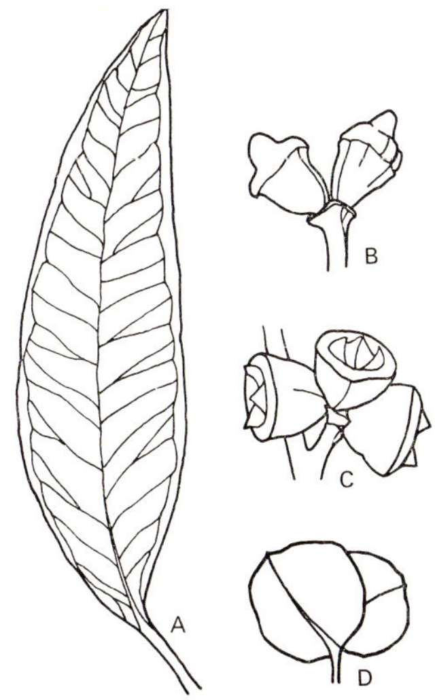


FIG. 8 *Eucalyptus johnstonii*: A, mature leaf; B, buds; C, fruits; D, sucker leaves. (All actual size except D × 1/3.)

On the fringe of the subalpine woodland other conifers are found. The pencil pine, *Athrotaxis cupressoides*, a most graceful and dignified tree, frequents the fringes of the tarns and lakes with its closely appressed leaves and spherical cones; the celery-top pine, *Phyllocladus aspleniifolius*, one of the quaintest and most unpinelike, carries its cones at the tips of the "leaves", which are really phyllodes; *Microstrobus niphophilus*, also hugging the waterways, appears as a dense, branching small shrub with tiny leaves and pendulous cones. Then there is *Diselma archeri*, a much-branched pine which attains shrub height in the woodland; it has small, thick, blunt leaves closely pressed against the stem, and solitary, terminal cones. Two prostrate pines are found scrambling among the boulders here, too—the alpine pine, *Podocarpus lawrencii*, also occurring on the mainland, and the creeping pine, *Microcachrys tetragona*, endemic to Tasmania.

Associated with these trees in the subalpine woodland is the Tasmanian waratah, *Telopea truncata*, growing to the height of a tall shrub or almost a small tree, with thick leaves and brilliant red flowers in dense terminal heads, making a grand show in summer. It is a widespread plant, occurring also in the temperate rainforest but very common in the subalpine woodland. White and yellow-flowering forms are occasionally found.

Other members of the family common in the region are *Hakea microcarpa* and several species of *Orites*. The waxberry, *Gaultheria hispida*, with white flowers enhanced by the bright red of the new young foliage and followed by attractive waxy white "berries", is a feature and the cheeseberry, *Cyathodes parvifolia*, with bunches of small white flowers followed by great masses of brilliant red berries, is common, with other species of the same genus.

Common here, too, is the widespread dogrose, *Bauera rubioides*, with white or pink open flowers and dainty foliage. In this habitat it is a scrambling shrub.

In the lower altitudes forests of the quaint pandani or giant grasstree, *Richea pandanifolia*, are found. This grows 3 to 9 metres high, usually with a bare trunk bearing a crown of long, broad, leathery leaves commonly 30 to 120 centimetres long, hanging down and protecting the lower ones—a feature particularly welcome to bushwalkers as the basis of a fire, sometimes difficult to light in these wet parts. Out of the upper branches rise the dense heads of creamy flowers. At higher elevation pandani is rather stunted; in the sheltered temperate rainforests it is taller and more robust. A smaller and rather less spectacular relative, *R. dracophylla*, is found in the drier parts. This grows 1 to 1.5 metres high, rather like a miniature pandani but bearing long heads of most attractive pink or cream flowers, each flower subtended by delicately tinted, striated, transparent bracts.

A characteristic group of smaller plants makes an interesting understorey in the Tasmanian subalpine forests. *Pittosporum bicolor*, a quaint little stunted tree or tall shrub, is found in exposed situations. It has slender, tubular flowers, dark crimson or yellow on the outside, crimson or purple inside. Two species of bushy heath, both growing into compact shrubs with delicate prickly foliage and soft pink flowers, are common in the shelter of taller plants. These are *Archeria comberi* and *A. eriocarpa*. The former has blunt, crowded leaves; the latter has more distant and sharply pointed leaves and occurs also in the rainforest. Many berry-bearing epacrids such as *Trochocarpa, Leucopogon* and *Monotoca* of several kinds are common here too. In full summer their variously coloured berries make a beautiful sight.

Rocks at this elevation are used for protection and support by scrambling plants. In their lee several kinds of *Orites*, *Richea*, *Cyathodes*, *Coprosma, Baeckea,* as well as *Tasmannia lanceolata*, take shelter. It is an area rich in species.

### Temperate Rainforests of Tasmania

At subalpine and montane elevations in Tasmania in places where the annual rainfall exceeds 50 inches and valleys provide protection, the vegetation is luxuriant and resembles in general the tropical rainforest of the east coast on the mainland. This is temperate rainforest, differing in species from any that occur in other parts of Australia. Large beech and myrtle trees grow thickly together, mosses and lichens encrust and hang from the tree trunks; there is a rich fern flora among the rotting timber on the ground. The southern sassafras, *Atherosperma moschatum*, occurring also in the mainland rainforests, is common. It is a fragrant tree with large leaves paler on the undersurface and small white flowers. *Pittosporum bicolor*, also found in the more open woodland, is met here, too.

The deciduous beech, *Nothofagus gunnii*, which intrudes to the highest elevations of Mount La Perouse, Mount Field and Cradle Mountain as a rather stunted tree 1.5 to 2 metres high grows much taller in the rainforest, often with tangled branches.

*Nothofagus cunninghamii* (Fig. 262), commonly known as myrtle, is a more widely dispersed plant, penetrating to sea level and occurring also on the mainland, but not reaching the highest altitudes. It is a dignified much-branched tall tree, sometimes exceeding 30 metres in height but stunted at higher elevations.

Dense stands of these myrtles occur throughout the rainforest. Our myrtles and beeches, which are related to the true beeches, *Fagus*, have particularly interested botanists because they are not closely related to any other elements in the Australian flora. Some students think they migrated northwards from Antarctica along a continuous land-mass with a cool moist climate which united the present northern continents with Antarctica. This, they believe, occurred before the Antarctic flora was exterminated by Pleistocene glaciation. Fossil evidence has certainly established the previous existence of a flora in northern Antarctica which was suited to a more temeprate climate than it enjoys today and which included some kinds of *Nothofagus*. Others claim that these plants had their origin in North America and reached Australia along the western shore of the Pacific. However it is, the beeches and myrtles, confined as they are to isolated patches of country on the mainland and more extensively in Tasmania, have a fascination for the botanist on account of their peculiar lack of affinity with other elements of the Australian flora and for others because of their peculiar beauty.

In the most characteristic temperate rainforests of Tasmania the myrtle canopy is fairly open and there is a distinct and peculiar understorey.

One of the most interesting plants of this understorey is the terrifying horizontal, *Anodopetalum biglandulosum* (Fig. 136), which loves the lush gullies beneath the myrtles. It commences life with a slender, erect stem which soon bends horizontally, sending out shoots which follow the same pattern, resulting in a mass of tangled branches and trunks which form a platform some distance from the ground. This can be an almost impenetrable barrier, making horizontal one of the bushwalker's dreads.

Another common understorey plant of the more poorly drained areas is Tasmanian blackwood, *Acacia melanoxylon*, which despite its common name also occurs on the mainland. It is a handsome, shapely tree with rough bark and dense, blunt foliage, and produces heads of creamy-yellow flowers. It is handsome in fruit, too, its brown pods curling over as they open, revealing shiny black seeds to each of which is attached a bright-red aril, a structure which aids in water absorption prior to and during germination.

Among the smaller trees of the Tasmanian rainforests are four outstandingly handsome kinds: Tasmanian laurel (*Anopterus glandulosus*), leatherwood (*Eucryphia lucida*), *Cenarrhenes nitida*, and *Agastachys odorata*. Tasmanian laurel is an elegant small tree or tall shrub with glossy leaves and delicate white or blush-pink flowers borne in long clusters. Leatherwood is one of the most beautiful blossom trees. Seen from the air a leatherwood tree in full bloom might be housing flocks of white birds. The flowers are large and showy and it is a prolific honey-tree. Indigenous Tasmanians and acclimatized mainlanders claim superior qualities for the honey. One form of this plant, *Eucryphia lucida* var. *milliganii*, occurs as a shrub, prostrate or up to 3 metres in height, with smaller leaves and flowers and more tolerant of exposed situations and higher altitudes. The other two plants mentioned, *C. nitida* and *A. odorata*, are both members of the Waratah family. *C. nitida* is an erect, spreading tree, or a shrub sometimes found scrambling over rock surfaces. *A. odorata* is a most handsome shrub reaching 4 to 6 metres but often much smaller, enjoying the protection of the forest giants.

The temperate rainforest of Tasmania is rich in attractive flowering trees and shrubs, but none can compare with the vivid majesty of the climbing heath, *Prionotes cerinthoides*. This is an epiphyte, using trees and taller shrubs to support its long, slender, much-branched stems as they reach for the light through the dense forest growth. It may cover the trunks of myrtle and other forest trees to a height of 10 metres or more.

Five species of conifers also occur in this formation, including celery-top pine (*Phyllocladus aspleniifolius*), pencil pine (*Athrotaxis cupressoides*), King William pine (*A. selaginoides*)—these three have already been described —*A. laxifolia*, and Huon pine (*Dacrydium franklinii*). The last-named was a valuable boatbuilding timber very common at altitudes of

2,500 feet fringing the rivers and swampy flats and famous in the picturesque south-west of the State at Huon, Macquarie Harbour, and Port Davey. These centres are rich with anecdotes of the tough old "piners" but now Huon pine is very rare. It grows to a height of 20 to 30 metres with spreading branches and a dense pyramidal crown. The leaves are very small, closely overlapping with a rounded end, the outer surface covered with scattered pores.

*A. laxifolia* is closely related to pencil pine, but it is a shorter tree and is distinguished mainly by the fact that the leaves, although overlapping, are also slightly spreading, with the apex incurved. It is not very common, usually occurring in association with pencil pine or King Billy pine.

At these higher altitudes the rainforest is confined to the gullies, the more exposed surrounding parts being occupied by *Eucalyptus coccifera* and *E. vernicosa* forest. Very often the stands of myrtle give way to pockets of King Billy or other pines or deciduous beech at the head of the valley occupied by the rainforest.

Tree ferns are common through the gully, and the ubiquitous *Bauera rubioides* grows to a considerable height, almost ranking with the horizontal plant in its impenetrability.

Below the taller plants of the understorey an interesting mixture of smaller plants occurs. Heaths or epacrids are common, among them the two most widespread species of *Richea*, namely *R. sprengelioides* and *R. scoparia*, and several species of *Cyathodes*, *Monotoca*, *Trochocarpa* and *Archeria*.

Other shrubs of the rainforest include *Tasmannia lanceolata*, species of *Baeckea* and many others. It is a rich domain.

# MONTANE COUNTRY

Below the subalpine tract and descending to about 3,000 feet on the mainland and to 2,000 feet in Tasmania is the area classified as montane country. This is characterized by steep and precipitous slopes and deeply entrenched creeks and rivers.

Strong winds are encountered in many parts of this area, but temperatures are generally higher than at the higher elevations and there is less permanent snow. Summer dews are heavy and autumn frosts can be quite severe. Precipitation is considerably less than at higher altitudes, for the subalpine tract captures much of the rain cloud coming from the alpine regions.

Grassland, heathland and scrubland occupy considerable areas of montane country and carry a number of species which penetrate to higher altitudes but also a number which will not tolerate even subalpine habitats. New associations here are savannah woodland, wet mallee and schlerophyll forest, more extensive on the tablelands at lower elevations but also significant in the montane.

## Heathland and Scrubland

Montane heathland and scrubland occur in considerable stretches on the slopes of Mount Kosciusko, in the Australian Capital Territory, on the eastern side of the Blue Mountains, on parts of the New England tableland in New South Wales and on sloping land adjoining the subalpine areas in Victoria and Tasmania, in the latter mainly on the marshy flats. In much of this association the soil is poorly aerated and precipitous slopes are the exception. In these poorly drained areas of the mainland, species of subalpine epacrids and eyebright (*Euphrasia*) as well as willow herbs (*Epilobium*) and

*Kunzea muelleri* persist; trigger plants (*Stylidium*) and *Brachycome* and other composites are common; *Comesperma retusum* is more common than at higher altitudes, and one of the tea-trees, *Leptospermum myrtifolium*, is widespread.

On the windswept peaks and plateau areas the dwarf she-oak, *Casuarina nana*, is common. Grasses and cypresses make up much of the understorey, among which are many plants of the Lily family such as *Dianella tasmanica* and *Stypandra glauca* as well as the purple iris, *Patersonia sericea.* There is a form of *Grevillea victoriae*, also two species of *Banksia*—*B. integrifolia* and *B. marginata*—as well as several kinds of *Hakea* and *Persoonia.* Among the most common of the taller shrubs are *Lomatia fraseri, Acacia siculiformis* and *A. kettlewelliae.*

A number of plants commonly identified with subalpine habitats persist. Among the most conspicuous of these are *Oxylobium ellipticum* and *Bossiaea foliosa.* A handsome yellow-flowering bottlebrush, *Callistemon pallidus*, replaces *C. sieberi* of higher altitudes, although an occasional stand of the latter may still persist.

Fringing the rivers and streams the heathland merges into scrub with three species of tea-tree, *Leptospermum lanigerum*, *L. myrsinoides*, and *L. phylicoides*, while a tall form of *Baeckea gunniana* often forms dense stands. There are a number of taller wattles, including the handsome silver wattle, *Acacia dealbata*, as well as the smaller *A. buxifolia* and *A. pravissima.*

A very fine form of *Correa lawrenciana* var. *rubra* with red bells, belonging to the Boronia family, occurs, also the dainty lilac-flowered *Prostanthera phylicifolia* and the green-flowering *Prostanthera walteri*, the two former on the sandy flats, the latter among the denser scrub. *Cassinia aculeata*, one of the composites, is plentiful.

On moister parts of the plateaux *Scaevola hookeri* makes a carpet of rich green dotted with dainty, pale-lilac, fan-shaped flowers, and *Lotus corniculatus* with bright-yellow pea flowers also favours these moister spots. The delicate pale-pink flowering *L. australis* prefers the dry stretches.

Interspersed with heathland are stretches of grassland which may be conveniently divided into sod-tussock grassland and wet-tussock grassland. The former occurs on damp, level country carrying small shrubs including several species of *Epacris*, *Pultenaea fasciculata*, many kinds of grasses and small ground orchids. In wet-tussock grassland are found a variety of small plants such as *Veronica gracilis*, *Geranium antrorsum* and many grasses and sedges. *Boronia algida*, a prostrate plant with tiny leaves and dainty pink flowers, is another alpine that likes the moist pockets of the montane zone.

The drier stretches also carry a number of species which extend from the coastal slopes, including such plants as *Calytrix tetragona, Isopogon anemonifolius, Gompholobium huegelii, Dillwynia floribunda, Epacris impressa* and, among the lilies, *Stypandra caespitosa* and *Laxmannia gracilis.* Descending from higher elevations are such species as *Helichrysum bracteatum, Phebalium ozothamnoides, Westringia rubiifolia, Coprosma hirtella* and *Parahebe perfoliata.*

In Tasmania the montane heath is quite extensive. It is interrupted by patches of bog or swamp in many parts, large stretches of which are dominated by button grass, *Gymnoschoenus sphaerocephalus*, in acid quartzite areas and by wiry rush, *Restio australis*, in dolerite and less acid situations. It is particularly noticeable here that many plants or close relatives of plants from higher altitudes mingle with others which are commonly found in sea-coast habitats.

Pea-flowers are common. These include *Pultenaea gunnii*, *P. subumbellata, P. fasciculata, P. juniperina,* two attractive prostrate bossiaeas, *B. cordigera* and *B. riparia,* and a form of *Hovea longifolia. Bauera rubioides*, found in almost every habitat in Tasmania, is a medium-sized, rather bushy shrub here. *Correa lawrenciana* is represented, but the flowers are greenish-yellow, not red as they are in the mainland variety at this elevation. *Stackhousia pulvinaris*, a lovely little prostrate plant restricted to higher altitudes on the mainland, is sometimes found.

The beautiful Tasmanian laurel, *Anopterus glandulosus*, common in the temperate rainforest, reaches the height of a small tree here.

A handsome bottlebrush with greenish-yellow spikes of flowers, *Callistemon viridiflorus*, is a common inhabitant of the wetter dolerite areas. It is a small tree of attractive bushy habit.

The area is rich, too, in several species of buttercups, including *Ranunculus lappaceus* and its close relative *R. pascuineus* (which also extends to subalpine habitats), *R. scapigerus* (related to the mainland *R. hirsutus*), *R. sessiliflorus*, *R. pumilo*, *R. collinus* and, in the moister parts, *R. pimpinellifolius* and *R. decurvis*.

## Wet Mallee

On deep rocky soils in windswept areas of the montane habitat, stands of dwarf many-stemmed mallees are a conspicuous feature. This is particularly characteristic of the mainland. The most obvious association is that of *Eucalyptus aggregata* and *E. cypellocarpa* (two of the species described under the heading of Savannah Woodland). Other species occurring are *E. glaucescens*, *E. moorei* and its variety *latiuscula*, *E. kybeanensis*, *E. radiata* and *E. perriniana*, which occur mainly on the higher parts of the subalpine tract. *E. perriniana* is described under the heading of Subalpine Woodland.

Tingiringa gum, *E. glaucescens* (Fig. 9), is a rather ragged-barked tree, the bark stripping off in reddish-brown flakes, leaving a white stem. It somewhat resembles *E. perriniana* in the latter's adult form. The sucker leaves are opposite, stalkless, grey-green and rounded; the mature leaves are grey, alternate, 7 to 12 centimetres long. The fruit is without a stalk, in groups of three, rounded, mealy-white or bluish, 5 to 6 millimetres in diameter, the valves enclosed.

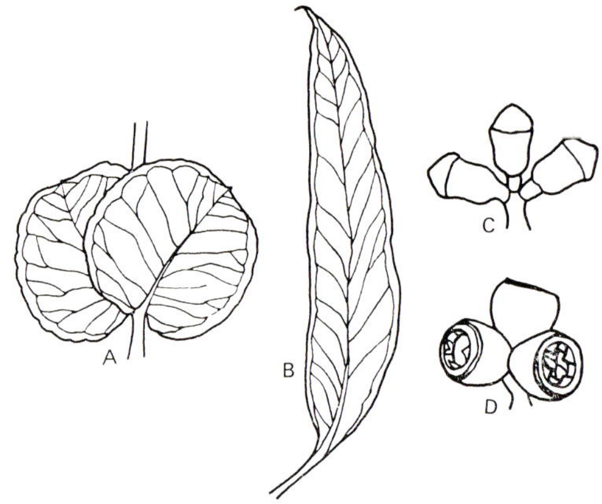


FIG. 9 *Eucalyptus glaucescens*: A, sucker leaves (actual size); B, mature leaf (× $\frac{1}{2}$); C, buds (actual size); D, fruits (× $\frac{1}{2}$).

Narrow-leafed sally, *E. moorei* (Fig. 10), is a small tree, often forming thickets, with smooth bark, peeling in ribbons, and a little rough bark at the butt. The sucker leaves are opposite, slightly grey-green, without stalks, rather narrow, 2 to 3 centimetres long; the mature leaves are alternate, narrow, 4 to 7 centimetres long, stalked, pale-green, thick and with a hooked point. The fruits are without stalks, in groups of seven to thirteen, globular, 3 to 4 millimetres in diameter, the valves enclosed.

*E. moorei* var. *latiuscula* (Fig. 11) has a grey or greyish-green bark. It is often called little sallee and rather resembles black sallee, *E. stellulata*. The sucker leaves are broader and the mature leaves slightly larger than those of *E. moorei*, very like those of *E. stellulata* though somewhat smaller. It may be a hybrid of *E. stellulata* x *E. moorei*.

FIG. 10 *Eucalyptus moorei*: A, mature leaves; B, sucker leaves; C, buds; D, fruits. (All actual size except D × $1\frac{1}{2}$).

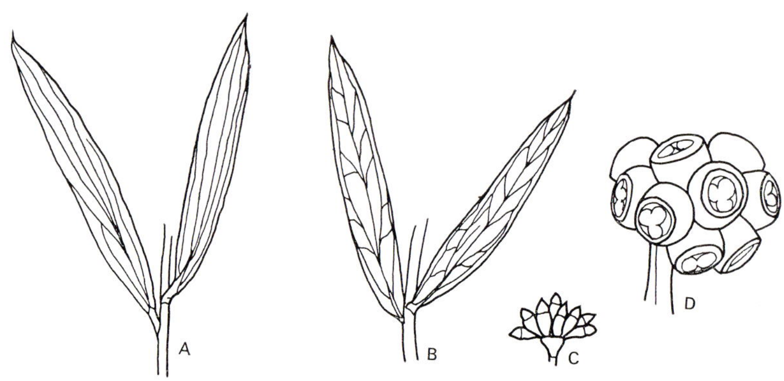

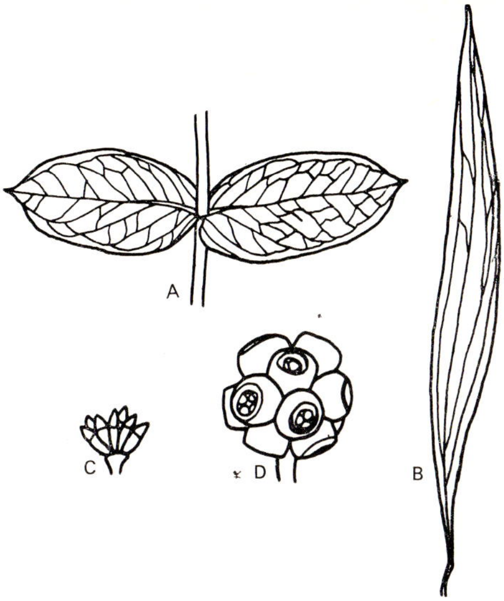


FIG. 11 *Eucalyptus moorei* var. *latiuscula*: A, sucker leaves; B, mature leaf; C, buds; D, fruits. (All actual size).

FIG. 13 *Eucalyptus radiata*: A, mature leaves (× $\frac{1}{2}$); B, sucker leaves (× $\frac{1}{2}$); C, buds (actual size); D, fruits (actual size).

*E. kybeanensis* (Fig. 12) is a slender mallee with smooth, greenish stems and opposite, shortly stalked, pale-green, oblong sucker leaves; the mature leaves are alternate, long, pointed, dark-green and rather thick. The flowers are in groups of three to seven. The fruit is without a stalk, hemispherical, 7 to 8 millimetres in diameter, the valves slightly protruding.

Narrow-leafed peppermint, *E. radiata* (Fig. 13), is a medium-sized tree, rarely assuming mallee form, with peppermint bark on all but the smaller branches, which are smooth. The sucker leaves are stalkless, opposite, and narrow; the mature leaves are alternate and stalked, but otherwise rather similar to the sucker leaves, with a strong peppermint scent. The flowers are in groups of eight to fifteen. The fruit is stalked, hemispherical, about 4 to 5 millimetres in diameter, the valves enclosed.

The understorey of the wet mallee formation is poor in species as well as in individuals. Some fairly large shrubs occur, such as *Banksia*

FIG. 12 *Eucalyptus kybeanensis*: A, sucker leaves (actual size); B, mature leaves (actual size); C, fruits (× $1\frac{1}{2}$); D, buds (× $1\frac{1}{2}$).

*integrifolia*, *Grevillea victoriae*, *Eriostemon myoporoides*, *Phebalium ozothamnoides*, *Callistemon pallidus*, *Correa lawrenciana*, and *Prostanthera phylicifolia*. A few epacrids, including *Leucopogon* and *Epacris*, also occur, and a few scrambling plants including *Boronia algida* and *Kunzea muelleri*. A few kinds of grasses occupy the open spaces. *Calytrix tetragona*, common at lower altitudes, also occurs.

In Tasmania the wet mallee carries such species as *Eucalyptus aggregata*, *E. ovata* (both described under the heading of Savannah Woodland), *E. perriniana* (see heading Subalpine Woodland), and *E. gunnii* (see heading Wet Sclerophyll). *E. aggregata* reaches elevations of 3,000 feet, *E. ovata* 4,000 feet, and *E. gunnii* elevations of 2,000 to 3,000 feet in poorly drained woodland.

**Savannah Woodland**

On poorly drained soils in the cooler parts of the montane habitat on the mainland a savannah woodland vegetation carrying a number of taller eucalypts covers quite extensive areas. These include *Eucalyptus aggregata*, *E. albens*, *E. bridgesiana*, *E. camphora*, *E. cypellocarpa* (syn. *E. goniocalyx*), *E. dalrympleana*, *E. dives*, *E. ovata*, *E. parvifolia*, *E. pauciflora*, *E. perriniana*, *E. robertsonii*, *E. rubida*, *E. stellulata* and *E. viminalis*. *E. pauciflora*, *E. perriniana* and *E. stellulata* have already been described under the heading Subalpine Woodland.

Manna gum, *E. viminalis* (Fig. 14), is a widespread and handsome tree with ribbons of bark hanging commonly from the limbs and with a smooth white bole. It is easy to confuse this with candlebark, *E. rubida* (Fig. 15), with which it is commonly associated in the dry schlerophyll forest. Manna gum is somewhat more spreading and pendulous, however, and is more common in the valleys though it occurs also in association with candlebark on more elevated land. The stems of candlebark tend to

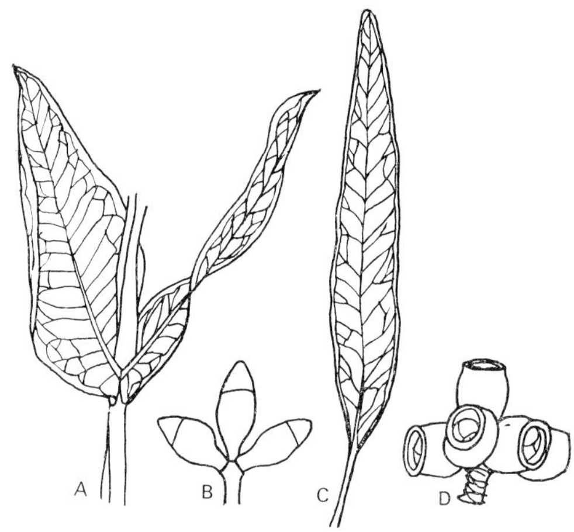


FIG. 15 *Eucalyptus rubida*: A, sucker leaves; B, buds; C, mature leaf; D, fruits. (All actual size except C × ½.)

become more colourful, with streaks and blotches of red and orange, though this is not a definite distinction. The two, however, are clearly distinguished by the sucker leaves which

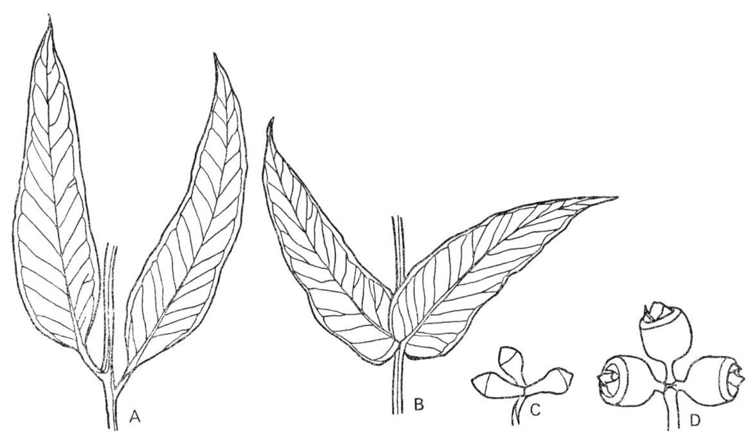


FIG. 14 *Eucalyptus viminalis*: A, mature leaves (× ½); B, sucker leaves (× ½); C, buds (actual size); D, fruits (actual size).

in manna gum are without stalks, opposite, narrow, pale-green; in candlebark they are also stalkless and opposite, but round and grey-green. The flowers of both species are usually in groups of three. The fruit of manna gum is bright-green, stalkless or almost stalkless, spherical, 5 to 6 millimetres in diameter, with protruding valves; that of candlebark is similar but usually grey-green.

Mountain gum, *E. dalrympleana* (Fig. 16), is also very easily confused with these last two species, particularly with candlebark, and it may be that there is no real distinction between the two. Mountain gum is a magnificent tree, occurring frequently in this formation in large stands and often showing rich colourings in reds, pinks, orange and green on the smooth bark. A general description of the characteristics of candlebark applies to this species in relation to both juvenile and mature leaves and flowers. There is a slight difference in the fruit and buds. The buds of candlebark are grey-green, those of mountain gum bright-green; the fruit of candlebark is sometimes slightly stalked and usually grey-green, that of mountain gum is stalkless and not grey-green.

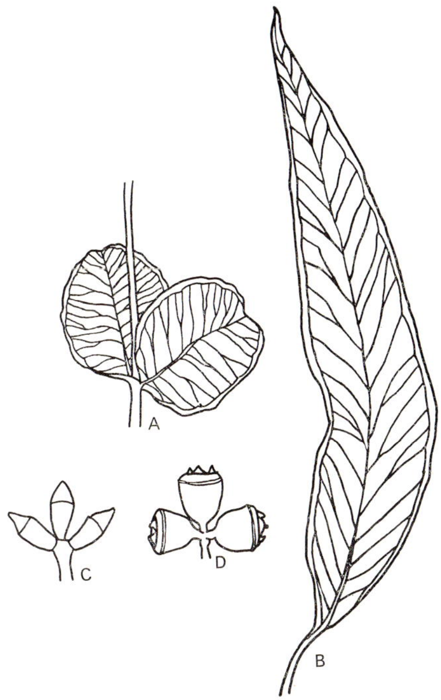


FIG. 16 *Eucalyptus dalrympleana*: A, sucker leaves (× ½); B, mature leaf (× ½); C, buds (actual size); D, fruits (actual size).

All three species occur also in Tasmania, *E. viminalis* from sea-level to 1,500 feet, usually in moist sheltered valleys; *E. rubida* from 500 to 2,000 feet, replacing *E. viminalis* in the colder, drier areas; and *E. dalrympleana* between 1,500 and 3,000 feet on well-drained soils in the wetter areas particularly.

Black gum or woollybutt, *E. aggregata* (Fig. 17), is a spreading tree, often tall, with a dark-coloured, flaky bark. It grows commonly along watercourses and on alluvial flats. The sucker leaves are opposite, rounded and shortly stalked; the mature leaves are alternate, stalked, rather narrow, wavy in outline, 7 to 12 centimetres in diameter. The fruit is shortly stalked, nearly globular, about 4 millimetres in diameter, with protruding valves.

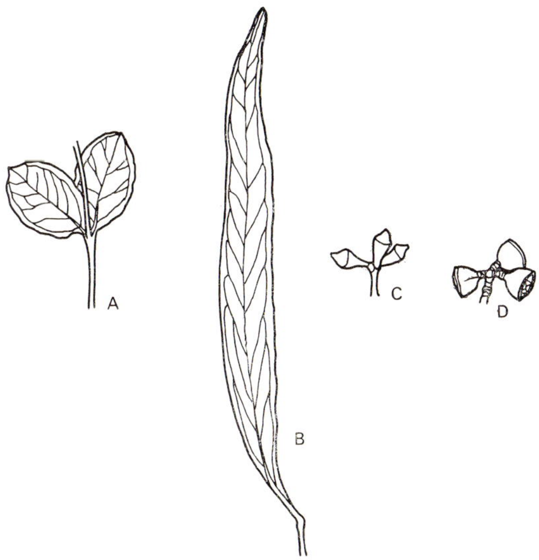


FIG. 17 *Eucalyptus aggregata*: A, sucker leaves (× ½); B, mature leaf (× ½); C, buds (actual size); D, fruits (actual size).

White box, *E. albens* (Fig. 18), is a very handsome twisted tree with a mealy-white appearance, the bark finely fibrous on the trunk, the limbs smooth. The sucker leaves are opposite, shortly stalked, grey-green, broad, 4 to 7 centimetres long, the mature leaves are alternate,

FIG. 18 *Eucalyptus albens*: A, sucker leaves; B, mature leaf; C, buds; D, fruits. (All actual size except B $\times \frac{1}{2}$.)

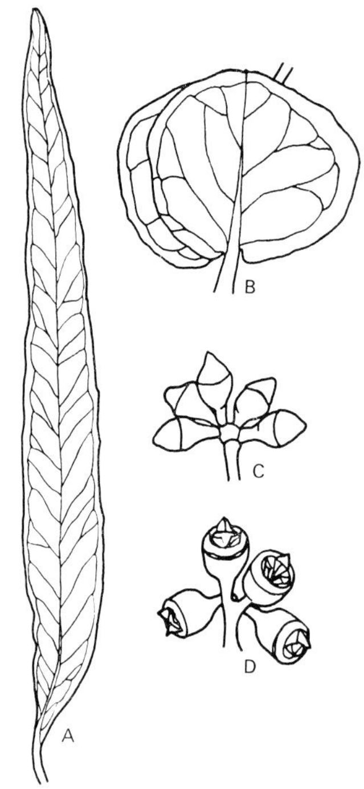


FIG. 19 *Eucalyptus bridgesiana*: A, mature leaf ($\times \frac{1}{2}$); B, sucker leaf ($\times \frac{1}{2}$); C, buds (actual size); D, fruits (actual size).

stalkless, broad, leathery, glaucous, 10 to 15 centimetres long. The fruit is usually without a stalk, three to seven in a group, grey-green, rather barrel-shaped, 9 to 12 millimetres in diameter, the valves deeply enclosed.

But-but or apple, *E. bridgesiana* (Fig. 19), is a fairly tall tree with a spreading crown and pendulous branches with soft whitish bark almost to the tips. The sucker leaves are opposite, without stalks, stem-clasping, very grey-green, rounded, 5 to 6 centimetres long, the mature leaves narrow, stalked, wavy, pale-green, 12 to 25 centimetres long. The fruit is stalked, or rarely stalkless, in groups of four to seven, cone-shaped, 5 to 6 millimetres in diameter.

Broad-leafed sallee, *E. camphora* (Fig. 20), occurs fairly commonly along watercourses and on damp flats in this formation. It is a crooked little tree with a ragged, flaky bark for considerable distances up the stem, but the branches are smooth and dark. The sucker leaves are oppos-

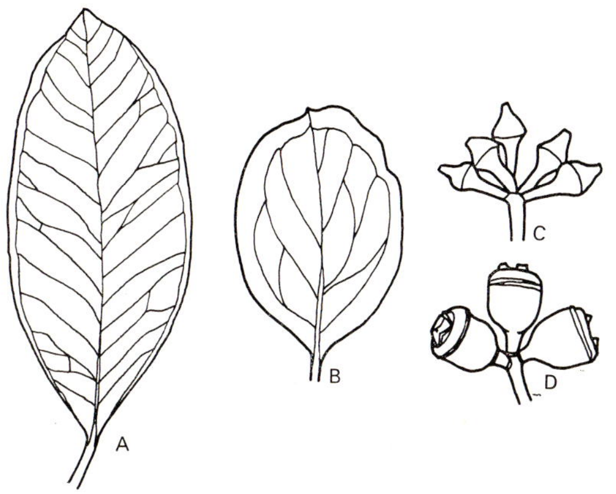


FIG. 20 *Eucalyptus camphora*: A, mature leaf (× ½); B, sucker leaf (× ½); C, buds (actual size); D, fruits (actual size).

ite, rounded, on short stalks; the mature leaves are alternate, rather broad, wavy, 8 to 14 centimetres long. The fruit is stalked, turban-shaped, 6 to 8 millimetres in diameter, the valves protruding.

Spotted mountain grey gum, *E. cypellocarpa* (syn. *E. goniocalyx*)—Fig. 21, is a medium-sized tree at these altitudes with a smooth, mottled stem, often with flaky bark on the butt. The sucker leaves are opposite, stalkless, broad, rounded at the base, 6 to 10 centimetres long and covered with a greyish bloom; the mature leaves are alternate, stalked, narrow, tapering, dark glossy-green, thick, 12 to 25 centimetres long. The fruit is shortly stalked, rounded or pear-shaped, in groups of about seven, 8 to 10 millimetres in diameter, the valves slightly protruding.

Broad-leafed peppermint, *E. dives* (Fig. 22), is a medium-sized spreading tree with rough bark on the trunk and branches. The sucker leaves are opposite, without stalks, stem-clasping, broad, leathery, very grey-green, strongly aromatic, 3 to 10 centimetres long; the mature leaves are alternate, stalked, broad, thick, 10 to 15 centimetres long. The fruit is stalked, in groups of six to ten, domed, 6 to 8 millimetres in diameter, the valves enclosed or slightly exserted.

FIG. 21 *Eucalyptus cypellocarpa* (syn. *E. goniocalyx*): A, sucker leaves (actual size); B, mature leaf (actual size); C, buds (× ½); D, fruits (× ½).

FIG. 22 *Eucalyptus dives*: A, sucker leaves; B, mature leaf; C, fruits; D, buds. (All actual size except B × ½).

Swamp gum, *E. ovata* (Fig. 23), is a fairly tall tree with rough bark at the butt, smooth and ribbony elsewhere. The sucker leaves are opposite, egg-shaped, wavy, shortly stalked, 4 to 8 centimetres long; the mature leaves are egg-shaped, leathery, often very glossy, 9 to 12 centimetres long. The fruit is usually stalked, in groups of four to eight, hemispherical, 5 to 9 millimetres in diameter, the valves enclosed or shortly protruding. This species occurs also in Tasmania from sea-level to 2,000 feet, often on badly drained soils.

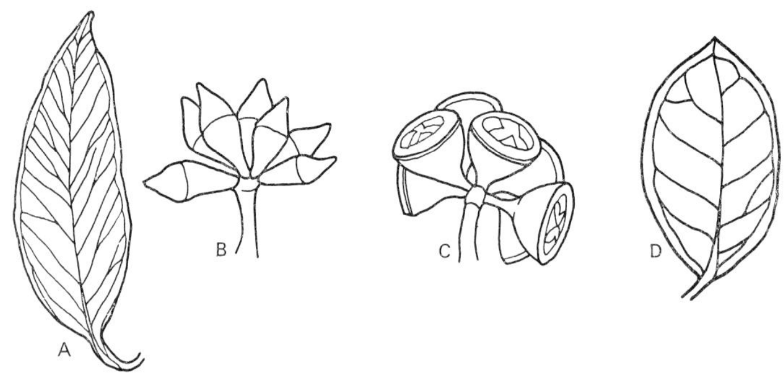


FIG. 23 *Eucalyptus ovata*: A, mature leaf (× ½); B, buds (actual size); C, fruits (actual size); D, sucker leaf (× ½).

Small-leafed gum, *E. parvifolia* (Fig. 24), is a small spreading tree with a smooth dull-grey bark, limited in distribution to an area near Nimmitabel in southern New South Wales and a few scattered patches in similar habitats. The sucker leaves are opposite, stalkless, grey-green,

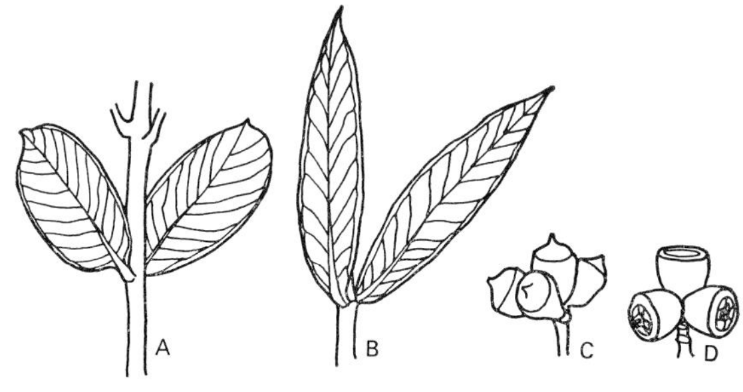


FIG. 24 *Eucalyptus parvifolia*: A, sucker leaves (× ½); B, mature leaves (× ½); C, buds (actual size); D, fruits (actual size).

rounded, paler on the underside; the mature leaves are alternate, opposite, 6.5 centimetres long. The flowers are in groups of four to seven. The fruit is stalkless, globular, about 5 millimetres in diameter, the valves very small, sometimes enclosed, sometimes protruding.

Fine-leafed peppermint, *E. robertsonii*, is a handsome straight tree sometimes reaching 50 metres, with a fine-textured but rough bark. The sucker leaves are opposite, stalkless, rather

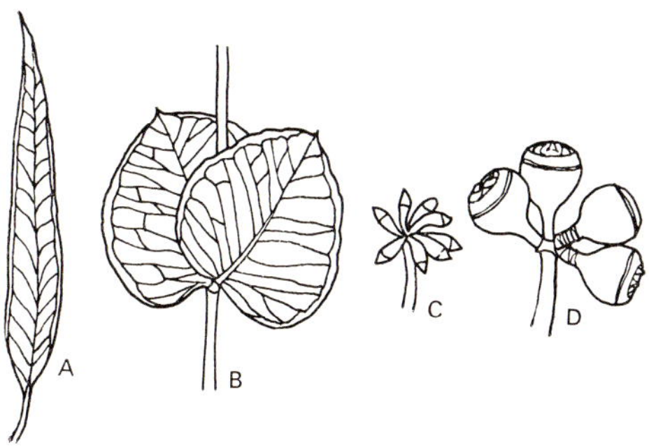


FIG. 25 *Eucalyptus robertsonii*: A, mature leaf (× ½); B, sucker leaves (actual size); C, buds (× ½); D, fruits (actual size).

narrow; the mature leaves are alternate, narrow to medium broad, finely veined, 7 to 17 centimetres long. The flowers are in groups of nine to twenty-one; the fruits are stalked, pear-shaped, about 5 millimetres in diameter, the valves enclosed.

Savannah woodland has a wide range of species in the understorey. It is rich in grasses and sedges and often presents a parklike appearance, with a wide sweep of grassland. Among the grasses are a considerable variety of small plants including many orchids, a number of small pea plants such as *Bossiaea riparia, Dillwynia prostrata, Glycine clandestina, Gompholobium huegelii, Hardenbergia violacea, Hovea heterophylla, Lotus australis, L. corniculatus, Oxylobium alpestre, O. ellipticum, O. procumbens, Psoralea adscendens and Swainsona oroboides.* Many small daisies are also common.

Other smaller plants found particularly in the grasslands of the area include bulbine lily (*Bulbine bulbosa*), fringed violet (*Thysanotus tuberosus*), *Drymophila cyanocarpa, Stypandra unbellata*, and several ground orchids. Other small plants include the ubiquitous *Stellaria pungens*, several cardamines, *Acaena anserinifolia, Geranium antrorsum,* the lovely little blue *Linum marginale, Tetratheca glandulosa, T. thymifolia, Poranthera microphylla, Hibbertia obtusifolia,* several small pimeleas, *Trachymene humilis*, and some species of bluebell (*Wahlenbergia*). Permanently wet areas house a variety of other small plants including *Drosera arcturi, D. peltata, Scleranthus biflorus, Ranunculus pimpinellifolius,* several kinds of native violets and the bladderwort, *Utricularia dichotoma.*

The area is rich in small leucopogons, including *Leucopogon hookeri, L. fraseri, L. lanceolatus,* and *L. microphyllus.*

Among the larger plants are many proteaceous kinds including *Banksia integrifolia, B. marginata, Grevillea lanigera, G. victoriae, Hakea dactyloides, H. eriantha, Lomatia fraseri, L. myricoides* and *Telopea oreades.*

The dwarf she-oak, *Casuarina nana*, occurs in this formation, and several species of *Leptospermum*, including *L. attenuatus, L. myrsinoides*, and *L. phylicoides*, are also present.

*Pomaderris elliptica* and *Coprosma quadrifida* are commonly met and, where moisture is more abundant, *Prostanthera lasianthos* and *P. walteri.*

Among the wattles here are blackwood (*Acacia melanoxylon*), sometimes associated with snow gum in this formation, silver wattle (*A. dealbata*), and *A. rubida* and *A. siculiformis.* It is one of the richest communities.

In Tasmania the savannah woodland is a somewhat indistinctly defined region of the central plateau and the north-east occurring at altitudes as low as 1,000 feet and as high as 2,500 feet and, rarely, to 3,000 feet. Winter snows occur occasionally but never last very long.

At the lower level *Eucalyptus pauciflora* is the dominant tree—and it is not characteristic of colder and higher areas as it is on the mainland. Among the undergrowth a few tall species occur, including *Bursaria spinosa, Exocarpos cupressiformis* and occasional specimens of *Banksia marginata* and *Dodonaea viscosa.* The smaller undershrub vegetation is sparse, consisting largely of heaths such as *Acrotriche serrulata*, *Epacris impressa*, *Astroloma humifusum;* daisies such as *Olearia stellulata*, *O. erubescens*, and occasionally *O. algida*; and a peaflower, *Lotus corniculatus*.

Frequently the open patches are gay with clumps of the dainty flag iris, *Diplarrena moroea*, along with scattered plants of *Hibbertia serpyllifolia*.

Where the savannah woodland extends to higher elevations in Tasmania an entirely different association occurs. Frosts are often severe here and snow may lie for many days. As this zone lies in the rain shadow of the nearby mountains, the rainfall is low. The chief trees are mountain ash (*Eucalyptus delegatensis*), yellow gum (*E. johnstonii*) and sometimes the low range form of Tasmanian snow gum (*E. coccifera*). The understorey is never very dense and includes such species as mountain pepper (*Tasmannia aromatica*), *Olearia algida, Helichrysum purpurascens*, *Pimelea drupacea* and, occasionally, patches of woolly tea-tree, *Leptospermum lanigerum*.

A similar vegetation zone is to be found in Tasmania in many parts of the central plateau at elevations of 1,000 to 2,000 feet. Here, particularly on rocky dolerite outcrops, stands of *E. delegatensis* and *E. coccifera* occupy considerable stretches, often to the exclusion of almost all other plants. In other parts the trees are more widely scattered and a variety of undershrubs, including several of those which occur also on the mainland, are associated with them.

### Wet Sclerophyll Forest

Wet sclerophyll forests are widely distributed in the montane tract, mainly in the cool, humid, less exposed parts. They occur on the slopes of Mount Kosciusko and of the Victorian Alps, and in the Australian Capital Territory and Tasmania. There is usually some snow in winter but it is ephemeral, disappearing within a day or two and only occurring at most over 2 to 3 weeks. Frosts are relatively light. A wide range of eucalypts occurs here, often in different associations. Among these are *Eucalyptus st. johnii*, *E. bridgesiana*, *E. chapmaniana*, *E. dalrympleana*, *E. delegatensis*, *E. dives*, *E. fastigata*, *E. fraxinoides*, *E. glaucescens*, *E. globoidea*, *E. macrorrhyncha*, *E. nitens*, *E. nortonii*, *E. obliqua*, *E. pauciflora*, *E. radiata*, *E. sieberi* and *E. viminalis*.

Of these *E. pauciflora*, *E. bridgesiana*, *E. dalrympleana*, *E. dives*, and *E. viminalis* have already been described (see Subalpine Woodland, Savannah Woodland).

One of the most interesting associations is that of mountain ash, *E. delegatensis*, and mountain gum, *E. dalrympleana*, which occur in the cooler, moister parts and form dense forests. These mountain-ash forests are majestic and quite different from any other forests in the area. With rough ragged bark on their butts, and smooth trunks, the mountain ashes tower against the sky. Sometimes they occur in pure stands, at other times the white stems of mountain gum are mixed with them. Unlike most eucalypts, mountain ash does not form a lignotuber and for this reason is more vulnerable to fire, not having the recuperative powers of the lignotuber species.

In the valleys occupied by these giant gums an association rather similar to the coastal subtropical rainforests prevails, with broad-leafed tall shrubs and trees and the giant tree fern, *Dicksonia antarctica*, as well as many smaller ferns. Here will be found the Southern sassafras, *Atherosperma moschatum*, with its gentle fragrance, and the large-leafed, tall composite *Bedfordia salicina* with its masses of yellow blossom, another large-leafed composite with white flowers, *Olearia argophylla*, two species of *Lomatia* (*L. myrsinoides* and *L. fraseri*) and, representing the wattles, the blackwood (*Acacia melanoxylon*) and silver wattle (*A. dealbata*). *Elaeocarpus holopetalus*, a characteristic rainforest type, also occurs. Composites—including some brachycomes, *Cassinia aculeata*, and *C. longifolia*, billy buttons (*Craspedia uniflora*), and several kinds of helichrysum, among them some paper daisies—are scattered throughout this area. Other composites, including helipterums, native dandelion (*Microseris scapigera*), a wide range of olearias and several species of *Senecio* are a significant feature of the vegetation.

Eucalypts listed for this area and not already described are detailed below.

Eurabbie, *Eucalyptus st. johnii* (syn. *E. bicostata*) (Fig. 26), is a majestic tree with rough bark low on the trunk, the upper parts and the limbs gleaming white. The sucker leaves are stalkless, stem-clasping, grey-green or almost mealy white, very large and broad, 4 to 14 centimetres long or even longer; the mature leaves are alternate, stalked, dark glossy-green, fairly narrow, falcate, 10 to 25 centimetres long or longer. The fruit is without a stalk, in groups of three, grey-green, turban-shaped or round, 2-ribbed, often slightly warty, 14 to 17 millimetres in diameter, the values protruding but sometimes concealed by the broad disc. This species occurs also on Flinders Island, off Tasmania, but not in a similar habitat.

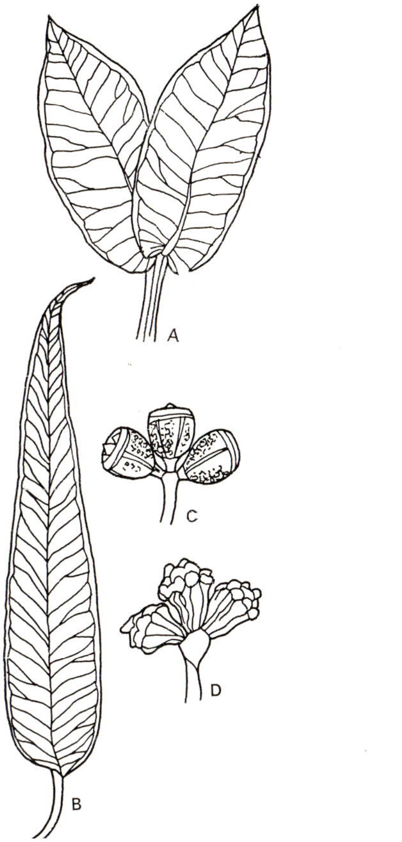


FIG. 26 *Eucalyptus st. johnii* (syn. *E. bicostata*): A, sucker leaves; B, mature leaf; C, fruits; D, buds. (All × ½ except B × ¼.)

Bogong gum, *E. chapmaniana* (Fig. 27), grows to a height of 25 to 30 metres with a rough, grey, finely fibred bark on the trunk, decorticating in ribbons from the smooth limbs. It occurs in Victoria. The sucker leaves are opposite, stalkless or with very short stalks, 15 to 20 millimetres long; the mature leaves are stalked, slightly grey-green, alternate, fairly narrow, 14 to 26 centimetres long. The buds are without stalks, grey-green. The fruit is grey-green, bell-shaped or egg-shaped, 11 to 13 millimetres long, the valves protruding.

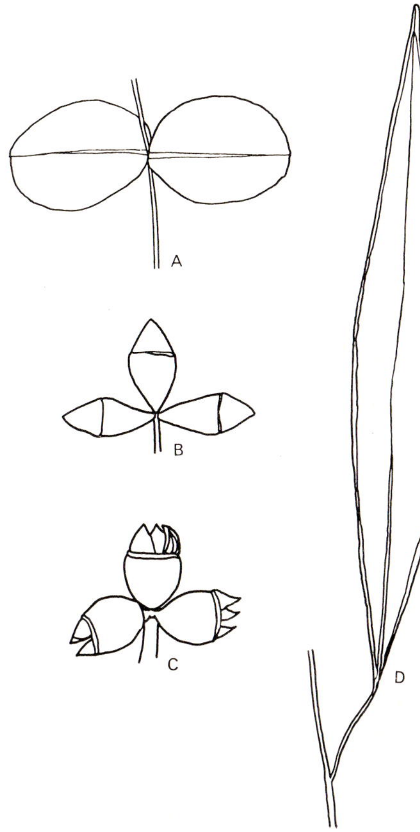


FIG. 27 *Eucalyptus chapmaniana*: A, sucker leaves; B, buds; C, fruits; D, mature leaf. (All × ½.)

Alpine ash, *E. delegatensis* (Fig. 28), a majestic tree, is one of the distinctive features of the montane wet sclerophyll in valleys at 3,000 to 4,000 feet on the mainland, and a valuable

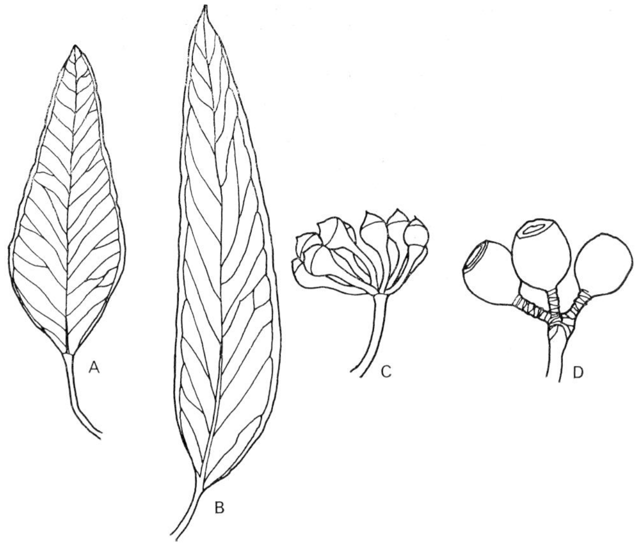


FIG. 28 *Eucalyptus delegatensis*: A, sucker leaf (× ½); B, mature leaf (× ½); C, buds (actual size); D, fruits (actual size).

timber tree. In Tasmania, where it is known as gum-topped stringybark or whitetop, it is abundant from 1,000 to 3,500 feet where is forms extensive pure stands and is an important source of timber, being marketed as Tasmanian oak (as is also *E. regnans* in that State). It is a tall, straight tree with a rough, fibrous bark extending for a considerable distance up the main stem and decorticating in ribbons. The upper part of the stem and the branches are gleaming white. The sucker leaves are opposite, stalked, thick, broad and greyish-green; the mature leaves are broad, leathery, strongly and pleasantly aromatic, 8 to 16 centimetres long. The flowers are in groups of seven to fifteen. The fruits are stalked, pear-shaped, 10 to 12 millimetres diameter, the valves enclosed.

Cut-tail, *E. fastigata* (Fig. 29), is a tall tree with a stringy bark on the trunk and branches. The sucker leaves are opposite, stalked, broad, distantly toothed, 3 to 6 centimetres long; the mature leaves are alternate, stalked, thin, fairly broad, 10 to 18 centimetres long. The fruit is stalked, seven to twelve in a group, pear-shaped or turban-shaped, 6 to 7 millimetres in diameter, the valves slightly protruding. This species can persist in the more exposed parts as well as in the sheltered gullies.

White mountain ash, *E. fraxinoides*, is a tall tree with a gleaming white stem conspicuously marked with insect scribbles and with rough bark at the butt. The sucker leaves are opposite,

FIG. 29 *Eucalyptus fastigata:* A, sucker leaf (× 2); B, mature leaf (× ½); C, buds (actual size); D, fruits (actual size).

Rocky outcrops, a feature of the Australian Alps, add to the picturesqueness of the scenery. Here *Aciphylla glacialis* shelters beside a granite boulder on the slopes of Lake Albina, Kosciusko

TOP: The gentle slopes of alpine mountains, showing evidence of glaciation and deeply dissected by streams

BOTTOM: The ranges of the Snowy Mountains, showing the flat surface of the raised peneplain

TOP: Billy buttons, *Craspedia uniflora,* flowering valiantly in summer snows

BOTTOM: Tussocky growth of *Helipterum albicans* is an adaptation against severe winds

TOP: Tall alpine herbfield with snowgrass and a variety of shrubs

BOTTOM: Short alpine herbfield on an exposed site, showing a stream bordered by *Brachycome stolonifera*

TOP: An alpine stream with marsh marigold, *Caltha introloba*, held against the swiftly running water by boulders

BOTTOM: *Caltha introloba* in flower

TOP: Fjaeldmark, windswept areas of the high tops characterized by mosses and lichens with low-growing shrubs in the depressions

BOTTOM LEFT: Cushion bushes of Tasmania, *Pterygopappus lawrencei,* intermingled with *Abrotanella forsterioides*

BOTTOM RIGHT: *Dracophyllum minimum,* another Tasmanian cushion bush. Cushion bushes are characteristic of Tasmanian fjaeldmark or fell field

TOP: Subalpine heathland with *Bossiaea foliosa* in the foreground and mixed shrubs behind

CENTRE: *Acaena anserinifolia*, a small scrambling plant common on heathlands

BOTTOM: *Parahebe perfoliata*, with brilliant blue flowers, conspicuous on the heathlands

Black sallee, *Eucalyptus stellulata,* a small, compact tree of the subalpine woodlands, has smooth, satiny, olive green bark with brown blotches

Spinning top or whistling gum, *Eucalyptus perriniana,* of subalpine and montane areas of Tasmania and New South Wales

TOP: Temperate rainforest of Tasmania with *Richea pandani-folia* in the foreground

BOTTOM: Myrtle or beech *Nothofagus cunninghamii* occurring in Tasmanian and mainland rainforest

P: A montane swamp with *ptospermum lanigerum* and *Baeckea gunniana*

TTOM: Jindabyne Valley (now owned), a typical montane :ubland, with scattered trees *Eucalyptus rubida* and *E. viminalis*

*Eucalyptus bridgesiana,* a picturesque, rough barked tree common in exposed situations on savannah woodland

Mountain ash, *Eucalyptus delegatensis,* a valuable timber tree of wet sclerophyll habitats

Silvertop, *Eucalyptus nitens,* a magnificent towering tree of the wet sclerophyll forests

*Dicksonia antarctica*, a common tree fern of the wet sclerophyll valleys

Giant grass-trees, *Xanthorrhoea australis,* are conspicuous among the eucalypts of the dry sclerophyll forests

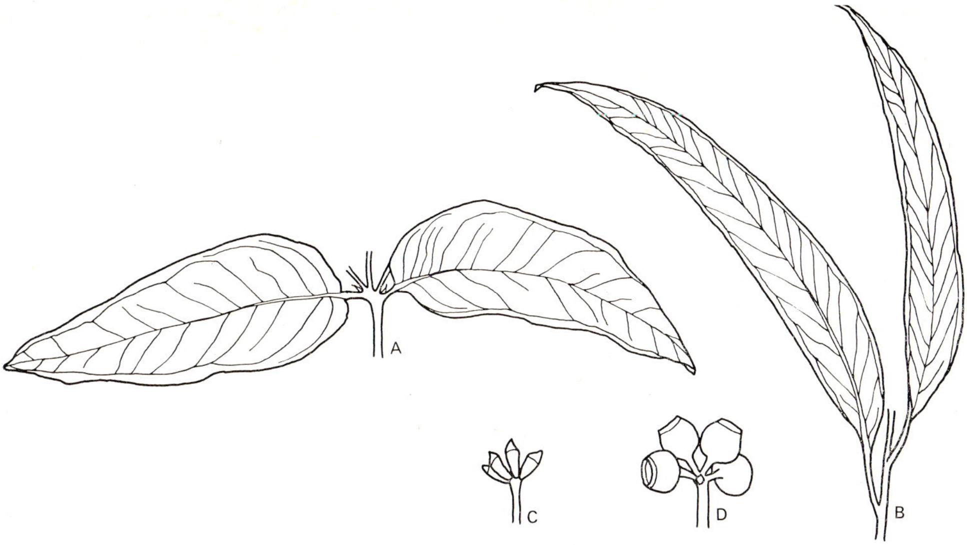


FIG. 30 *Eucalyptus fraxinoides*: A, sucker leaves (× ½); B, mature leaves (× ½); C, buds (slightly reduced); D, fruits (slightly reduced).

shortly stalked, narrow, slightly grey-green, 4 to 10 centimetres long; the mature leaves are alternate, stalked, narrow, rather thick, 8 to 15 centimetres long. The fruit is in groups of five to eight, stalked, more or less globular, 9 to 10 millimetres in diameter.

Stringybark, *E. macrorrhyncha* (Fig. 31), is a small to medium sized tree with persistent stringy bark on the stem and the larger branches. The sucker leaves are opposite, shortly stalked, rounded or broad, more or less toothed, 3 to 5 centimetres long; the mature leaves are alternate, petiolate, rather narrow, 7 to 12 centimetres long. The fruit is stalked, six to twelve in a group, globular, 7 to 10 millimetres in diameter, the valves protruding prominently. This species

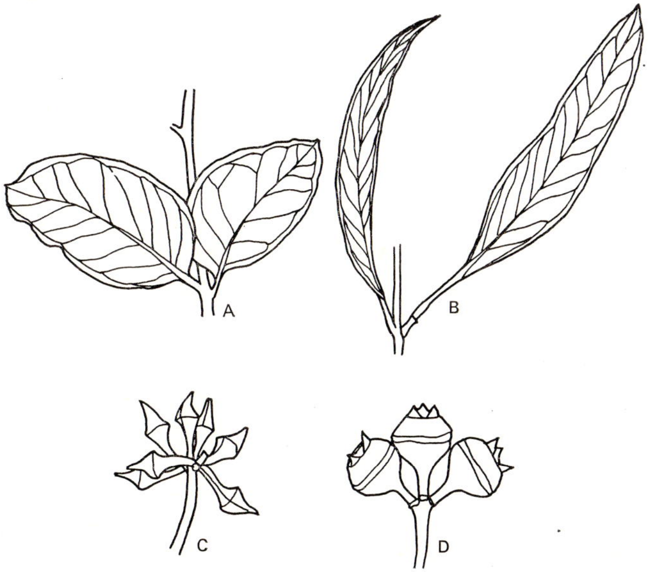


FIG. 31 *Eucalyptus macrorrhyncha*: A, sucker leaves (× ½); B, mature leaves (× ½); C, buds (actual size); D, fruits (actual size).

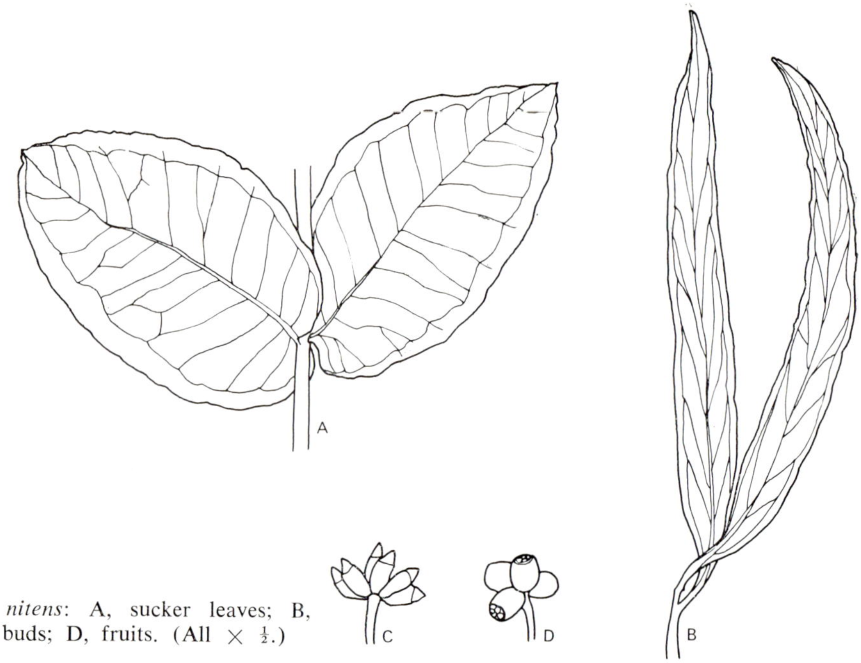


FIG. 32 *Eucalyptus nitens*: A, sucker leaves; B, mature leaves; C, buds; D, fruits. (All $\times \frac{1}{2}$.)

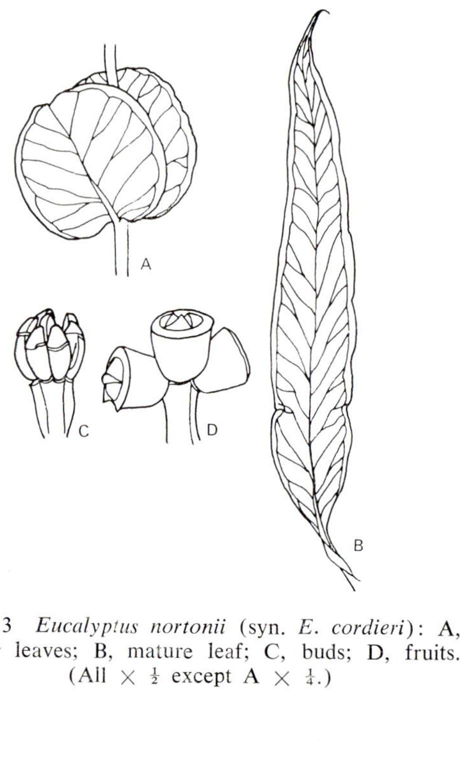


FIG. 33 *Eucalyptus nortonii* (syn. *E. cordieri*): A, sucker leaves; B, mature leaf; C, buds; D, fruits. (All $\times \frac{1}{2}$ except A $\times \frac{1}{4}$.)

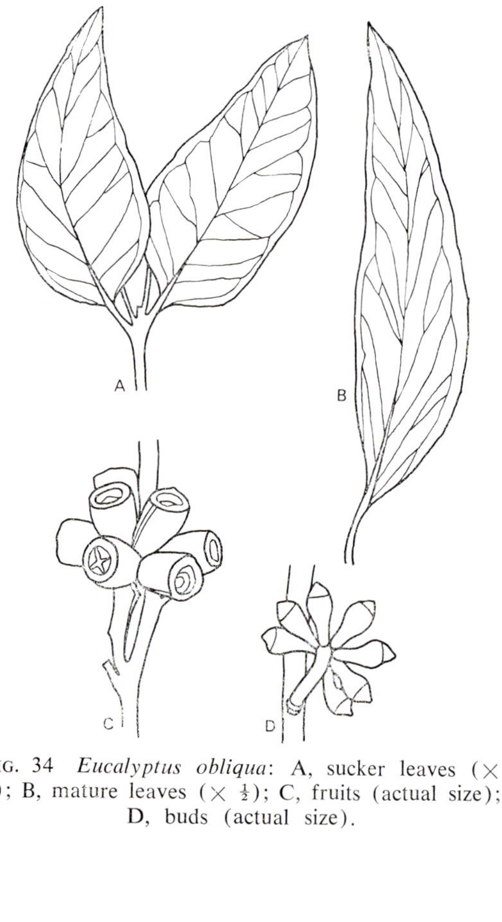


FIG. 34 *Eucalyptus obliqua*: A, sucker leaves ($\times \frac{1}{2}$); B, mature leaves ($\times \frac{1}{2}$); C, fruits (actual size); D, buds (actual size).

occurs in exposed places as well as in the more sheltered areas.

Silvertop, *E. nitens* (Fig. 32), is a majestic tall, straight tree with smooth, shining bark, often with ribbons hanging on the branches and more or less rough at the butt. The sucker leaves are opposite, stalkless, often stem-clasping, grey-green, broad, 7 to 10 centimetres long; the mature leaves are alternate, stalked, smooth and shining, rather narrow, remotely toothed, 10 to 30 centimetres long. The fruit is without a stalk, in groups of four to eight, shining, egg-shaped, about 6 millimetres in diameter, the valves very small, enclosed.

Mealy bundy, *E. nortonii* (syn. *E. cordieri*) —Fig. 33, is a tall tree with a rough, dark bark persistent to the small branches. The sucker leaves are opposite, stalkless, rounded or heart-shaped, 8 to 10 centimetres long; the mature leaves are alternate, stalked, thick, narrow, up to 21 centimetres long. The fruit is hemispherial or egg-shaped, in groups of three to seven, 8 to 10 millimetres in diameter, the valves enclosed.

Messmate, *E. obliqua* (Fig. 34) is a large tree with deeply furrowed bark persistent even to the small branches. The sucker leaves are opposite, stalked, broad, slightly toothed, 6 to 8 centi-

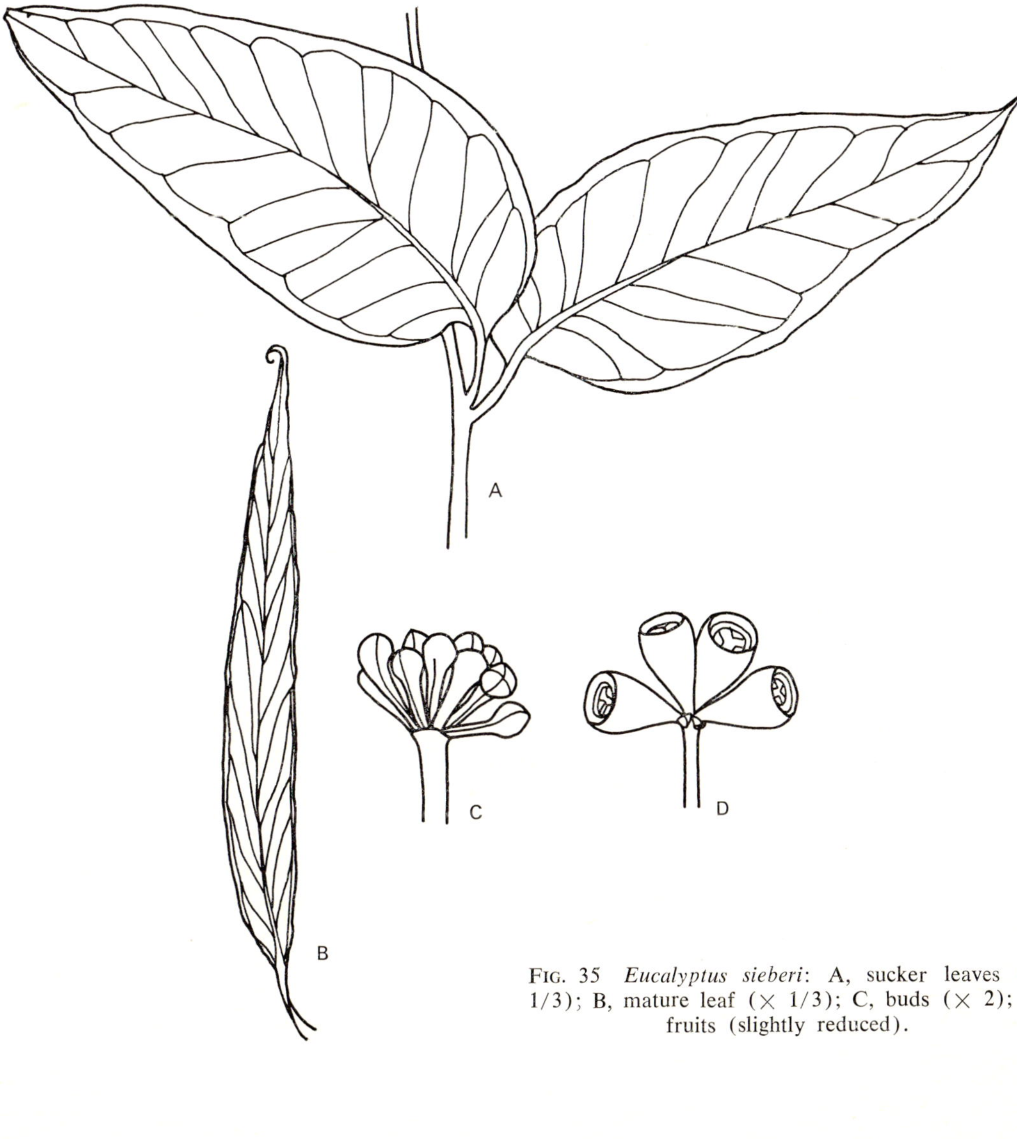


FIG. 35 *Eucalyptus sieberi*: A, sucker leaves (× 1/3); B, mature leaf (× 1/3); C, buds (× 2); D, fruits (slightly reduced).

metres long; the mature leaves are stalked, alternate, fairly broad, oblique, 10 to 16 centimetres long. The fruit is stalked, seven to fifteen in a group, pear-shaped, 8 to 9 millimetres in diameter, the valves deeply enclosed. This species occurs also in Tasmania from sea-level to 2,000 feet, often in exposed situations, but reaching its best development in sheltered situations.

Mountain ash, *E. sieberi* (Fig. 35), is a medium-sized tree with deeply furrowed bark on the stem but with smooth branches. The sucker leaves are opposite, stalkless or slightly stalked, elliptical, thick, slightly grey-green, 8 to 12 centimetres long; the mature leaves are alternate, stalked, 9 to 18 centimetres long. The fruit is stalked, in groups of 5 to 15, pear-shaped, 6 to 11 millimetres in diameter, the valves enclosed. This species reaches its best development in the sheltered gullies but occurs also in exposed situations.

below, 6 to 12 centimetres long. The fruit is shortly stalked, in groups of five to twelve, egg-shaped, rather thick, sometimes rough, 4 to 7 millimetres in diameter, the valves small, enclosed. This species occurs in the somewhat more exposed parts but sometimes descends into the more sheltered valleys.

As might be expected in a milder environment there is a wide range of plants in the understorey, ranging from fairly tall plants including a few of the rainforest type such as *Bedfordia salicina* and *Elaeocarpus holopetalus* but with a much wider range tending towards dry sclerophyllous types on the one hand and wet mallee on the other. Of the taller plants many acacias occur, including *Acacia buxifolia*, *A. dealbata*, *A. kettlewelliae*, *A. pravissima*, *A. rubida*, *A. siculiformis* and *A. vomeriformis*. Several taller shrubby pea flowers are common, including *Daviesia corymbosa*, *D. ulicina*, *Goodia lotifolia*, *Mirbelia oxyloboides*, *Indigofera australis*,

FIG. 36 *Eucalyptus globoidea*: A, intermediate leaves (× ½); B, mature leaves (actual size); C, buds (actual size); D, fruits (× ½).

*E. globoidea* (Fig. 36), is a medium-sized tree with a fibrous bark persistent to the branches. The sucker leaves are opposite, elliptical, softly hairy when very young, paler underneath, fragrant, 3.5 to 5.5 centimetres long; the mature leaves are alternate, stalked, fairly broad, often oblique, thin, shining on top, paler

*Oxylobium trilobatum*, *Bossiaea buxifolia*, *B. foliosa*, and, among the smaller pea-flowers, *Lotus corniculatus* and *L. australis*.

*Eucalyptus viminalis* and *E. rubida* also occur in Tasmania in similar situations.

Some additional species are also found in Tasmania in this habitat which do not occur on the mainland. These include *E. gunnii* and *E. cordata* as well as *E. johnsonii*, *E. pulchella*, *E. regnans*, *E. nitida* and *E. tasmanica*.

Silver gum, *E. cordata* (Fig. 37), is a most handsome small tree or dense shrub, silvery-grey all over with a smooth, white bark blotched

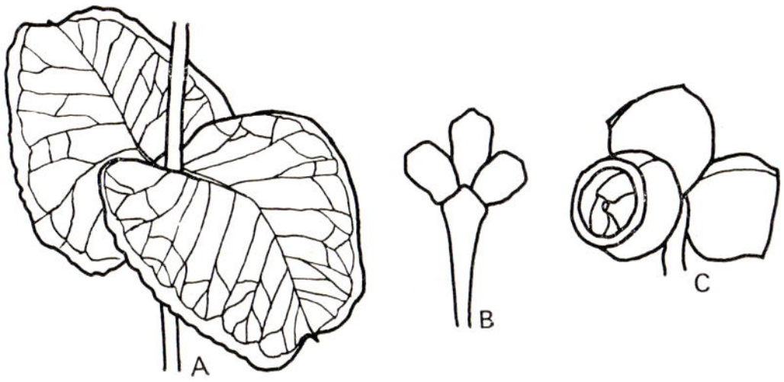


FIG. 37 *Eucalyptus cordata*: A, mature leaf ($\times \frac{1}{2}$); B, buds ($\times \frac{1}{2}$); C, fruit (actual size).

with green and purple. Sucker and mature leaves are similar, opposite, stalkless, grey-green, round, blunt, stem-clasping, wavy, 4 to 9 centimetres long. The fruit is grey-green, in groups of three, round or turban-shaped, 10 to 12 millimetres in diameter, the valves deeply enclosed.

Cider gum, *E. gunnii* (Fig. 38), is an endemic Tasmanian occurring in stands in marshy situations on the central plateau at 2,000 to 3,500

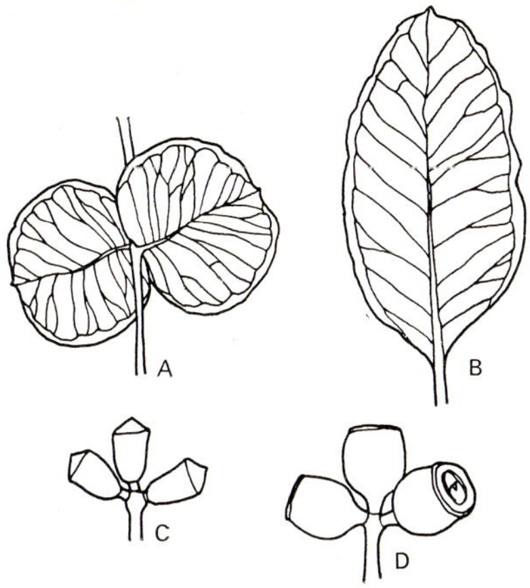


FIG. 38 *Eucalyptus gunnii*: A, sucker leaves ($\times \frac{1}{2}$); B, mature leaf ($\times \frac{1}{2}$); C, buds (actual size); D, fruits (actual size).

feet. It is a fairly tall tree with smooth limbs and upper trunk but scaly and persistent at the butt. The limbs are often coloured green or orange. The sucker leaves are round, opposite, stalkless, 2.5 to 4.5 centimetres long; the mature leaves are 4 to 8 centimetres long, fairly broad, stalked and pointed. The fruit is round or bell-shaped, usually stalkless, in groups of three, often grey-green, 6 to 9 millimetres in diameter, the valves enclosed or slightly protruding.

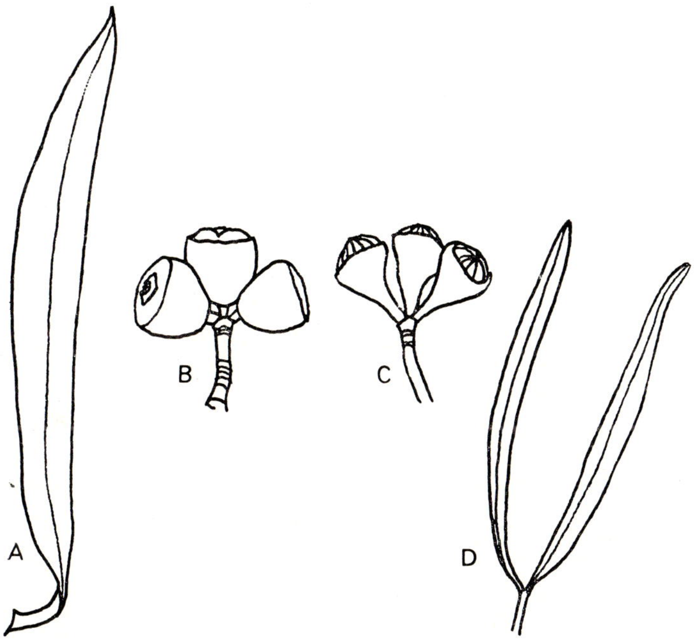


FIG. 39 *Eucalyptus pulchella* (syn. *E. linearis*): A, mature leaf (actual size); B, fruit ($\times$ 2); C, buds ($\times$ 2); D, sucker leaves (actual size).

White peppermint, *E. pulchella* (syn. *E. linearis*) (Fig. 39), is an extremely graceful tree with a smooth white trunk throughout, often with bluish streaks. The sucker leaves are opposite, rather narrow, and fairly long; the mature leaves are similar in shape but alternate, 5 to 12 centimetres long. The fruit is in groups of five to thirteen, pear-shaped, 7 to 12 millimetres in diameter, the valves enclosed or level. This species reaches an elevation of about 2,500 feet but also occurs at lower elevations.

Swamp or stringbark gum, *E. regnans* (Fig. 40), is very abundant and widely distributed in Tasmania and is important economically. It occupies sheltered valleys up to 2,500 feet and supplies the timber marketed as Tasmanian oak. It occurs also on the mainland at lower elevations. It is a tall tree with a shaft-like trunk, buttressed at the base in big trees. The stem is smooth, white streaked with grey with a considerable amount of rough bark at the butt. The upper bark peels off in ribbons. The sucker leaves are opposite, pointed, 4 to 8 centimetres long, stalked, rather broad; the mature leaves are alternate, stalked, broad with a sharp point, asymmetrical, glossy, 10 to 16 centimetres long. The fruit is bell-shaped, in groups of seven to sixteen, and measures 5 to 6 millimetres in diameter; the valves are enclosed.

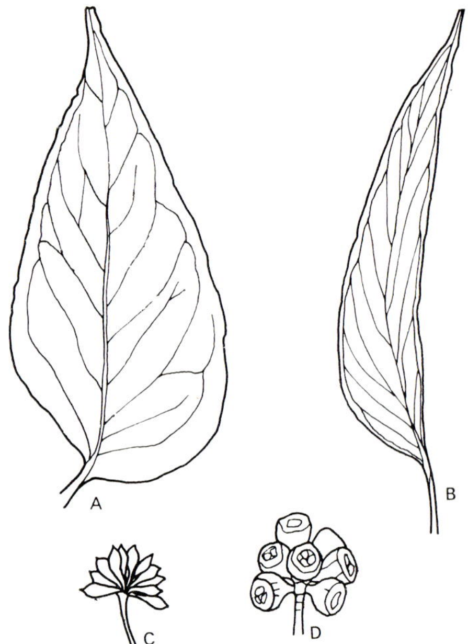


FIG. 40 *Eucalyptus regnans*: A, sucker leaf; B, mature leaf; C, buds; D, fruits. (All × ½ except A, actual size).

Smithton peppermint, *E. nitida* (syn. *E. simmondsii*) (Fig. 41), is a rough-barked tree with smooth branches, widespread but reaching only about 1,500 feet on the lower montane slopes of Tasmania. The sucker leaves are opposite, usually stalkless, broad with a sharp point, the base often heart-shaped, grey-green or dark-green, 3 to 10 centimetres long: the mature leaves are alternate, stalked, leathery, narrow with a sharp point, 6 to 15 centimetres long. The fruit is turban-shaped in groups of seven to twenty-three, and measures 10 to 12 millimetres in diameter; the valves are level.

Silver peppermint, *E. tenuiramis* (syn. *E. tasmanica*) (Fig. 42), is a tree of variable height, often quite small but very attractive, grey-green all over, often with pendulous branches and a greyish-white, smooth trunk. The sucker leaves are stalkless, opposite, clasping the stem, broad

FIG. 41 *Eucalyptus nitida*: A, buds (actual size); B, fruits (actual size); C, sucker leaf (× ½); D, mature leaf (× ½).

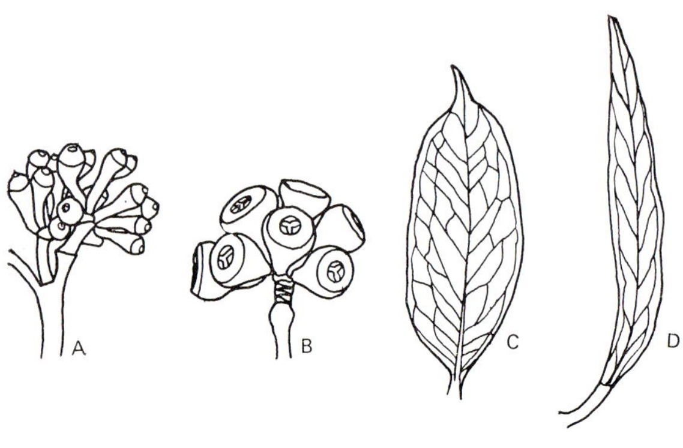

FIG. 42 *Eucalyptus tenuiramis*: A, sucker leaves (× ½); B, mature leaves (× ½); C, buds (actual size); D, fruits (actual size).

with a sharp point, markedly grey-green, 3 to 5 centimetres long; the mature leaves are alternate and stalked, somewhat elliptical, 5 to 16 centimetres long. The fruit is pear-shaped or rounded, in groups of seven to twenty-three, and measures 7 to 11 millimetres in diameter; the valves are level or enclosed.

The understorey of the wet sclerophyll region in Tasmania resembles its counterpart on the mainland in general. The area is rich in members of the Waratah family. The Tasmanian waratah, *Telopea truncata*, is well represented here as in other parts of the island, and *Banksia marginata* is common. Other members on the mainland include *Grevillea australis* persisting from alpine habitats, *G. lanigera* and *G. victoriae*. *Hakea lissosperma* and *H. eriantha* from subalpine parts also appear. *Lomatia fraseri* and *L. longifolia* come from slightly higher altitudes. *Persoonia chamepeuce* from other habitats in the montane occurs, as well as a number of other species of *Persoonia* not represented in other divisions.

Pea flowers and acacias are well represented: several species also found in the dry sclerophyll are represented here as well as *Acacia alpina* from subalpine habitats and blackwood, *A. melanoxylon*, common in Tasmania as well as widespread on the mainland as a tall umbrageous tree.

Heaths are common with a number of species of *Leucopogon*, *Brachyloma*, *Epacris* and *Acrotriche*. Daisies and daisy bushes are very common.

**Dry Sclerophyll Forest**

Although more typical of areas of lower elevation, dry sclerophyll forests are represented in the more arid stretches of the mainland montane. A fairly wide range of eucalypt species includes *Eucalyptus bridgesiana*, *E. dives*, *E. elaeophora*, *E. macrorrhyncha*, *E. pauciflora*, *E. rubida*, *E. sieberi* and *E. stellulata*. Of these, *E. pauciflora* (Subalpine), *E. bridgesiana* and *E. dives* (Savannah Woodland) and *E. macrorrhyncha*, *E. rubida* and *E. sieberi* (Wet Sclerophyll) have already been described.

The area carries a wide range of other plants as well. It is a low rainfall area with light but not persistent winter snow. The terrain is mainly hilly and steep, poorly aerated, level ground being uncommon. *E. dives* forms distinct associations in the colder parts with *E. pauciflora,* and in other parts with *E. bridgesiana*, *E. rubida* or *E. elaeophora*. On warm ridges *E. dives* forms associations with *E. sieberi*, while *E. macrorrhyncha* forms associations with *E. elaeophora* in some parts and with *E. rubida* in others.

Transition to savannah woodland is clearly marked through the association of *E. pauciflora* and *E. stellulata*.

Bundy, *E. elaeophora* (Fig. 43), is usually a small, crooked tree with pendulous branches and a rough, whitish bark. The sucker leaves are opposite, stalkless, rounded, 4 to 13 centimetres long; the mature leaves are alternate, with stalks, rather broad, pointed, 10 to 22 centimetres long. The fruit is stalkless, three to

FIG. 43 *Eucalyptus elaeophora*: A, sucker leaf (actual size); B, mature leaf (× ½); C, buds (× ½); D, fruits (× ½.)

eight in a head, egg-shaped or turban-shaped, 5 to 6 millimetres in diameter, the valves either enclosed or protruding.

A wide variety of shrubs occur throughout the formation.

Many of the plants occurring in the savannah woodland and the wet sclerophyll forest are found here also. The grasses and other small plants of these formations, including the ground orchids, also occur here but are not commonly so abundant.

Among the taller shrubs a wider range of acacias is found including *Acacia penninervis* and *A. falciformis*. Many species of *Dillwynia* and *Pultenaea* are common as well as *Hardenbergia violacea* and several oxylobiums.

*Pomaderris apetala* and *Hymenanthera dentata* both grow fairly tall in this habitat, and there are a number of pimeleas, including *Pimelea axiflora* and *P. curviflora*. In the moister parts *Prostanthera phylicifolia* and *Leptospermum phylicoides* appear and, among the epacrids, *Brachyloma daphnoides*, *Epacris microphylla*, and *E. impressa*. *Ajuga australis* makes a gay show in dry parts. Among the composites *Gnaphalium collinum*, *G. luteoalbum*, *Helichrysum apiculatum*, *H. scorpioides*, *Cassinia aculeata*, *Cymbonotus lawsonianus*, and *Brachycome marginata* are presentative.

# II

# Descriptive List of Alpine Species

IN THIS part the plant families and the individual species occurring in the ecological divisions outlined in Part 1 are described in some detail. It has been necessary to use a certain number of technical words, though they have been kept to a minimum: a glossary appears at the back of the book.

The arrangement of families follows the system of Bentham and Hooker, with some slight variations. Genera and species are arranged in alphabetical order for the most part. Where a type genus, such as *Epacris* (family Epacridaceae), occurs the genus is given precedence.

The genus *Eucalyptus* has been omitted from family Myrtaceae in this part of the book, where it rightly belongs, and the species of the genus are described in Part 1 in their ecological habitats. This is done because the tree flora, being the dominant feature of a habitat, defines it, and a description of the habitat is thus largely a description of the tree species.

# DICOTYLEDONS

Flowering plants are divided into two large groups, dicotyledons and monocotyledons. Of these, the dicotyledons comprise by far the larger group, containing the majority of trees and shrubs and some herbs.

Dicotyledons have two cotyledons or seed leaves in the embryo. The stem has its vascular bundles—conducting tissues—arranged in a single ring; in trees and shrubs these bundles unite with growth to form a ring of woody tissue. The leaf venation is usually netted, except in a few members (including those of the family Epacridaceae), in which the venation is parallel. The parts of the flower are typically in fours or fives or multiples of these numbers.

### Family RANUNCULACEAE

Buttercups and their relatives are found all over the world. In Australia the alpine buttercups are a more conspicuous feature of the vegetation than they are in any other part of the continent, both in number of kinds and profusion of individuals. So prolific are they that a whole hillside is a blaze of gold or a mass of delicate white when they are in full bloom.

Members of the family are not difficult to recognize. The plants are all comparatively small herbs, usually with conspicuous yellow to cream or white open flowers. The flowers have 4, 5 or more free sepals and petals (the petals occasionally absent altogether), a large number of stamens inserted above the other flower parts and a number of separate carpels ripening into individual one-seeded fruits, either non-splitting achenes or splitting follicles. The common alpine genera are *Anemone*, *Caltha*, *Clematis* and *Ranunculus*. In the first three there are no petals, but the sepals are petal-like. In *Ranunculus* there are both sepals and petals but the latter are often deciduous.

*ANEMONE* (from Gk *anemos*, the wind, probably referring to the exposed habitat of some species). Fig. 44.

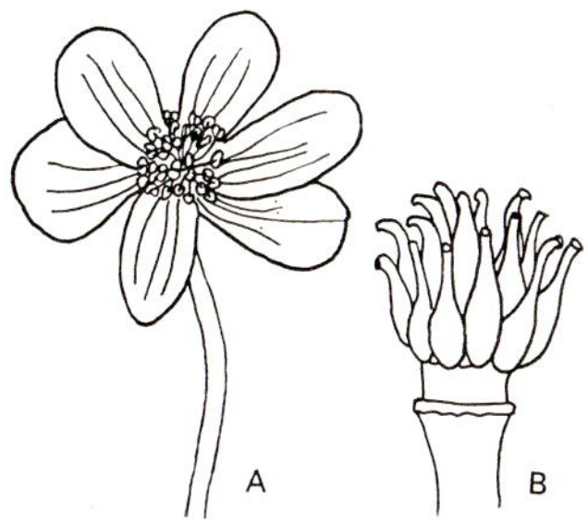


FIG. 44 *Anemone*: A, flower; B, fruits. (Both actual size.)

*A. crassifolia*. White. Alpine, subalpine. W. Tas. The only species of *Anemone* in Australia, this is a small plant with a rosette of divided leaves, each with 3–4 segments, thick and almost succulent, often purple on the underside. Flowers 1–3 on erect stalks with 3 leaves forming a ring just below the flower, a characteristic that makes an attractive background to the open flower; flower with 6–8 free petal-like sepals about 1.5 millimetres long, soft and rounded at the tip; no petals; numerous bright-yellow stamens enhance the delicate beauty of the open flower; carpels numerous, smooth, developing into achenes.

*CALTHA* (ancient Roman name for a plant, possibly marigold). Fig. 45. Represented in Australia by two species, both small rosette plants which enjoy the swiftly flowing streams of the alpine meadows and slopes or wet flushes on scree. The leaves are a rich light green, heart-shaped at the base and stalked, each with a characteristic wing-like appendage formed by the upturned bases of the leaves, making it easy

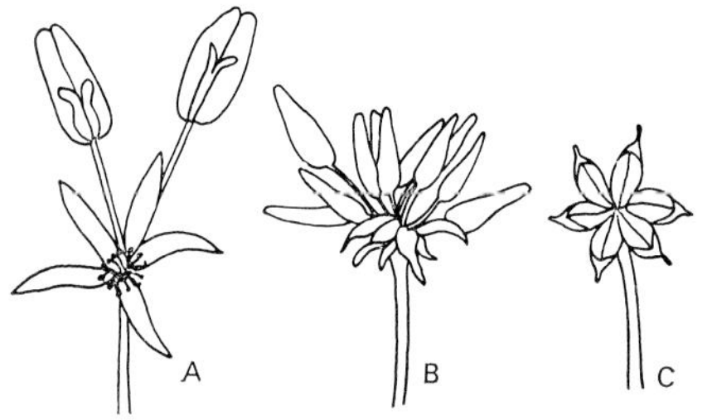


FIG. 45 *Caltha*: A, flower (actual size); B, young fruit (× ¾); C, opened fruit (actual size).

to identify even without flowers. These plants often occupy considerable areas and in early summer make an attractive display. After the blossom come stalked fruits rising to a much greater height than the flowers. These stiff little brown heads of follicles are quaint and attractive.

*C. introloba.* Alpine Marsh Marigold. White, cream or pink. Alpine, subalpine. N.S.W., Vic.

A tufted plant 1.5–2 cm long. Leaves stalked, spear-shaped or rounded. Flowers with 4–8 narrow, free, petal-like sepals, 1.5 cm long or slightly less; no petals; stamens several, numerous, with fairly long reddish stalks; carpels several, fruit splitting.

*C. phylloptera.* Tasmanian Marsh Marigold. Yellow. Alpine, subalpine. Tas.

Plant similar to *C. introloba*. Flowers with 5–8 narrow sepals, 8–10 mm long; no petals; stamens few and less conspicuous than in former species; carpels 5–8.

*CLEMATIS* (from Gk *klematis,* name of a plant, derived from *klema,* a twig, perhaps because of the twiggy branches). Fig 46. A widespread genus of few species, all of which are woody climbers or small shrubs. There are 4 or more petal-like sepals, no petals and numerous stamens and carpels. The flowers are succeeded by most attractively plumed fruits (achenes with feathery styles), giving the plant the common name of Old Man's Beard. Three species are found in the montane areas but all have a wide distribution at lower elevations also. These are *C. aristata*, *C. gentianoides* and *C. microphylla.*

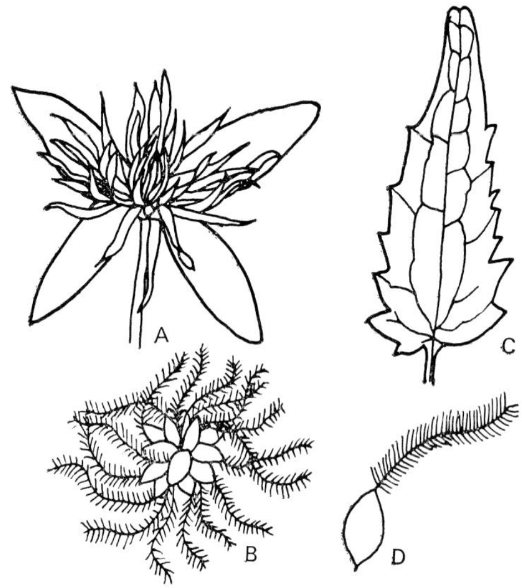


FIG. 46 *Clematis*: A, flower (× ¾); B, fruiting head (actual size); C, leaf (× ¼); D, single fruit (× 2).

*C. aristata.* Creamy white. Montane. N.S.W., Vic., Tas.

A robust woody climber. Leaves smooth with 3 oval or heart-shaped leaflets, often toothed and on long, often twisted stalks. Flowers 3–6 cm in diameter, on long stalks in the upper leaf axils; sepals 4, narrow.

*C. gentianoides.* White, pink flush. Montane. Tas.

A tufted plant 15–50 cm high. Leaves simple, pointed, elongated, commonly toothed, on short stalks. Flowers solitary, delicately perfumed, terminal, up to 7 cm across; sepals 4–8.

*C. microphylla.* White, cream. Montane. N.S.W., Vic., Tas.

A robust woody climber. Leaves smooth, stalked with 5 leaflets. Flowers rather broad, about 5 cm in diameter in upper leaf axils.

*RANUNCULUS* (from Latin *rana*, a frog, many species inhabiting marshy places where frogs abound). Fig. 47. A fairly large genus of wide distribution, with several species in the alpine parts of Australia. Several are restricted to elevations over 4,000 feet, while others may occur from the coast to subalpine habitats. Most

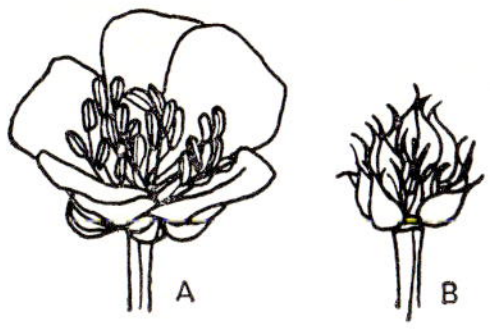


FIG. 47 *Ranunculus*: A, flower; B, fruit. (Both slightly enlarged.)

species have glossy yellow flowers with 5–8 sepals, often small and deciduous; 5 to many petals, though rarely more than 8, the outer surface often brown, red or purplish, especially at higher altitudes and with a little pocket-like nectary gland at the base of each petal. The stamens are usually very numerous and attractive. The carpels ripen into non-splitting achenes.

**Alpine Buttercups of the Mainland**

Among the mainland buttercups almost entirely restricted to the highest altitudes are *R. anemoneus*, *R. gunnianus* and *R. niphophilus*.

*R. anemoneus*. Alpine Windflower. White. Alpine, subalpine. N.S.W., Vic.

The largest of all the buttercups and, to me, the loveliest. Growing as it does along the line of melting snow among the harsh granite boulders of Mount Kosciusko and nearby peaks and occasionally in Victoria at the highest altitudes but descending to subalpine levels, it never ceases to amaze with its apparent frailty but its capacity to withstand the harshest conditions. As the snow melts, the flower buds push their naked way up into the frosty air and, as they burst from the discarded sepals the backs of the soft petals are often seen tinged with a most delicate rose-pink. When they open, however, the many narrow petals are displayed in all their dazzling whiteness, each up to 2.5 cm in length surrounding a ring of numerous golden-yellow stamens.

As the flowers open, the very attractive leaves begin to uncurl from an underground stem. These are much-dissected, rounded in outline and on rather fleshy stalks which are 5–10 cm long. At this stage a robust plant will have many flowers, 1–3 in a head, on stalks up to 30 cm in length, making a sight never to be forgotten. In the breezes so common on the mountainside the anemone windflowers, often the dominant note over a wide expanse of territory, dance continually to an ever-changing theme, sometimes slow and steady, often with wild abandon but always graceful and breathtakingly beautiful.

This buttercup is one which has made a spectacular recovery after severe depredations during the long period when summer grazing was permitted on the high moors. The flowers were quickly and effectively trampled in soon after they had pushed their way through the melting snow and the leaves rarely had a chance to appear at all, so that this lovely plant was in imminent danger of disappearing altogether. Now it is to be seen gaily dancing its way during the summer months from the very roof of Australia down to about 5,000 feet.

*R. gunnianus*. Golden yellow, glossy. Alpine. N.S.W., Vic., Tas.

A very handsome species from wet habitats readily distinguished from most other buttercups by its long, stalked leaves finely dissected into very narrow segments, basal, smooth, 1–2 cm long on stalks 2–4 cm long. Flowers solitary, large, with 5 small petals, 5–12 (commonly 5) narrow petals 1–1.5 cm long, often tinged with purple on the back; stamens and carpels very numerous, carpels smooth with a straight beak.

*R. niphophilus*. Golden yellow, glossy. Alpine. N.S.W.

A lover of streams and boggy places below permanent snow, often spreading extensively and making a wonderful carpet of gold attractively displayed against the grey granite boulders, this is a small tufted plant with basal leaves, mostly smooth, divided into 3 rounded lobes again divided, variable in size, the lobes about as long as broad and on stalks up to 16 cm long but usually much shorter. Flowers usually solitary, 12–30 mm in diameter on stems 1.5–15 mm long or rarely more, with 5 sepals, 5 egg-shaped petals 7–14 mm long, 30–35 stamens and 30–35 carpels, each ending in a short, straight style.

A number of other buttercup species are found in alpine habitats but extend also to subalpine and montane areas. These include *R. clivicola*, *R. dissectifolius*, *R. inundatus*, *R. lappaceus*, *R. muelleri* and *R. productus*.

*R. clivicola.* Golden yellow, glossy. Alpine, subalpine, montane. N.S.W.

A recently discovered species from Mount Kosciusko which may have a wider range, rather similar to *R. niphophilus* and occupying similar habitats, but sometimes descending to lower altitudes. Leaves usually with 3–5 segments again divided. Flowers 18–35 mm in diameter 2–4 together on stems 12–50 mm high; style arching towards the centre of the flower.

*R. dissectifolius.* Golden yellow, glossy. Alpine, subalpine. N.S.W.

Closely related to *R. clivicola*, but occurs more commonly in sphagnum bogs and wet grasslands rather than near running water. Leaves often hairy, 10–45 mm long with 3 major lobes again divided. Flowers solitary, 16–30 mm in diameter, on stems 6–25 mm long; style with a recurved tip.

*R. inundatus.* Pale golden yellow. Alpine, subalpine, montane. N.S.W., Vic., Tas.

A smooth, creeping plant with rosettes of leaves 1–3 cm long, divided into narrow segments, either floating or submerged. The flowering stalks are slender, 5–30 cm long with 1–3 flowers. Flowers small, petals 5–12. Fruit rather inflated with irregular warty ridges and with a short, reflexed beak.

*R. lappaceus.* Golden yellow, glossy. Alpine, subalpine, montane. N.S.W., Vic., Tas.

A small tufted plant, widespread in Australia in a variety of habitats from sea-level to 5,000 feet, common in grasslands. Leaves basal, numerous, hairy, egg-shaped, 2–3 times dissected or lobed with toothed margins, 5 cm broad on stems 5–20 cm long. Flowers solitary or few together, 15–40 mm in diameter on erect stems longer than the leaves, variable in length from 4–70 cm; 5 small sepals and 5 rounded petals variable in size, up to 2.5 cm long, sometimes bronze or purple on the back, especially at higher elevations; stamens 4–110; carpels 20–50, with straight styles ending in recoiled tips.

*R. muelleri.* Golden yellow, intensely glossy. Alpine, subalpine. N.S.W., Vic.

One of the buttercups less likely to be recognized when it is not in flower, this is a beautiful little plant, often nestling in the crevices where the water lies, found at altitudes of 6,000 feet in the grasslands and herbfields and in wetter situations down to 5,500 feet and, where it can find the shelter of a friendly boulder, extending to the fjaeldmark at 6,700 feet. The fjaeldmark form differs from the normal species in having considerably longer flower stalks and shorter leaf stalks and is regarded as var. *brevicaulis.*

Normally a small tufted plant with basal leaves 8–27 mm long, not dissected though sometimes lobed or indented, pointed, elliptical or egg-shaped, covered with close hairs giving a rough feel and appearance, often tinged with purple on the back, on stems 0.5–9 mm long. Flowers solitary, 18–35 mm in diameter on stems 2–16 cm long; sepals 5, small; petals 5, egg-shaped, 7–22 mm long; stamens 15–70; carpels 30–120, each ending in a straight style with a recurved tip.

*R. productus.* Pale yellow. Alpine, subalpine. N.S.W.

A recently described species from wet grasslands on sloping ground. Leaves mostly basal, very hairy, divided, with 3–7 egg-shaped segments, very similar to those of *R. niphophilus* but with narrower, more hairy segments. Flowers small, 3–4 mm in diameter, on stems up to 20 cm long but usually under 10 cm. There are 5 elliptical petals 1.5–2.5 mm long; stamens 15–20; carpels 24–45; styles ending in short recurved tips.

### WATER-LOVING SPECIES

Most of the buttercups from higher elevations like moist situations and many can tolerate an abundance of moisture. The three outstanding alpine species found constantly in wet places, sometimes completely submerged, are *R. millanii*, *R. pimpinellifolius* and *R. tricophyllus.*

*R. millanii.* Cream. Alpine, subalpine. N.S.W., Vic.

A most attractive denizen of the swamps and fens. It loves depressions subject to flooding and may spend the whole of its flowering period completely submerged. The spreading habit of these attractive little plants with their wide open flowers make a good cover, often over considerable areas.

Leaves basal, smooth, dissected into 3–5 narrow segments again divided, the whole leaf

4–20 mm long. Flowers solitary, 6–7 mm in diameter on stalks 5–80 mm long; sepals green, 5 (rarely 6); petals 5 (occasionally 6), not shining, egg-shaped, 3.5–8 mm long; stamens 8–35; carpels 9–25, each with a stout, straight style with a recurved tip.

*R. pimpinellifolius.* Golden yellow, glossy. Alpine, subalpine, montane. N.S.W., Vic., Tas.

A small species making bright patches in fens and swamps. Leaves basal with a few along the erect stems at times, hairy or almost smooth on the upper surface, dissected, with 3–4 segments, each somewhat egg-shaped, the whole leaf 0.5–3 cm long. Flowering stems 2–2.5 cm high, each with 2–5 small flowers, 5–15 mm in diameter; sepals 5, green; petals 5, elliptical or narrow egg-shaped, 4–9 mm long; stamens 20–25; carpels 8–40, each with a stout, straight style ending in a recurved beak.

*R. trichophyllus.* Pale yellow. Subalpine, montane. N.S.W., Vic., Tas.

A smooth plant, frequently completely submerged, sometimes creeping in mud. Leaves 1–4 cm in diameter in basal rosettes, submerged ones finely divided, floating or aerial ones with a few rounded lobes. Flowers solitary, 1–4 cm in diameter on stalks 5–30 cm long. Fruit somewhat inflated, hairy, with transverse ribs, and a short beak. Some authors prefer to classify this as *Batrachium tricophyllum*, distinguishing the genus from *Ranunculus* by the ribbed fruit, the light-coloured petals, the greenish tinge on the sepals and the finely dissected leaves.

**Tasmanian Buttercups**

Four species of alpine buttercups are endemic to Tasmania—namely *R. collinus*, *R. decurvus*, *R. pascuinus* and *R. setaceus.*

*R. collinus.* Golden yellow. Subalpine. Vic., Tas.

A small creeping plant found in places where there is water seepage. Leaves basal, broad, deeply lobed or divided into 3 broad segments each usually 3-toothed, dark-green and shining above. Flowers small, 8–12 mm in diameter, solitary on stalks longer than the leaves; sepals 5, small; petals 6–11; carpels with tips of styles recurved.

*R. decurvus.* Pale yellow. Montane. Tas.

A small-flowered hairy plant of wet situations. Leaves numerous, basal, on long stems with 3–5 wedge-shaped, sharp-pointed leaflets. Flowers 3–4 mm in diameter, solitary or in small heads; sepals 5, petals 5, narrow, blunt; style short and recurved at tip.

*R. pascuinus.* Golden yellow. Alpine, subalpine. Tas.

Closely related to *R. lappaceus*, a small, erect species with basal leaves, simple or divided, on stalks 1–10 cm long. Flowers solitary, up to 2.5 cm in diameter, very similar to *R. lappaceus*, often flushed with purple on the back.

*R. setaceus.* Cream. Alpine. Tas.

A waterways species closely related to *R. millanii* of the mainland. Leaves entire or with one pair of leaflets only. Flowers solitary, about 0.8 mm in diameter on stalks shorter than the leaves; petals smaller and narrower than those of *R. millanii*, often flushed purple on the back.

**Buttercups from Lower Altitudes**

A number of buttercup species which do not ascend to higher altitudes appear at subalpine and montane levels, including *R. eichleranus*, *R. graniticola*, *R. plebeius*, *R. pumilo* and *R. scapigerus.*

*R. eichleranus.* Yellow. Subalpine. Vic.

A very small tufted plant of the grasslands and woodlands, recently discovered near Mount Hotham. Basal leaves with narrow hairy segments, twice divided. Flowers small, 16–20 mm in diameter, solitary, on stalks 5–20 cm long; sepals 5, green; petals 5, egg-shaped; stamens 35–65; carpels 35–65, each with a short, slender, straight style ending in a recurved tip.

*R. graniticola.* Golden yellow, glossy. Subalpine. N.S.W., Vic.

This buttercup may be easily confused with *R. lappaceus* upon a cursory glance. It is a somewhat smaller plant of the grasslands and swamps. Leaves basal, hairy, simple with coarsely toothed margins or sometimes dissected or lobed with toothed margins, 1–4 cm long, on stems 9–10 cm long. Flowers 10–36 mm in diameter, 2–3 in a head on stalks longer than the leaves; sepals 5, green, hairy, deciduous; petals 5 (occasionally as many as 8), sometimes tinged with purple on the back, egg-shaped, 0.5–2 cm long with an oblong or tapering nectary (wedge-shaped in *R. lappaceus*); stamens 30–70; carpels about 20, each ending in a short, stout style with recurved tip (in *R. lappaceus* the style is slender).

As the carpels ripen to form the fruits they frequently turn a dark-brown or almost black colour giving a characteristic black-eyed appearance to this dainty little buttercup. This, however, is not a sufficiently reliable point to be used in identification.

*R. plebeius*. Golden yellow. Montane. N.S.W., Vic.

An erect straggling plant 10–80 cm high. Leaves mainly basal, hairy, stalked, egg-shaped, 1–7 cm long with 3 pointed segments. Flowers 7–16 mm in diameter on fairly long stems; petals 5, egg-shaped; stamens 9–25; carpels 25-60. Fruit rugose with a slender, strongly recurved beak.

*R. pumilio*. Yellow. Montane (sea-level to 3,000 feet). N.S.W., Vic., Tas.

A widely distributed, slender, hairy plant with basal and stem leaves. Leaves round in outline, deeply divided, 5–15 mm in diameter. Flowers very small, with stalks; carpels each with a short style recurved at the tip. Fruit hairy or with small tubercules, sometimes smooth, pale-brown with a green margin.

*R. scapigerus*. Golden yellow. Subalpine. N.S.W., Vic., Tas.

A fairly widely distributed species of grasslands and woodlands, once included with *R. lappaceus*. Leaves mostly basal, hairy, egg-shaped, deeply lobed. Flowers 8–16 mm in diameter, on stems 25–26 mm long with 2–4 flowers on each stem; sepals 5, green; petals 5, elliptical or egg-shaped, often tinged with purple on the back; stamens 15–35; carpels 20–50 with straight styles with recoiled tips.

### Hybrid Buttercups

*R. victoriensis*. Golden yellow. Subalpine. Vic. (Mount Bogong High Plains and Mount Hotham).

A grassland species regarded as a stabilized hybrid of *R. eichleranus* x *R. muelleri*. Very variable with variously divided leaves from almost simple and notched to doubly dissected, the whole leaf 12–50 mm long. Flowers 16–30 mm in diameter; sepals 5, green; petals 5, broad, 5–7 mm long; carpels in a globular head each with a stout style ending in a recurved tip.

Another lovely little hybrid ascribed to R. *muelleri* x *R. niphophilus* is fairly common near the Blue Lake at Mount Kosciusko. The leaf is very similar to that of *R. muelleri*, 1–1.5 cm long on stalks 2–3 cm in length. Flowers large, solitary; sepals 5, longer than the petals; petals 5, in shape similar to those of *R. niphophilus*. The habit and general appearance approach *R. muelleri*.

A hybrid of *R. graniticola* x *R. millanii has* also been reported.

### Family DILLENIACEAE

Guinea flowers or hibbertias are closely related to buttercups, and this is clearly shown in the structure of the flower. There are 5 sepals and 5 free petals, the number being consistent and not variable as it is in many of the buttercups; the petals are usually yellow, thin and often crinkled, not glossy and frequently with large bracts at the base. The stamens are numerous, inserted above the other flower parts; carpels few, ripening into follicles or berry-like fruits.

Guinea flowers are shrubs found in a wide range of habitats; only one ascends higher than montane altitudes.

*HIBBERTIA* (after George Hibbert, London merchant and patron of botany, d. 1837 or 1838). Fig. 48.

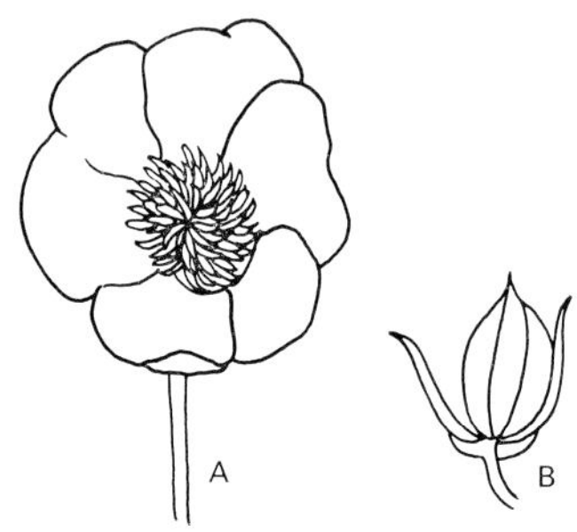


FIG. 48 *Hibbertia*: A, flower; B, fruit. (Both × ½.)

*H. fasciculata*. Yellow. Montane. N.S.W., Vic., Tas.

A common coastal plant on the mainland also, and sometimes ascending to 3,000 feet in Tasmania. Prostrate, with spreading branches. Leaves narrow, clustered, thick, 4–12 mm long. Flowers numerous, solitary; petals heart-shaped at top, about 8 mm long; stamens 8–12 in 3 groups round the numerous carpels.

*H. linearis* var. *obtusifolia*. Yellow. Montane. N.S.W., Vic.
Semi-prostrate, leaves flat, broad, 8–20 mm long. Flowers shortly stalked, fairly large; stamens and carpels numerous.

*H. pedunculata*. Yellow. Montane. N.S.W., Vic.
Usually prostrate. Leaves narrow, stiff, blunt, 3–10 mm long. Flowers stalked, about 12 mm in diameter; stamens 12–20, arranged all round the numerous woolly carpels.

*H. procumbens*. Deep yellow. Subalpine, montane. Vic., Tas.
A lovely little prostrate plant with remarkable range from sea-level to 4,000 feet in Tasmania. Branches spreading. Leaves flat, narrow, sharp, usually smooth, 5–12 mm long. Flowers abundant, most attractive, solitary, about 2.5 cm in diameter; petals blunt; stamens 20–25 in groups round the carpels. The knobbly brown flower buds enrich the golden beauty of the flowers.

*H. serpyllifolia*. Yellow. Montane. N.S.W., Vic., Tas.
A smooth, low shrub, spreading 30–50 cm and widely distributed in other habitats also. Leaves narrow, blunt, the edges rolled under, 3–6 mm long. Flowers solitary on short stalks; petals heart-shaped at the top, 8–10 mm long; stamens about 12, arranged all round the hairy carpels.

**Family WINTERACEAE**

This group of trees and small shrubs is quite a small one, represented in these habitats by one genus, *Tasmannia*, of two species; both are, however, widely distributed and a rather conspicuous feature of the vegetation.

*TASMANNIA* (syn. *DRIMYS*). Fig. 49.

*T. lanceolata*. Mountain Pepper. Creamy, inconspicuous. Subalpine, montane. N.S.W., Vic., Tas.
A tall shrub or small tree, extending to sea-level in Tasmania. Stems characteristically bright-red and rather fleshy-looking when young. Leaves aromatic with transparent oil dots, bright-green, elliptical, on short stalks, varying in length from 1.5 cm at higher altitudes to 13 cm at sea-level. Flowers in small clusters at the ends of the branches, either staminate or pistillate on individual plants with 2–8 petals, 20–25 stamens inserted above the other flower parts, and 1 carpel. Fruit black, berry-like, about 5 mm in diameter, more conspicuous than the flowers.

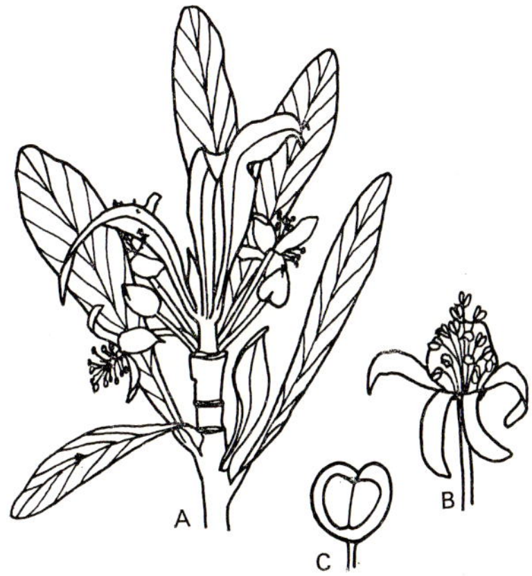

FIG. 49 *Tasmannia*: A, flowerhead (actual size); B, flower (× 2); C, fruit (× 2).

Another species, *T. vickeriana*, occurs in New South Wales at montane and subalpine levels.

The bright-red stems and the equally bright-green leaves, together with the spreading habit of these plants, make them conspicuous among the commonly more sombre colouring of their associates.

**Family CRUCIFERAE**

A group of soft plants to which the well-known garden stock, honesty and alyssum belong. The flowers have 4 sepals and 4 petals, and there are 6, sometimes fewer, stamens inserted above the other flower parts. The fruit is a specialized capsule frequently elongated but in some members short and broad, always opening by 2 valves separating from the base and revealing a persistent central partition to which the seeds are attached.

A considerable number of small plants of this family are scattered throughout alpine habitats, including a number of introduced forms.

*BARBAREA* (after St. Barbara, early Christian martyr). Fig. 50.

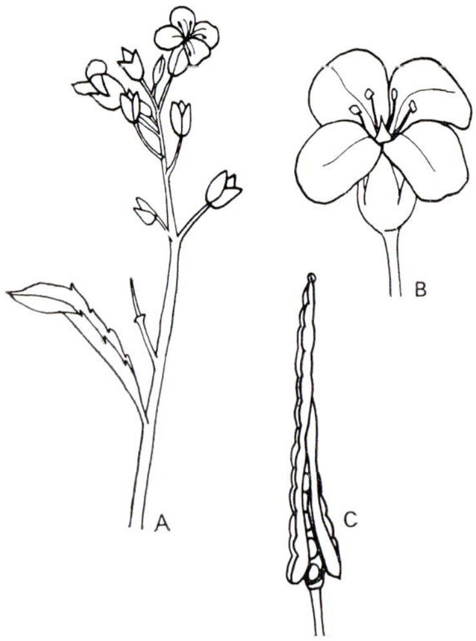


FIG. 50 *Barbarea*: A, flowerhead (actual size); B, flower (× 3); C, fruit (× 2).

*B. australis*. Yellow. Subalpine, montane. N.S.W., Vic., Tas.

An erect herb about 50 cm high, scarcely branched. Basal leaves stalked, divided, 8–10 cm long with 2–3 pairs of small egg-shaped lateral lobes and a large terminal one; upper leaves simple and wavy with wings, stem-clasping. Flowers in long slender heads. Fruit long, 3–5 cm, narrow, crowded.

*CARDAMINE* (from the Gk name for a cress-like plant). A genus of alpine to montane species occurring in New South Wales, Victoria and Tasmania with rather conspicuous heads of white or purple flowers. One species grows in colonies up to a height of 120 cm and makes a fine display in full bloom, but most of the species are small and rather inconspicuous. The genus is under review for classification.

*CHEESEMANIA* (after T. F. Cheeseman, 20th-century botanical author). Fig. 51. *C. radicata* is a semi-prostrate Tasmanian species from alpine and subalpine areas common in rock crevices at about 4,000 feet. The branches are cylindrical, brittle and bare, ending in rosettes of rather thick spoon-shaped leaves 4–9 mm long. The flowers are borne in leafy heads at the ends of the branches.

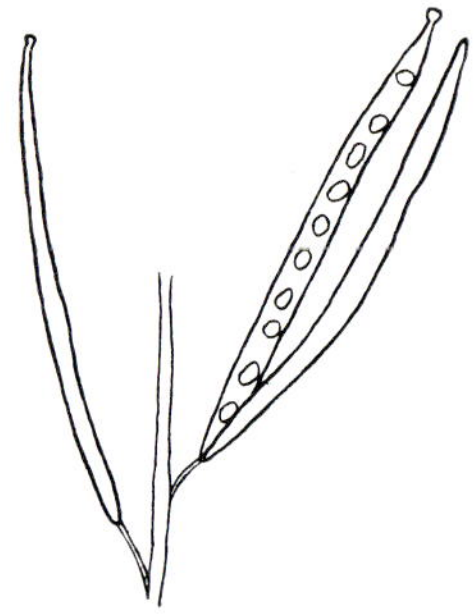

FIG. 51 *Cheesemania*: pod (actual size).

*DRABASTRUM* (from *drabe*, Gk name for a kind of cress, and *aster*, a star). Fig. 52. *D. alpestre,* mountain cress, is a small montane species 12–15 cm high from the mainland, with small pink or white flowers often tinged with purple. The stalked basal leaves are up to 5 cm long, coarsely toothed or lobed. The fruit is 6 mm long, on spreading stalks.

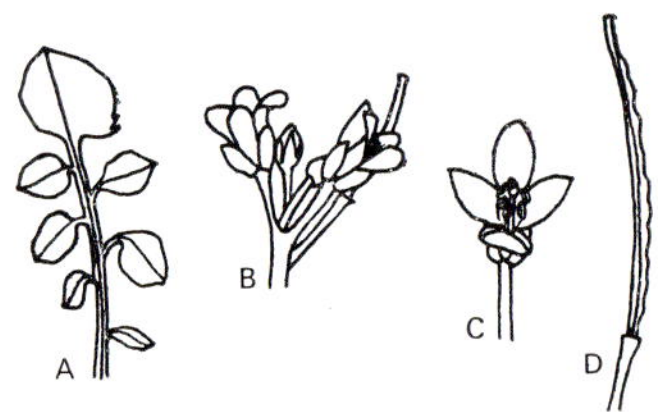


FIG. 52 *Drabastrum*: A, leaves (actual size); B, opening flower (× 2); C, flower (× 3); D, fruit (× 5).

**Family VIOLACEAE**

Violets, well-known plants of many habitats, are a feature of the alpine areas, particularly of the wetter parts of the moors and fens. They are small, soft plants with irregular, solitary flowers on relatively long stalks which carry 2 leaf-like bracts; the 5 sepals produced into short appendages; the petals are spreading, the lowest of the 5 is usually large and spurred, acting as an alighting platform for pollinators; there are 5 stamens inserted above the other flower parts and 3 fused carpels which develop into a characteristic 3-chambered splitting capsule.

*HYMENANTHERA* (from Gk *hymen,* a membrane and *anthera,* an anther). Fig. 53.

FIG. 53 *Hymenanthera*: branch with fruit and flowers (actual size).

*H. dentata.* Tree Violet. Yellow. Montane. N.S.W., Vic.
A stiff, much-branched shrub up to 2 metres high, stunted at higher elevations, often thorny. Leaves alternate, blunt, smooth, leathery, 6–30 mm long, without stalks, or narrowed into a short stalk about 4 mm long. Flowers regular with 5 equal petals, almost stalkless, rather fleshy; sepals and petals 5, the tips of the 5 stamens united in a tube round the carpels. Fruit a purple berry.

*VIOLA* (ancient Roman name for violet). Fig. 54.

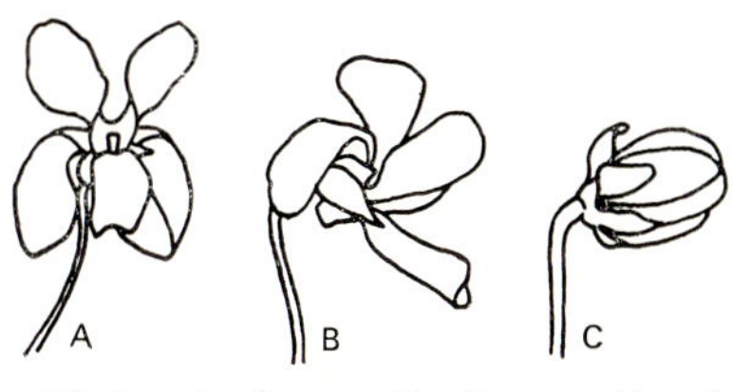


FIG. 54 *Viola*: A, flower; B, flower side view; C, fruit. (All actual size.)

*V. betonicifolia.* Mountain Violet. Deep purple. Alpine, subalpine, montane. N.S.W., Vic., Tas.
A small tufted plant of damp, grassy patches with basal, oblong leaves 2–5 cm long, stalked. Petals 12–18 mm long.

*V. caleyana.* White. Montane. N.S.W., Vic., Tas.
A straggling, rather weak plant sometimes ascending in height 5–25 cm. Leaves triangular or almost round, heart-shaped at the base and with slightly wavy margins, stalked 2–3.5 cm long. Petals 6–15 mm long.

*V. cunninghamii.* Pale violet, white. Alpine, subalpine, Tas.
A small plant confined to the western highlands of Tasmania. Leaves basal, tufted, round or triangular, 10–20 mm in diameter. Petals 6–15 mm long.

*V. hederacea.* Lilac, white. Subalpine, montane. N.S.W., Vic., Tas.
A much more common species occurring at lower altitudes also. Leaves kidney-shaped, coarsely toothed. Petals 8–12 mm long.

A form with rhomboid leaves 5–10 mm long and smaller flowers is now regarded as a distinct species, *V. sieberana.*

### Family PITTOSPORACEAE

Plants of this family are shrubs, climbers or trees with an aromatic odour and simple leaves. The flowers have 5 sepals, 5 petals and 5 stamens inserted above the other flower parts. The fruit often has seeds embedded in a scarlet pulp.

*BILLARDIERA* (after Jacques Julien Houten de la Billardière, French botanist, 1755-1834, who visited W.A. and Tas. and described many new plants). Fig. 55.

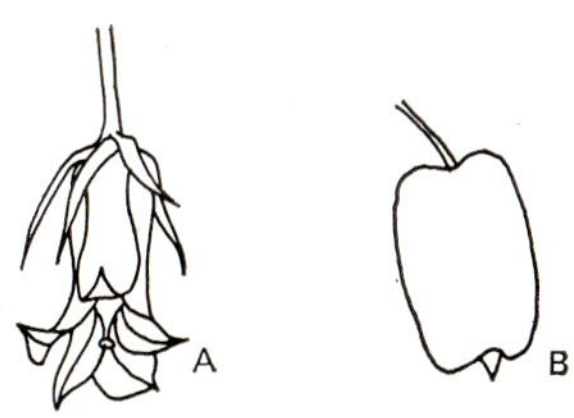


FIG. 55 *Billardiera*: A, flower; B, fruit. (Both actual size.)

*B. longiflora.* Climbing Blueberry. Greenish, sometimes yellow, changing to purple; berries white, purple, red. Subalpine, montane. N.S.W., Vic., Tas.
A twining plant with narrow, stout, alternate leaves. Flowers solitary, bell-shaped, expanded at the tip, 1.5–3.5 cm long; style protruding. Berries 2 cm long, most attractive.

*BURSARIA* (from Lat. *bursa*, a pouch, referring to the fruit). Fig. 56.

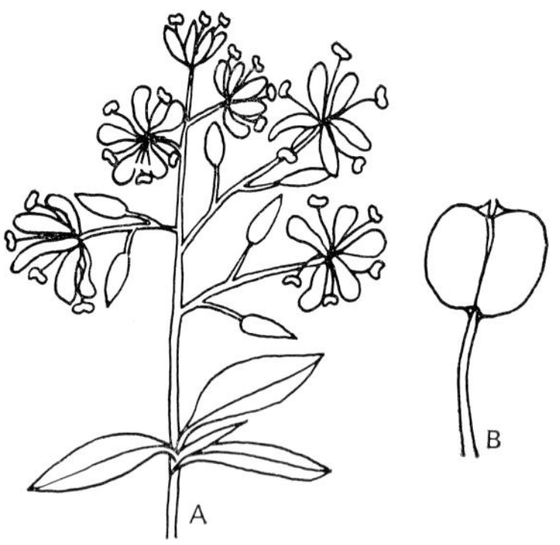


FIG. 56 *Bursaria*: A, flowerhead (actual size); B, fruit (× 2).

*B. spinosa.* Australian Blackthorn. White. Montane. N.S.W., Vic., Tas.
Common in a variety of other habitats also, this likes dry grassland or sclerophyll forest. A rather slender shrub or small tree to 13 metres high with spiny branches and blunt leaves about 1.5 cm long, often clustered. Petals spreading, about 5 cm long, forming open flowers. Fruit small, brown, flat and dry.

*MARIANTHUS* (from Gk *Maria* and *anthos*, a flower—flower of the Virgin Mary). Fig. 57.

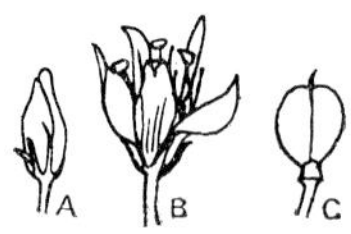


FIG. 57 *Marianthus*: A, bud; B, flower; C, fruit. (All actual size.)

*M. procumbens.* White, sometimes tinged with purple. Subalpine, montane. N.S.W., Vic., Tas.
A small, slender plant often spreading 10–60 cm. Leaves alternate, crowded, without stalks, 6–12 mm long, pointed. Flowers solitary or 2–4 together, on stalks 2–4 mm long; flowers shortly tubular, spreading at the top into narrow petals 5–8 mm long. Fruit a somewhat inflated capsule, blunt at the top, about 5 mm long.

*PITTOSPORUM* (from Gk *pitta,* pitch, and *sporos*, a seed, referring to the sticky pulp round the seeds). Fig. 58.

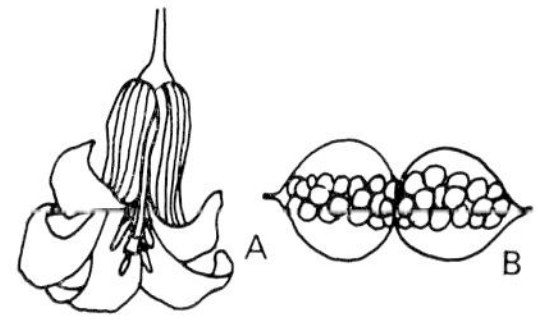


FIG. 58 *Pittosporum*: A, flower (× 1½); B, fruit (actual size).

*P. bicolor.* Maroon, yellow; outer crimson or purple. Alpine, subalpine. N.S.W., Vic., Tas.
An attractive slender plant, common at about 4,000 feet, growing to 14 metres in protected areas but stunted in exposed situations. Leaves crowded, entire, narrow, blunt and leathery, smooth on top, hairy and often brown below, 2–6 cm long. Flowers in small heads; petals joined to form a short tube with the tips curled back. Fruit woody, splitting to reveal numerous seeds embedded in a sticky red pulp.

## Family TREMANDRACEAE

The black-eyed susans are heathy shrubs with solitary, rather large, open-petalled flowers with 4–5 sepals, 4–5 spreading petals and 9–10 stamens in a single ring opening in terminal pores to shed their pollen and inserted above the other flower parts. The fruit is a flattened capsule. There are 4 montane species and 1 occurring at higher elevations.

*TETRATHECA* (from Gk *tetra*, four, and *theka*, a box, referring to the 4-celled anthers). Fig. 59.

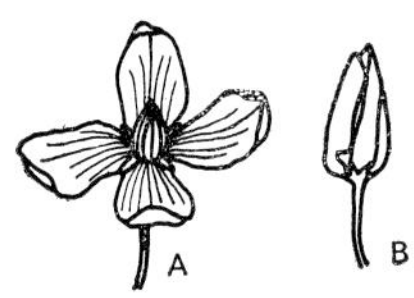


FIG. 59 *Tetratheca*: A, flower; B, fruit. (Both actual size.)

*T. ciliata.* Lilac, purple. Montane. N.S.W., Vic., Tas.
A small, slender shrub up to 1 metre high. Leaves in rings of 3–4, rounded, about 10 mm long. Flowers solitary on long slender stems in the leaf axils; petals 4, broad, about 10 mm long.

*T. glandulosa*. Purple, white. Subalpine, montane. Tas.

A spreading undershrub with branches up to 80 cm long, also found at lower altitudes. Leaves alternate, clustered, glandular, rather broad, 6–10 mm long, the margins often toothed. Flowers solitary in upper leaf axils; petals 4, each 8–10 mm long; stamens 8.

*T. glandulosa* var. *orbifolia* occurs in similar habitats on the mainland. It has almost round leaves with almost entire margins.

*T. subaphylla*. Deep purple. Montane. N.S.W., Vic.

A shrub 30–60 cm high almost leafless or with a few small scattered leaves. Flowers small, solitary on very short stalks; petals 4, each 6–8 mm long.

*T. thymifolia*. Purple, rarely white. Montane. N.S.W.

A heathy shrub 30–50 cm high, common in heathlands at lower elevations. Leaves narrow, hairy, whorled, or sometimes opposite, 3–15 mm long. Petals 4, spreading, 7–10 mm long; stamens 8.

**Family POLYGALACEAE**

A small family of small plants with alternate, entire leaves and irregular flowers. The flowers have 5 sepals, the 2 inner ones (called wings) larger and petal-like. There are usually 5 petals; 3 joined to form a keel in the front enclose 8 stamens united in a split sheath. The ovary has 2 parts and develops into a compressed capsule.

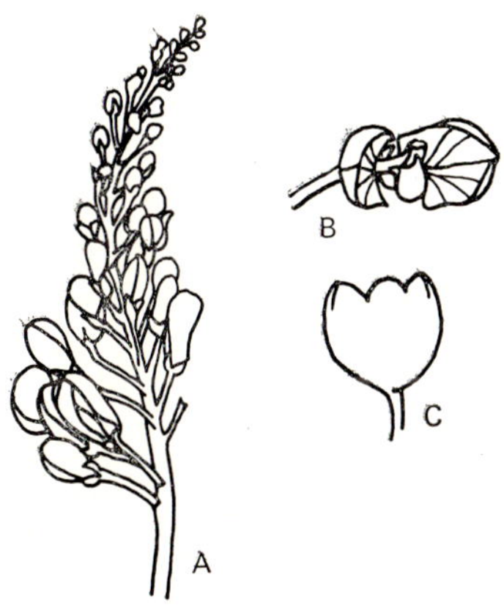


FIG. 60 *Comesperma*: A, flowerhead (actual size); B, flower (× 2); C, fruit (× 2).

*COMESPERMA* (from Gk *come*, hair, and *sperma*, a seed). Fig. 60.

*C. retusum*. Matches. Purple. Subalpine, montane. N.S.W., Vic., Tas.

A common inhabitant of the bogs and fens with brilliant flowers in long heads. Leaves small, alternate, crowded, thick, blunt, 3–4 mm long. Flowers superficially resembling pea-flowers on account of the 2 large sepals called wings which look like petals and spread out on either side of 3 fairly small petals; sepals 5 altogether; petals 5, the 3 lower ones forming a keel enclosing 8 stamens and ovary; upper 2 petals free. Fruit a small capsule.

*POLYGALA* (Lat. form of Gk *polygalon*: *polus*, much, and *gola*, milk, the ancients believing that eating these plants caused animals to produce more milk).

*P. japonica*. Pink. Montane. N.S.W., Vic.

A small wiry shrub up to 12 cm high, with a woody base. Leaves alternate, broad, 5–15 cm long. Flowers very small, the lowest petal crested; in small narrow heads produced along the stem.

**Family CARYOPHYLLACEAE**

The members of this small family are mainly herbs with thickened stems jointed at the nodes. The leaves are opposite or apparently whorled. The flowers are regular, with 4–5 sepals, 4–5 petals, sometimes missing, and 10 or fewer stamens, all inserted below the seedbox or on a rim round it. The fruit is a dry, splitting capsule or a non-splitting nut.

*COLOBANTHUS* (from Gk *kolobos*, stunted and *anthos*, a flower). Another small genus in this family, with 2 species in these habitats. *C. apetalus* occurs from alpine to montane habitats in New South Wales, Victoria and Tasmania. It is a small, densely tufted plant with opposite, overlapping, narrow leaves sheathing at the base, 1.5–2.5 cm long, and solitary terminal flowers on short stalks, with 5 leathery green sepals and no petals; there are 4–5 stamens inserted round the ovary. The capsule is globular. *C. benthamianus* is confined to alpine and subalpine habitats on the mainland. It is a tufted, moss-like plant 2–3 cm in height with densely set, small, narrow, pointed leaves, channelled above. The flowers are solitary on short stalks; the flowers and fruit are similar to those of *C. apetalus*.

*GYPSOPHILA* (from Gk *gypsos*, plaster of Paris, and *philos*, loving, some species prefering limy soil). This is a small genus, and one of the species, *G. australis*, is a small herb from montane habitats in New South Wales, Victoria and Tasmania: it has fairly broad stalkless leaves, heart-shaped at the base, and solitary red flowers with 5 tubular sepals and 5 free petals, the latter about 5 mm long; the capsule is rounded.

*SCLERANTHUS* (from Gk *skleros*, hard, and *anthos*, a flower, referring to the hard fruiting calyx). Fig. 61. Small, attractive mat plants

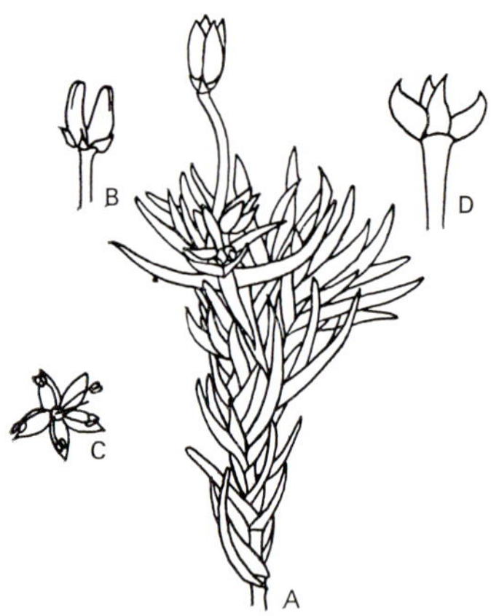


FIG. 61 *Scleranthus*: A, branch with flowerhead; B, young fruit; C, flower; D, opened fruit. (All × 4.)

forming bright-green, moss-like hummocks in wet and marshy land. The flowers are tiny, solitary or a few together or sometimes crowded. There are 5 sepals inserted on a rim round the 1–10 stamens; the petals are missing. The fruit is a small, single-seeded, non-splitting nut enclosed in the hardened sepals. There are 2 common species extending over a wide range of habitats at these altitudes.

*S. biflorus*. Green. Alpine, subalpine. N.S.W., Vic., Tas.

A tufted plant with crowded, narrow leaves 5–10 mm long. Flowers in pairs, without stalks and held within 4 bracts; stamen 1.

*S. singuliflorus* (syn. *S. minarioides*). Green. Alpine. N.S.W., Vic., Tas.

A tufted plant with stiff, pointed leaves 6–10 mm long. Flowers similar to those of *S. biflorus* but produced singly, each between 2 bracts.

*STELLARIA* (from Lat. *stella*, a star, referring to the open flowers). Fig. 62. A genus of small

FIG. 62 *Stellaria*: A, flower; B, young fruit; C, mature fruit. (All × 2.)

plants with starry white flowers. There are 5 sepals, 5 free bifid petals, sometimes absent, with dry or papery margins, 10 stamens inserted below the other flower parts, and 3 styles.

*S. pungens*. Common Starwort. White. Alpine, subalpine, montane. N.S.W., Vic., Tas.

A small spreading plant extending well below montane habitats and sometimes growing so luxuriantly as to be regarded as a pest, though very attractive in full bloom. Branches 4-angled, almost winged. Leaves smooth, stiff, almost stalkless, broad with sharp points, 3–10 mm long. Flowers solitary, starry, 10–15 mm in diameter; petals deeply cleft.

*S. angustifolia*, *S. flaccida*, *S. multiflora* and *S. palustris* are also common in New South Wales, Victoria and Tasmania, the first two at alpine and subalpine levels, the others at subalpine and montane levels. *S. angustifolia* is a moisture-loving species with smooth, narrow, blunt leaves and long, weak branches. *S. flaccida* reaches 70 cm in height with rounded leaves and 5 sepals and 5 fairly large petals on long stalks. *S. multiflora* grows about 15 cm high and has narrow lance-shaped leaves and white flowers with 5 large petal-like pointed sepals and no petals. *S. palustris* grows to 15 cm or more in height with narrow leaves and stalked flowers with 5 sepals and 5 petals.

**Family PORTULACACEAE**

The high moors are renowned for the exquisite beauty of their cover plants, which have adopted this mode of growth to meet the exigencies of a severe climate. Among the most widespread and charming of these is *Neopaxia*. Until quite recently this genus was described as two separate genera, *Montia* and *Claytonia*, the former with united petals split on one side and with 3–5 stamens inserted at the summit of the corolla tube, the latter with free petals or petals joined at the base only and 5 stamens opposite the petals.

*NEOPAXIA* (name replaces *Paxia*, after Ferdinand Pax, 1858-1942, German botanist). Fig. 63.

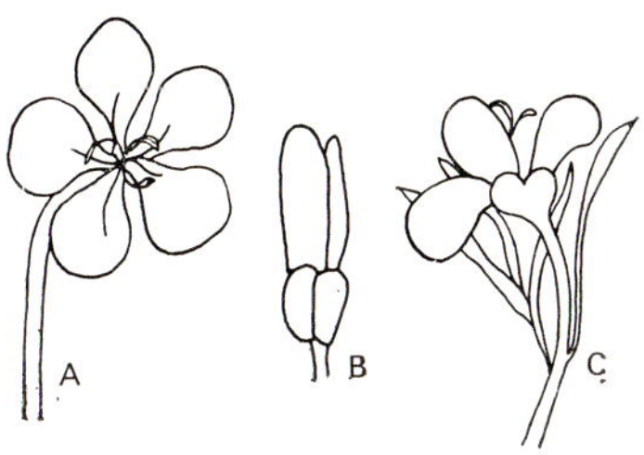


FIG. 63 *Neopaxia*: A, flower; B, fruit; C, flower side view. (All × 1½.)

*N. australasica*. White or pink. Alpine, subalpine, montane. N.S.W., Vic., Tas.
A small creeping plant preferring swampy depressed areas or moist crevices, one form truly aquatic and opening its flowers beneath the thin layer of snow at the periphery of snowdrifts. Leaves variable, erect, alternate, rather fleshy, forming a basal rosette but occurring along the stem also, compressed, cylindrical, bright- or grey-green, smooth, 2–9 cm long. Flowers conspicuous, starry, rather waxy, on long slender stalks, usually several together; sepals 2, petals 5, 5–8 mm long, the latter united and expanded but split down one side or free but united at the base with broad, free slightly unequal lobes, 3–8 mm long; stamens 5 opposite the 3 smaller petals. Fruit a small capsule.

A related plant, *Montia fontana*, with white flowers occurs in similar situations in Tasmania.

**Family MALVACEAE**

Represented by two genera only in subalpine and montane areas, being mainly a family of tropical herbs, shrubs or trees.

*GYNATRIX* (syn. *PLAGIANTHUS*) (from Gk *gyne*, woman, and *thrix, trikos,* a hair, referring to the hairy ovary). A small tree or shrub with soft leaves, small flowers with 5 sepals and 5 petals, and many stamens joined to form a tube.

*G. pulchella*. Greenish. Subalpine, montane. N.S.W., Vic., Tas.
Leaves smooth, oval, heart-shaped, 5–8 cm long. Flowers clustered in leaf axils, male and female flowers on separate plants; stamens 20–30. Fruit of 5 separate parts.

*ASTEROTRICHION* (from Gk *aster*, star, and *thrix*, *trikos*, hair).

*A. discolor*. Creamish-white. Montane. Tas.
Leaves stalked, broad, pointed, toothed, 5–15 cm long. Male and female flowers clustered in leaf axils and on separate plants. Fruit of 2 separate parts.

**Family STERCULIACEAE**

A small family of woody plants represented in these parts by a single species.

*BRACHYCHITON* (from Gk *brakhys*, short, and *khiton*, an undergarment, referring to the loose outer covering of the seeds). Fig. 64.

*B. populneum*. Kurrajong. Cream, blotched pink. Montane. N.S.W.
A handsome tree of characteristic pyramidal shape growing to a height of 6–18 metres occurring at a few localities at this altitude. Leaves broad or 3–5 lobed, about 4 cm long, on slender stalks. Sepals petal-like, forming a broad, short tube with expanded tips, 1.8–2 cm in diameter; petals absent; stamens 10–15 joined

FIG. 64 *Brachychiton*: A, leaf; B, flower; C, fruit. (All × ½.)

to form a tube inserted below the other flower parts; flowers male or female or sometimes both; carpels 5, surmounted by a style opening out into 5 stigmas, each forming a single dry splitting follicle.

## Family HYPERICACEAE

St John's wort or hypericums are small plants with stalkless leaves and yellow flowers with 5 sepals and 5 free petals, overlapping or contorted in bud. The stamens are indefinite, in 3–5 bundles, inserted below the other flower parts. The fruit is a capsule.

*HYPERICUM* (from Greco-Latin *hypericon*, name of some plant of uncertain identity). Fig. 65.

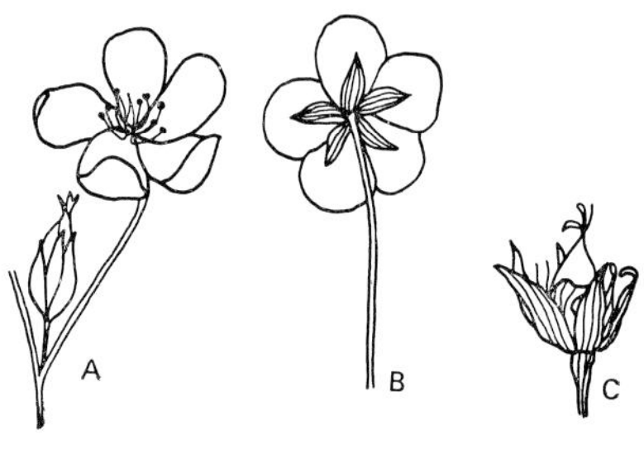


FIG. 65 *Hypericum*: A, flower; B, flower rear view; C, fruit. (All × 2.)

*H. gramineum.* Yellow. Montane. N.S.W., Vic., Tas.

A slender, smooth little plant with 4-angled stems 10–40 cm high. Leaves oblong, blunt, heart-shaped at the base, 5–20 cm long, the margins rolled back. Flowers solitary or few at the ends of the branches; sepals pointed, 5–8 mm long; petals longer, twisted in bud.

*H. japonicum.* Yellow. Alpine, subalpine, montane. N.S.W., Vic., Tas.

A slender shrub with stems 5–15 cm long. Leaves egg-shaped, flat, 4–8 mm long. Flowers few or solitary at the ends of the branches; sepals pointed, 5–8 mm long; petals longer, twisted in bud.

## Family ELAEOCARPACEAE

A small family of trees or shrubs. The flowers are regular with 4–5 sepals and 4–5 petals, the latter often fringed. The stamens are indefinite, inserted below the other flower parts.

*ARISTOTELIA* (after Aristotle, Greek philosopher). Fig. 66.

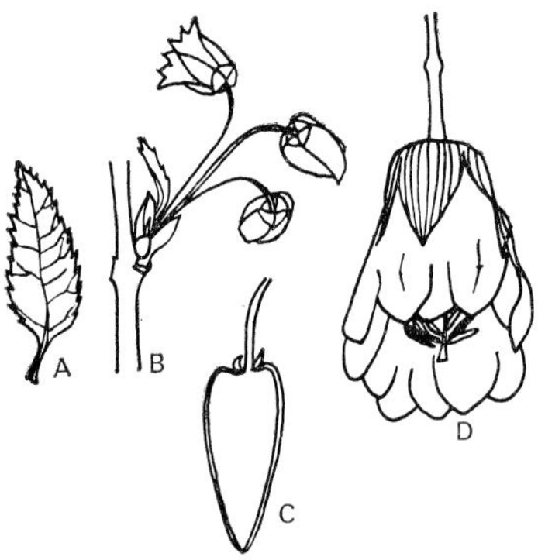


FIG. 66 *Aristotelia*: A, leaf (× ½); B, flowerhead (actual size); C, fruit (actual size); D, flower (× 4).

*A. peduncularis.* Heart Berry. White; fruit purplish-pink or red. Subalpine, montane. Tas.

A straggling plant 2–4 metres high. Leaves irregularly opposite or in rings of 3, shortly stalked, smooth, variable in shape, toothed, 1.5–7 cm long. Flowers solitary or 2–3 together in leaf axils on long, slender stalks; sepals 4, each 6–8 mm long; petals 4–5, deeply 3-lobed, 10–12 mm long; stamens usually 12, in 4 groups. Fruit an attractive berry 1–1.5 cm long.

*ELAEOCARPUS* (from Gk *elaia*, an olive tree, and *karpos*, a fruit). Fig. 67.

FIG. 67 *Elaeocarpus*: A, leaf; B. flowerhead; C, flower; D, fruit. (All actual size except C × 3.)

*E. holopetalus*. Black Olive Berry. White. Montane. N.S.W., Vic.

A lovely tree, sometimes very tall. Leaves broad, pointed, 5–10 cm long Flowers few in loose clusters, cup-shaped; petals 5, slightly wavy, 4–5 mm long, surrounding 15–20 stamens. Fruit succulent, ovoid, black.

## Family LINACEAE

*LINUM* (Latin name of the cultivated flax). Fig. 68.

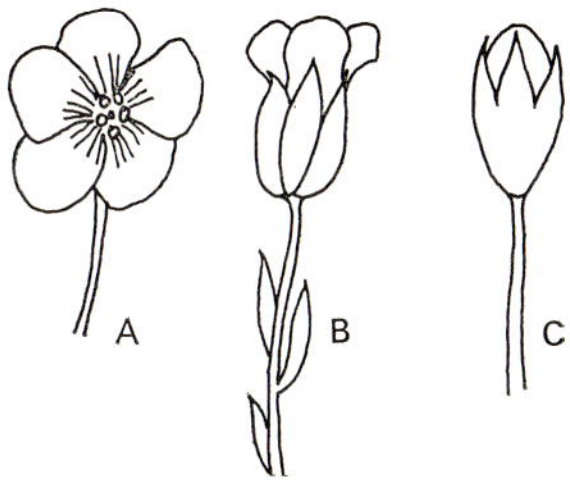


FIG. 68 *Linum*: A, flower; B, bud; C, fruit. (All × 2.)

*L. marginale*. Blue. Subalpine, montane. N.S.W., Vic., Tas.

This small, smooth plant grows 30–60 cm high with slender stems and branches and narrow, stalkless, pointed leaves 5–25 cm long. Flowers small, very delicate, in loose heads; sepals 5, broad, pointed, 4–6 mm long; petals 5, egg-shaped, translucent, free, fragile, falling soon after opening, 8–12 mm long; stamens 5, inserted below the other flower parts and alternating with 5 staminodes; ovary of 5 parts; stigmas 5. Fruit a capsule about 5 mm long, surrounded by the persistent sepals.

## Family GERANIACEAE

Soft-leafed plants with few Australian representatives. The flowers are regular or slightly irregular with 5 free sepals, 5 petals and 5, 10 or 15 stamens, usually in 2 rows, joined at the base and inserted below the other flower parts. The fruit is dry, elongated, separating into beaked or awned fruitlets.

*GERANIUM* (from Gk *peranos*, a crane, the fruit somewhat resembling a crane's head and bill). Fig. 69. This genus has large dissected or lobed leaves and regular flowers solitary or 2 together on equal stems, each with 10 stamens and a 5-lobed ovary with a 5-fid style. The fruit separates into 5 segments from the base.

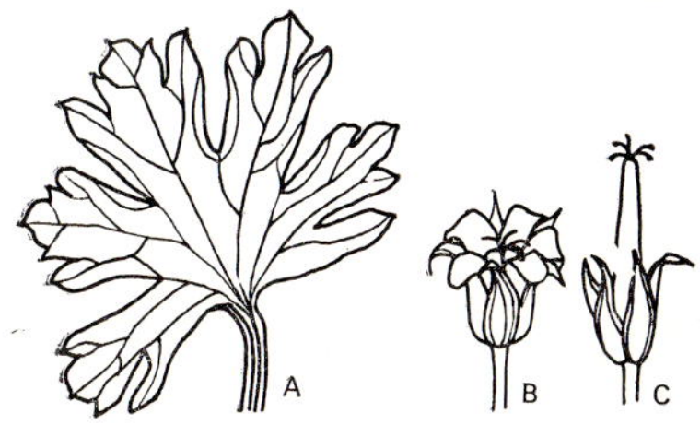


FIG. 69 *Geranium*: A, leaf; B, flower; C, fruit. (All slightly enlarged.)

*G. antrorsum* (syn. *G. potentilloides*). Pink, striped. Subalpine, montane. N.S.W., Vic., Tas.

A small, trailing, hairy plant. Leaves opposite, rounded in outline, deeply 5–7 lobed, often bluntly lobed again, 1–3 cm long, softly hairy, on slender stems. Flowers solitary with 2 bracts on the slender stem, softly hairy; petals egg-shaped, 5–6 cm long; stamens 10. Fruit hairy, 3.5 mm long; seeds brown.

Three closely related species are *G. obtusisepalum,* a subalpine and montane species from New South Wales which differs in having blunt petals and dark-brown seeds; *G. graniticola,* from montane habitats in New South Wales which differs in having white petals and the sepals diverging in the fruiting stage; and *G. solanderi,* a montane and lower elevations species from the mainland which differs in having twinned flowers and black or very dark-brown seeds.

*G. sessiliflorum.* Deep pink, white. Alpine, subalpine. N.S.W., Vic., Tas.
A small, compact, hairy plant with crowded, almost radical leaves, round or kidney-shaped in outline, 5–7 lobed, and again divided, hairy, 10–20 cm long. Flowers solitary on short, hairy stalks, the hairs pressed against the stem and turned backwards; sepals hairy, pointed, 4–8 mm long; petals narrow, 4–8 mm long; stamens 10. Fruit about 3 mm long; seeds black or very dark-brown.

*PELARGONIUM* (from Gk *pelargos*, a stork, referring to the fruit). Fig. 70. A genus of plants with slightly irregular flowers, the sepals produced into a short spur joined to the flower stalk. There are 10 stamens with 3–8 bearing anthers.

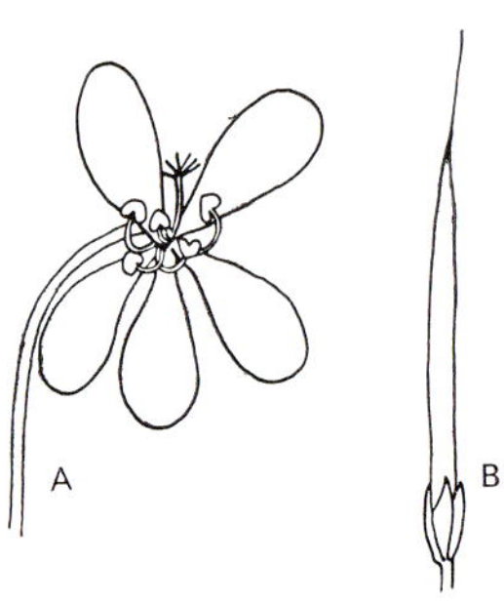


FIG. 70 *Pelargonium*: A, flower (actual size); B, fruit (× ¾).

*P. australe.* Pink. Alpine, subalpine. N.S.W., Vic., Tas.
A delightful, almost odourless small plant 15–60 cm high, common in other habitats also. Leaves opposite, rounded, with 5–7 usually shallow lobes, wavy, 2–9 cm long with broad stipules. Flowers in loose clusters of 4–12; sepals hairy, pointed, 4–6 mm long; petals egg-shaped, pointed, 8–11 mm long, the 2 upper ones broader than the others and striped; fertile stamens 7–8. Fruit splitting into beaked, stork-like sections.

*P. helmsii.* Deep pink. Subalpine. N.S.W.
Closely related to *P. inodorum*, this is a more strongly fragrant species. Flowers 5–12 in a head; sepals blunt, 1 mm long and covered with thick, short, closely packed hairs; petals about 8 mm long.

*P. inodorum.* Rose pink. Montane. N.S.W., Vic., Tas.
A small, slender, slightly hairy, faintly fragrant plant. Leaves opposite, egg-shaped, with 5–7 shallow lobes, wavy, 1.5 cm broad. Flowers 3–14 in a cluster, closely packed; sepals 3.5 mm long, pointed, covered with coarse, short, scattered hairs; petals 2–4 mm long, the lower ones shorter, striped; fertile stamens 3–5.

## Family OXALIDACEAE

*OXALIS* (Greco-Lat. name of some plant with sour leaves). Fig. 71.

FIG. 71 *Oxalis*: flower and fruit (actual size).

*O. lactea.* Wood Sorrel. White. Subalpine, montane. Vic., Tas.
A small creeping plant from damp places. The leaves are clover-like with 3 broad leaflets, 1–1.5 cm in diameter and in a basal rosette. Flowers solitary on long stalks; sepals 5, hairy, 3–5 mm long; petals 5, membranous, egg-shaped, 6–11 mm long; stamens 10, inserted below the other flower parts. Fruit a globular capsule.

## Family RUTACEAE

The boronias and their relatives are not very well represented in alpine areas, but there are a few attractive and interesting species. Members of the family have aromatic oil in their leaves. The flowers have 4–5 sepals and petals, free except in most correas, in which the petals are fused to form a tube. The flowers are commonly pink or white. There are the same number or twice the number of stamens as petals inserted below the other flower parts and the same number of carpels as petals. The fruit consists of a number of small non-splitting nuts.

*ASTEROLASIA* (from Gk *aster*, a star, and *lasios*, hairy, referring to the stellate hairs on these plants). Fig. 72. A genus of shrubs with simple, alternate leaves. The flowers have 5 very small sepals, 5 petals and 10 stamens, and the styles are united.

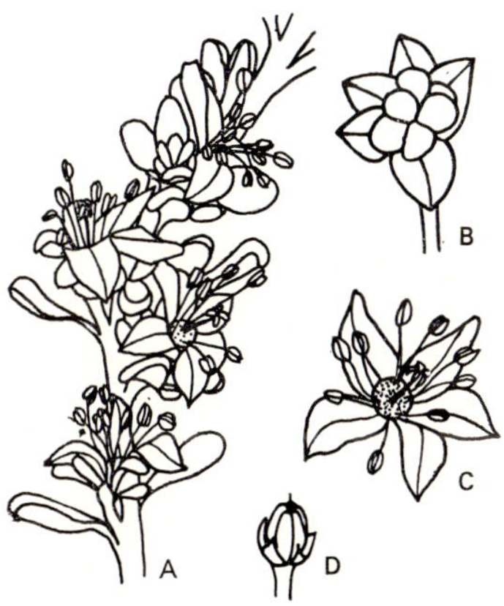


FIG. 72 *Asterolasia*: A, flowerhead (× 2); B, mature fruit (× 3); C, flower (× 3); D, young flower (× 2); C, fruit (× 2).

*A. asteriscophora.* Yellow. Subalpine, montane. N.S.W., Vic.
A slender, branching shrub 30–120 cm high. Leaves elliptical, rather large, hairy beneath. Flowers in small heads; sepals minute; stalks of stamens smooth; anthers blunt.

*A. trymalioides.* Yellow, cream. Montane, N.S.W., Vic.
A small rigid shrub with crowded, very small, smooth, egg-shaped or rounded leaves about 5–8 mm long. Flowers solitary or 2–3 together without stalks; petals without stalks, terminal, woolly outside; stigma conspicuous.

*BORONIA* (after Francesco Borone, botanical collector, d. 1794). Fig. 73. A genus of plants in which the flowers have 4 open petals and 8 stamens, and the leaves are usually opposite and divided.

FIG. 73 *Boronia*: A, flower; B, fruit. (Both × 1½.)

*B. algida.* White, pink. Subalpine, N.S.W., Vic.
A dainty little dwarf, often scrambling, with thick divided leaves; leaflets about 5, each 3–4 mm long. Flowers solitary at the ends of or in the fork of the branches, wide open and starry; sepals small and pointed; Petals somewhat pointed, 5–6 mm long.

*B. citriodora.* Pink, white. Subalpine, montane. Tas.
A lovely little plant of the moors, extending to the coast in Tasmania, this is a small, softly woolly shrub, sometimes quite prostrate but in protected positions reaching a height of 30–60 cm. Leaves with 3–7 narrow rather thick leaflets, each 5–15 mm long on stout stems with a fragrant lemon odour when crushed. Flowers solitary or 2–3 together; petals egg-shaped, 6–8 mm long.

*B. polygalifolia* (syn. *B. oppositifolia*). Pink. Subalpine, montane. N.S.W.

A low, smooth shrub. Leaves simple, 5–12 mm long, rather narrow. Flowers usually solitary in the leaf axils; petals 4–6 mm long. Stalks of stamens hairy; anthers with a conspicuous small abrupt point. Widely distributed as far as the coast, often in swampy land.

*B. rhomboidea*. Pink. Subalpine, montane. N.S.W., Tas.

A delightful little plant, completely prostrate at high elevations but reaching a height of 30–90 cm at lower elevations. Leaves without stalks, broadly rhomboid or egg-shaped, leathery, smooth, 5–12 mm long. Flowers solitary, or 2–3 together.

*CORREA* (after Jose Correa de Serra, Portuguese botanist, 1751–1823). Fig. 74. Plants with opposite, simple leaves, 4 sepals, 4 petals fused to form a comparatively long tube free at the tip, and 8 stamens.

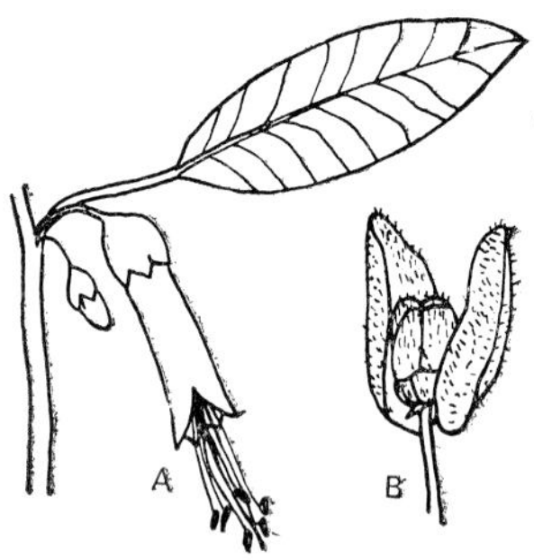


FIG. 74 *Correa*: A, flower; B, fruit. (Both actual size.)

*C. lawrenciana* var. *rubra*. Red. Subalpine. N.S.W., Tas. A lovely plant with an entirely different appearance from other members of the family because of the bell-shaped flowers. It is a robust shrub 1–4 metres high, often found near water. Leaves dark green, paler on the undersurface, shortly stalked, oblong, blunt, about 1.5–3 cm long. Flowers in leaf axils or terminal, 1–3 together on slender stalks, drooping, 5–25 cm long; the bells very attractive, in the alpine forms varying in colour from light-pink to deeply rusty-red on the same bush; some forms yellow. In Tasmania it extends to the coast.

*C. reflexa*. Red, yellow. Montane. N.S.W., Vic., Tas.

A variable rough shrub 1–5 metres high covering a wide range of lower habitats also. Leaves shortly stalked, egg-shaped with a heart-shaped base, scabrous on the upper surface, loosely woolly beneath. Flowers 1–3 together in the leaf axils, drooping, the tube 2–3.5 cm long with short, expanded tips.

*ERIOSTEMON* (from Gk *erion*, wool, and *stemon*, a stamen, referring to the woolly stalks of the stamens). Fig. 75. Plants with simple alternate leaves dotted with conspicuous oil glands. There are 5 sepals and 5 petals, and 8 or 10 stamens. The genus differs from *Phebalium* only in having free overlapping tips to the petals, which are not inflexed; and from *Asterolasia* in having a fairly conspicuous calyx and small free styles.

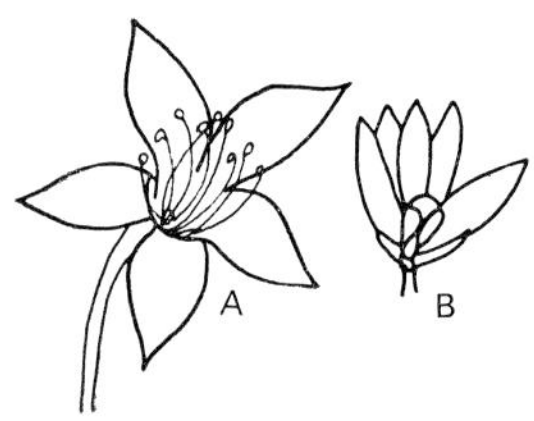


FIG. 75 *Eriostemon*: A, flower (× 1½); B, fruit (actual size).

*E. myoporoides*. White; buds pink. Montane. N.S.W., Vic.

Well known in cultivation, this is a handsome plant which extends to the mountain slopes. Very variable, commonly 90–150 cm high. Leaves dark-green, strongly aromatic, varying from egg-shaped to narrow, 2.5–10 cm long. Flowers open, 2–9 in a loose head; petals 6–10 cm long.

*E. trachyphyllus*. White. Montane. N.S.W., Vic. Extending to mountain slopes, this is a smooth shrub 60–90 cm high, conspicuously tuberculate on the stem. Leaves rather thin, more or less oblong with prominent midribs, 2.5–6 cm long. Flowers usually solitary; petals 7–8 mm long.

*PHEBALIUM* (apparently adapted from Gk name for an early fig-tree, *phebaleos*). Fig. 76. A genus of shrubs with alternate leaves. The flowers have 5 sepals and 5 petals slightly overlapping or just meeting one another in bud, the

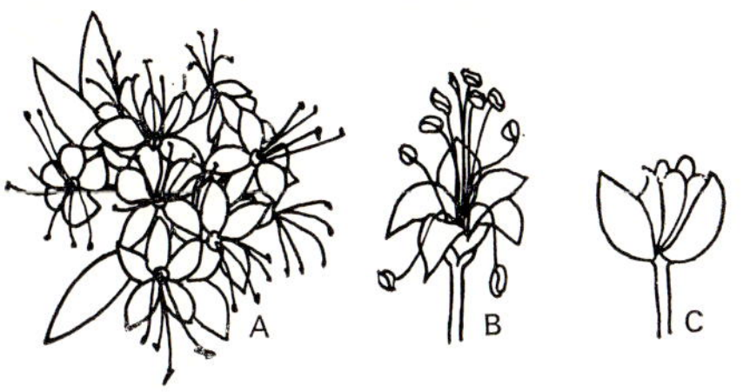


FIG. 76 *Phebalium*: A, flowerhead (× 1½); B, flower (× 2); C, fruit (× 2).

tips inflexed and strictly valvate, and 10 free stamens, usually protruding beyond the petals.

*P. lamprophyllum*. Yellow. Subalpine. N.S.W., Vic.
A small, smooth, bushy shrub with tubercles on the branches and fairly broad, shining, leathery leaves 10–12 mm long dotted with small glands. Flowers few, in small heads, with a small leaf at the base of each flower stalk.

*P. montanum*. White, pinkish. Alpine, subalpine. Tas.
A prostrate or almost completely prostrate shrub, the branches 15–50 cm long, the lower parts marked by prominent leaf scars. Leaves shortly stalked, thick or terete, with blunt points, crowded, about 1 cm long and with conspicuous glandular dots. Flowers solitary in the axils of the upper leaves, on short stalks, with 2 bracts just below the sepals; petals narrowly egg-shaped, 5–6 mm long.

*P. oldfieldii*. White, pinkish. Alpine, subalpine. Tas.
A shrub 30–90 cm in height, usually smooth, with narrow-oblong, crowded leaves, bluntly pointed, 10–12 mm long, the upper surface shining, the undersurface pale and covered with glandular dots. Flowers few, in clusters on short stalks; sepals small, egg-shaped, thick; petals rather narrow, 1-veined with several large dark glands on the back near the tip, 4–5 mm long.

*P. ovatifolium*. White; buds pink. Alpine, subalpine. N.S.W., Vic.
A very attractive shrub, commonly prostrate and scrambling over rocks but sometimes erect. Branches scurfy; leaves dark-green, stalked, broadly egg-shaped and blunt, about 13 mm in length, smooth and shining above, scurfy below. Flowers solitary in the leaf axils, produced in profusion to make a grand display at their best; petals 5–8 mm long.

*P. ozothamnoides*. Yellow. Subalpine, montane. N.S.W., Vic.
A stiff small shrub about 20–50 cm high with brown scurfy scales. Leaves egg-shaped, blunt, smooth above and covered with white hairs and scurfy scales below, about 10–12 mm long. Flowers small, crowded into small dense terminal clusters and covered with large scurfy scales which give the flowers an "everlasting" quality.

*P. phylicifolium*. Mountain Phebalium. Pale yellow. Subalpine, montane. N.S.W., Vic.
A dainty little hairy plant about 30–60 cm high. Leaves narrow, blunt, 8–12 mm long. Flowers rather large, crowded into small heads towards the ends of the branches.

*P. podocarpoides* (syn. *P. squamulosum* var. *alpinum*). Creamy yellow. Subalpine. N.S.W., Vic.
A fairly low-growing shrub with narrow, rounded leaves, recurved at the margins and scaly underneath, about 6–8 mm long. Flowers in small terminal heads.

*ZIERIA* (after John Zier, Polish botanist, d. 1796). Fig. 77. A genus of shrubs with opposite leaves, and flowers resembling those of *Boronia*, with 4 open petals, but only 4 stamens.

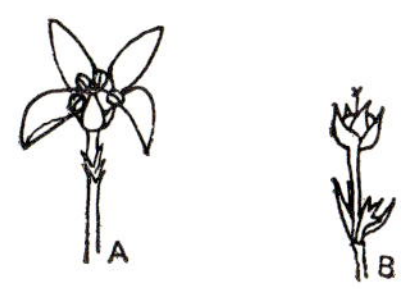


FIG. 77 *Zieria*: A, flower (actual size); B, fruit (× ½).

*Z. aspalathoides*. Pink. Montane. N.S.W., Vic.
A dainty, rather heath-like shrub with hairy branches. Leaves trifoliate, without stalks, leaflets narrow, 6–8 mm long. Flowers small; petals 3–6 mm in length.

### Family RHAMNACEAE

An interesting family not uncommon in Australia, with several representatives in montane areas. They are woody plants with alternate, simple, shortly stalked leaves. The flowers are small, regular, usually clustered. There are 4–5 valvate sepals, each with a prominent midrib. The petals number 4–5 or are absent, and often clawed and hood-shaped; there are 4–5 stamens, often enclosed by the petals and inserted above the other flower parts or on a rim round the ovary. The fruits are various.

*CRYPTANDRA* (from Gk *kryptos*, hidden, and *andros*, a man, the anthers being hidden by the petals ). Fig. 78.

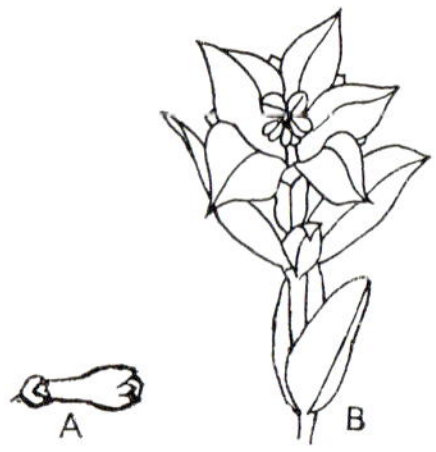


FIG. 78 *Cryptandra*: A, enclosed fruit (× ½); B, flowerhead (× 3).

*C. ericoides*. White. Montane. N.S.W., Vic.
A small spiny shrub with a wide distribution in other habitats also. Leaves narrow, hairy, 4–8 mm long. Flowers small, in terminal heads surrounded by persistent brown or reddish bracts; sepals 5; petals 5, hooded. Fruit a capsule enclosed in the persistent petals.

*DISCARIA* (from *Gk diskos*, round, and *aria*, a kind of oak). Fig. 79.

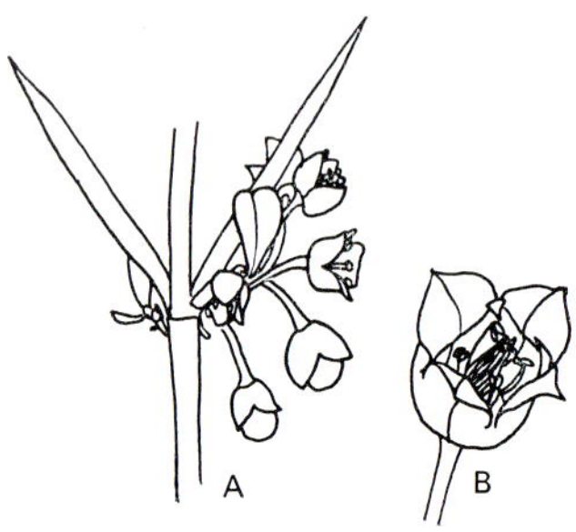


FIG. 79 *Discaria*: A, flowerhead, showing flowers and fruit; B, flower (× 4).

*D. pubescens*. Greenish. Montane. N.S.W., Vic., Tas.
A thorny shrub 30–90 cm high, the smaller branches reduced to spines. Leaves opposite but appearing to be clustered because of the shortness of the shoots. Flowers small, on short stalks; petals 4–5, hooded, falling soon after flowering; stamens 4–5. Fruit a round capsule.

*POMADERRIS* (from Gk *poma*, a lid, and *derris*, skin, referring to the membranous valve on the fruit of many species). Fig. 80. A genus

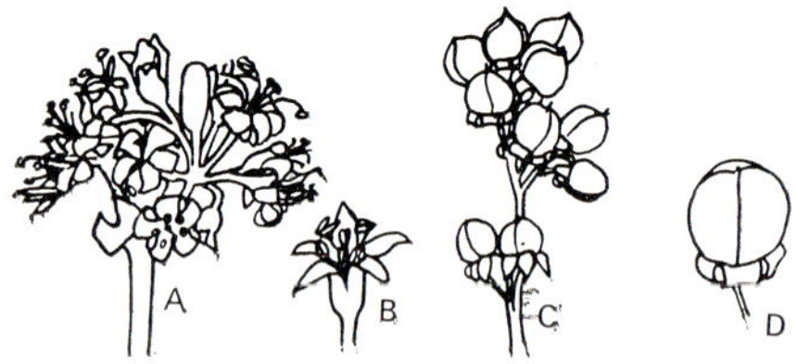


FIG. 80 *Pomaderris*: A, flowerhead (actual size); B, flower (× 2); C, fruits (× 1½); D, single fruit (× 3).

of hairy plants with 5 sepals, 5 petals (sometimes absent) and 5 free stamens (sometimes absent) not enclosed by the petals. Fruit a dry, splitting capsule.

*P. apetala*. Greenish-brown. Montane. N.S.W., Tas.
A tall shrub up to 10 metres high. Leaves broad, rough, wrinkled, 4–10 cm long. Flowers in small loose heads, much branched; sepals 5, each 2.5 cm in length; petals absent. Capsule 2–3 mm in diameter.

*P. betulina*. Brownish. Montane. N.S.W., Vic.
A slender shrub 4–5 metres high. Leaves rounded, rusty-white beneath, toothed, about 2.5 cm long. Flowers in dense globular heads surrounded by broad bracts; sepals 5; petals absent.

*P. pallida*. Cream. Montane. Vic.
Leaves narrow-elliptical, about 1 cm long, the upper surface grey-green, the lower slightly hairy. Flowers in small leafy heads, hairy.

*P. phylicifolia*. Brownish. Montane. N.S.W., Vic.
A shrub 1–3 metres high. Leaves narrow with stiff, short hairs above and a dense mat of hairs below, shortly stalked, narrow-oblong or broader on young plants, 6–12 mm long. Flowers in short, numerous heads; sepals 5, petals absent.

*SPYRIDIUM* (from Gk *spyridion*, a little basket, in reference to the flowerheads which are surrounded by leafy bracts). Fig. 81.

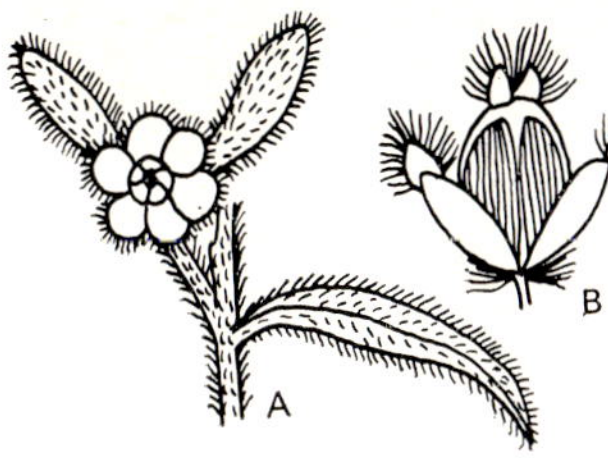


FIG. 81 *Spyridium*: A, flowerhead (× 2); B, fruit (× 4).

*S. parvifolium.* Dusty Miller. Silky white. Montane, subalpine. N.S.W., Vic.
A small shrub 1–3 metres high with egg-shaped or rounded leaves, 6–18 mm long, smooth above and hairy and white beneath. The flowers are in clusters at the ends of the stalks and surrounded by whitish and brown bracts; individual flowers small with 5 sepals, 5 petals and 5 stamens; ovary in 3 parts, developing into a small, dry, splitting capsule crowned by the persistent sepals.

**Family STACKHOUSIACEAE**

*STACKHOUSIA* (after John Stackhouse, Cornish botanist, 1742-1819). Fig. 82. A genus of small plants with alternate, entire leaves and

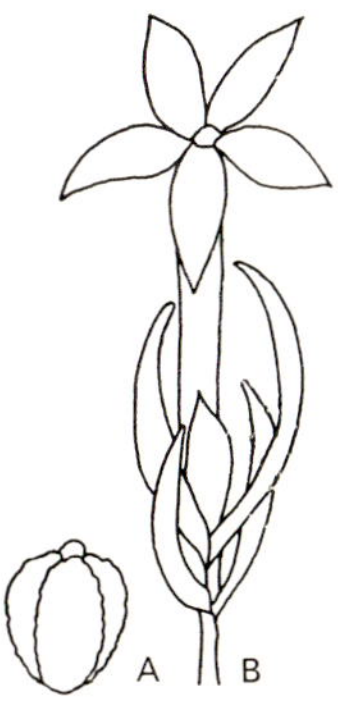


FIG. 82 *Stackhousia*: A, fruit; B, flowerhead. (Both × 4.)

fragrant flowers usually in terminal spikes, each flower with 5 sepals, 5 tubular petals and 5 stamens on a rim surrounding the carpels. The fruit consists of 3 separate non-splitting carpels.

*S. monogyna.* White, cream. Montane. N.S.W., Vic., Tas.
A soft plant 30–80 cm high with smooth, rather thick, narrow, sharply pointed leaves 1–4 cm long. Flowers in narrow heads, dense at first but later becoming more distant, sweetly scented, narrowly tubular, 5–7 mm long with narrow, spreading lobes.

*S. pulvinaris.* Soft rich yellow. Alpine, subalpine, N.S.W., Vic., Tas.
A charming little prostrate plant of the wet patches and depressions of the short alpine herbfields and along the watercourses of subalpine habitats, recognizable at a considerable distance by its fragrant and penetrating perfume. A tufted little plant with crowded, blunt, narrow, light-green leaves 5–8 mm long. Flowers solitary in leaf axils, crowded, tiny, about 3 mm long, expanding to form stars dotting its green cushion.

*S. viminea.* Yellow. Montane. N.S.W., Vic.
A variable plant 12–45 cm high with numerous stems. Leaves broad, blunt, 1.5–2.5 cm long. Flowers in rather distant clusters, sweetly scented, tubular, about 5 mm long.

**Family SAPINDACEAE**

*DODONAEA* (from Dodonaeus, Latinized name of Rembert Dodoens, botanist, 1517–1588). Fig. 83.

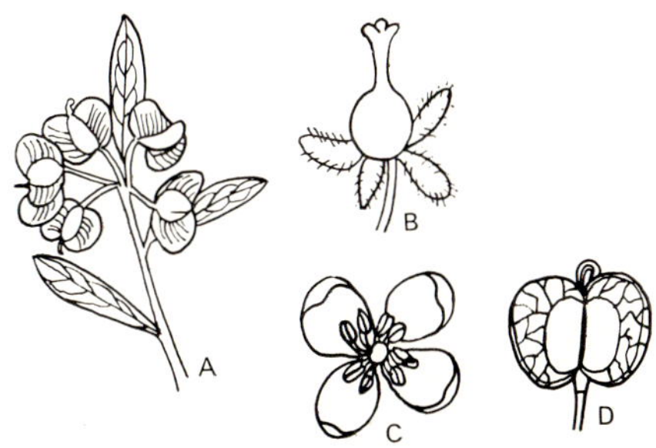


FIG. 83 *Dodonaea*: A, flowerhead; B, female flower; C, male flower; D, fruit. (All × 2 except D, actual size.)

The dodonaeas or hop-bushes are woody plants with inconspicuous flowers, either male or female, without petals and with angular winged fruits. There are usually 5 small sepals and 8 stamens inserted above the other parts of the flower. The ovary has 3 parts, with long styles which develop into single-seeded non-splitting nuts.

*D. procumbens.* Greenish. Montane. N.S.W., Vic.
A low, scrambling, smooth shrub with triangular or spoon-shaped, lobed or 3– or 5–toothed, crowded leaves, 1.2–2 cm long. The fruits have broad wings.

*D. viscosa.* Greenish. Montane. N.S.W., Vic.
A variable shrub, often tall, with a thin, reddish-brown bark and oblong, pointed, simple leaves 5–10 cm long. The fruits have broad wings.

## Family MIMOSACEAE

The wattles or acacias, with more than 600 species in Australia, are not very well represented in alpine areas, particularly at the higher elevations. They are closely related to the pea-flowers, Papilionaceae, and were at one time included with them in the same family, Leguminosae. Both groups have a legume as a fruit.

In *Acacia* the leaves are always divided or feathery, but in the majority of Australian kinds these true leaves are replaced in the adult stage by flattened leaf stalks (phyllodes) which assume a variety of forms, sometimes prickly, sometimes long and thin, sometimes broad, sometimes knife-shaped, wedge-shaped, narrow and so on. The flowers are small, without stalks, massed together in dense balls or rods. (See Fig. 84).

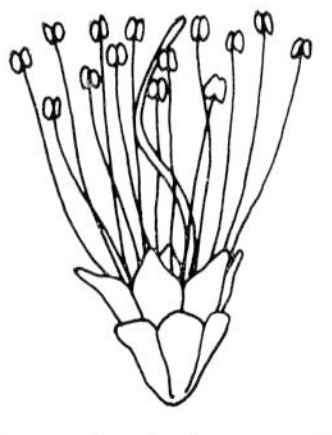

FIG. 84 A typical *Acacia* flower (× 2.)

Each flower has 4–5 sepals, 4–5 petals and a large number of attractive stamens inserted slightly above the other flower parts. Many of them are sweetly scented.

*ACACIA* (from Gk *akakia*, name of some prickly species growing in Egypt).

*A. alpina.* Yellow. Subalpine, montane. N.S.W., Vic. Fig. 85.
The only species of *Acacia* to reach beyond montane habitats. A low-growing shrub, sometimes

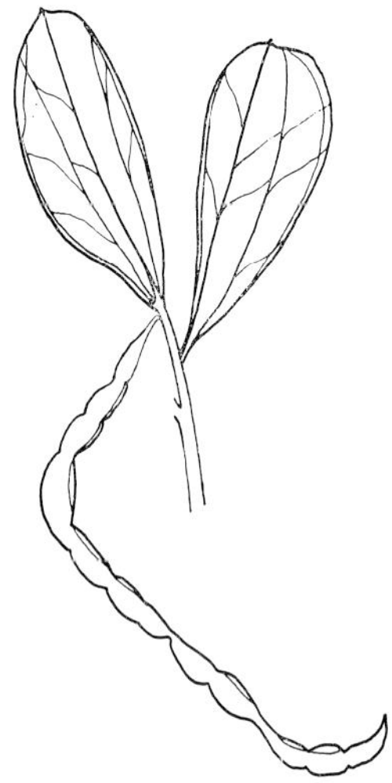

FIG. 85 *Acacia alpina*: pod and phyllode (actual size).

completely prostrate but growing erect in sheltered places. Branches angular, flattened. Phyllodes dark-green, egg-shaped, very blunt or occasionally with a short point, leathery, finely veined, 16–30 mm long. Flower spikes short with a few flowers in balls. Pod narrow.

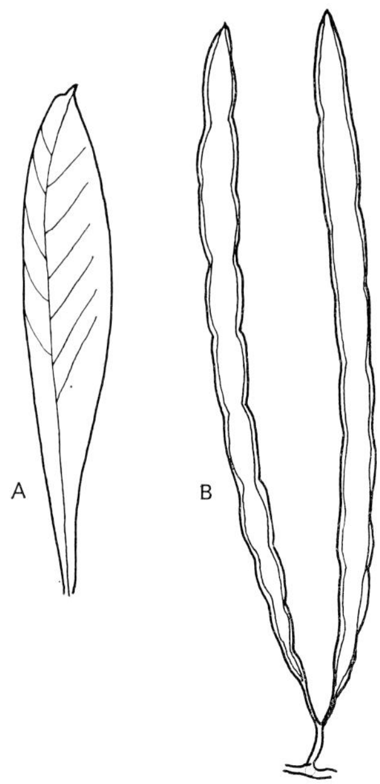


FIG. 86 *Acacia amoena*: A, phyllode; B, pods. (Both actual size.)

*A. amoena.* Yellow. Montane. N.S.W., Vic. Fig. 86.

A tall, smooth shrub. Phyllodes fairly broad, blunt, often oblique, 4–8 cm long, narrowed to the base with several veins. Flowers 8–12 together in long clusters of globular heads. Pod flat, up to 10 cm long.

*A. boormanii.* Golden. Montane. N.S.W. Fig. 87. A weak, upright shrub 1–2 metres high, with smooth, silvery stems. Phyllodes narrow, about

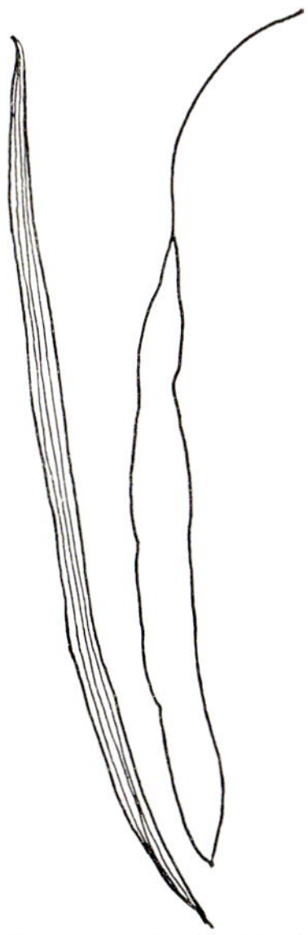

FIG. 87 *Acacia boormanii*: phyllode (× 2) and pod (actual size).

3 cm long, often hooked and with not very prominent veins. Flowers in globular heads in short clusters. Pod about 5 cm long, flat.

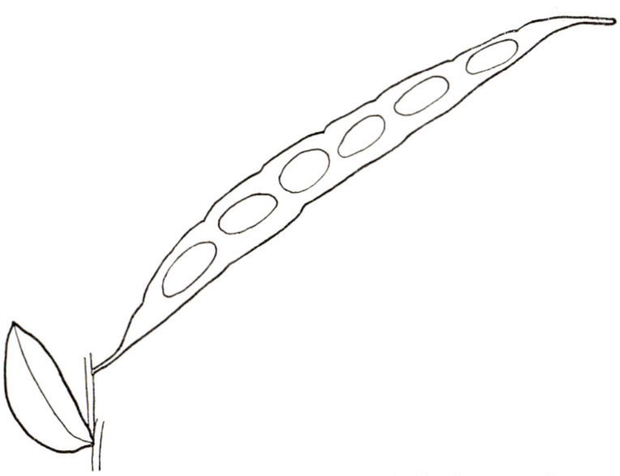

FIG. 88 *Acacia buxifolia*: phyllode and pod (actual size).

*A. buxifolia.* Golden yellow. Montane. N.S.W., Vic. Fig. 88.

A smooth shrub 3–6 metres high, rather angular, the branches often grey-green. Phyllodes oblong or egg-shaped, very oblique, leathery, 2.5 cm or less in length. Flowers 4–10, rather large individual blooms in balls, very attractive and sweet-scented. Pod flat and grey, 6–20 mm long.

*A. dallachiana.* Yellow. Montane. N.S.W., Vic. Fig. 89.

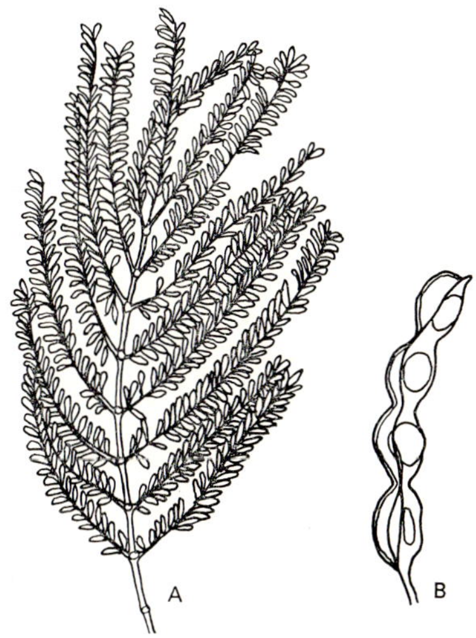


FIG. 89 *Acacia dallachiana*: A, leaf; B, pod. (Both actual size.)

A smooth, small tree with angular greyish branches. Phyllodes broad, curved, blunt, leathery, narrowed at the base, many-veined, 7.5–15 cm long. Flowers in elongated spikes. Fruit narrow, straight, about 10 cm long.

*A. dawsonii.* Yellow. Montane. N.S.W. Fig. 90. A twiggy shrub 1.5–2 metres high with erect, slightly flattened, terete branches. Phyllodes narrow, almost straight, blunt or with a small recurved point, about 10–12 cm long, stiff, many veined. Flowers in globular heads. Pod narrow, straight or slightly curved, smooth, black, about 3–5 cm long.

*A. dealbata.* Silver Wattle. Light yellow. Montane. N.S.W., Vic., Tas. Fig. 91.

One of the commonest species, often occurring in large stands, particularly in Tasmania, varying in habit and reaching a height of 30 metres or more with a slender stem and smooth, whitish

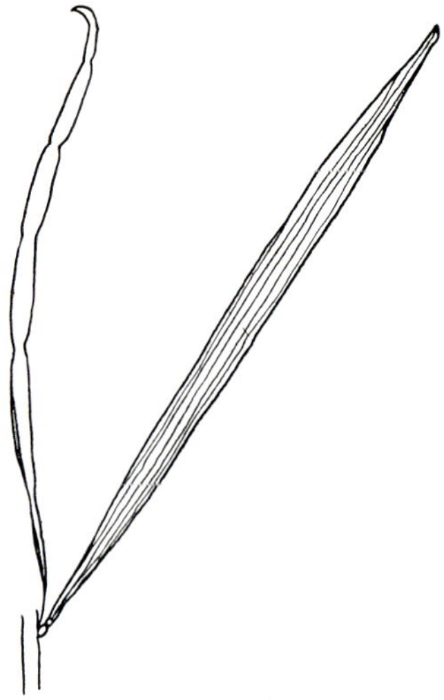

FIG. 90 *Acacia dawsonii*: phyllode and pod (actual size).

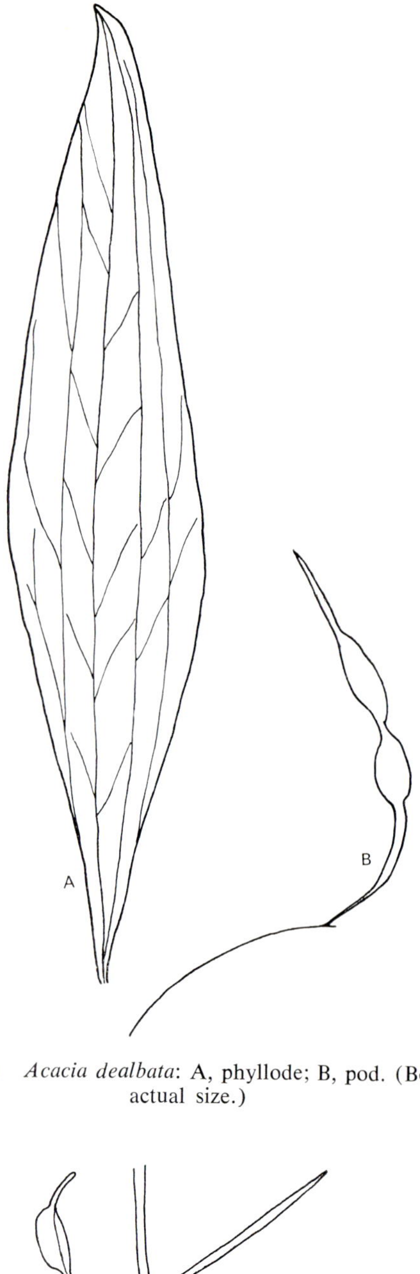


FIG. 91 *Acacia dealbata*: A, phyllode; B, pod. (Both actual size.)

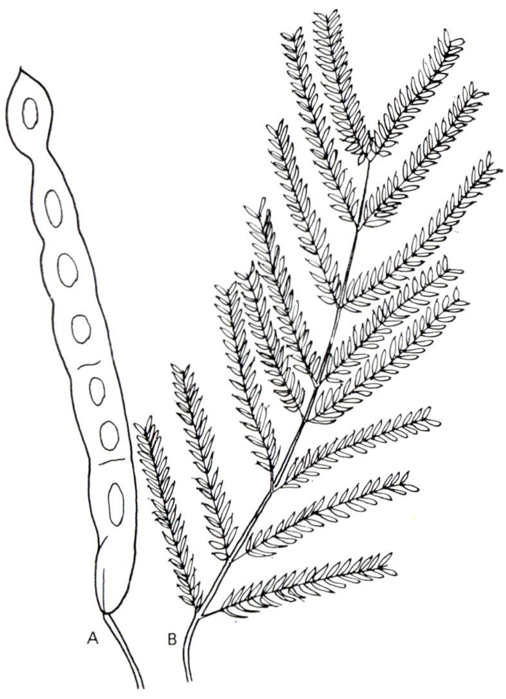


FIG. 92 *Acacia deccurrens*: A, pod; B, leaf. (Both actual size.)

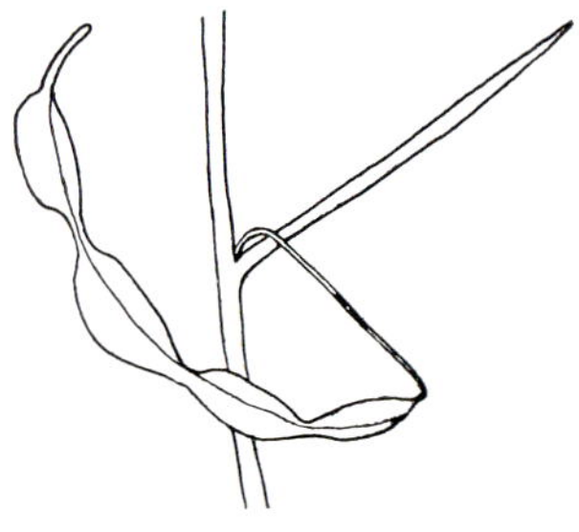

FIG. 93 *Acacia diffusa*: pod with phyllode (actual size).

bark. Leaves much-divided, feathery, delicately grey-green or silvery-white and very handsome. Flowers in small balls making a lovely picture against the soft foliage. Pod elongate, grey-green, 5–8 cm long.

J. H. Willis has described a closely related species, *A. nano-dealbata* from Victorian montane habitats, with almost smooth leaflets, 2.5 mm long or less, blunt and more closely set, the pod 4–6 cm long and broader than in *A. dealbata.*

*A. decurrens.* Yellow. Montane. Vic., N.S.W. Fig. 92.

A small tree with smooth, dark-green feathery leaves with 8–15 pairs of leaflets, pale beneath. Flowers 6–15 in globular heads in long clusters. Pod fairly narrow, 4.5–10.5 cm long.

*A. diffusa.* Pale yellow. Montane. N.S.W., Vic., Tas. Fig. 93.

A smooth, erect shrub up to 1.5 metres high with angular branches. Phyllodes narrow, thick, stiff, sharply pointed, 1-vcincd, 20–30 mm long. Flowers about 20 in a head, 2–3 heads clustered together. Pod flat, about 7.5–10 cm long.

*A. elongata.* Bright yellow. Montane. N.S.W., Vic. Fig. 94.

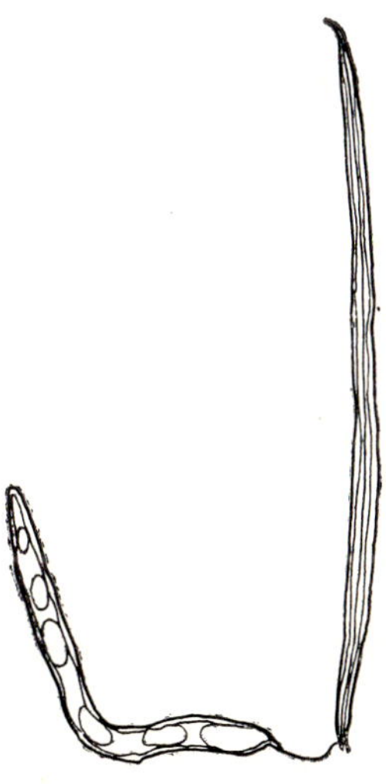

FIG. 94 *Acacia elongata*: pod with phyllode (actual size).

A spreading shrub. Phyllodes narrow with several veins. Flowers in globular heads, 1–2 together.

*A. falciformis.* Hickory Wattle. Golden yellow. Montane. N.S.W., Vic. Fig. 95.

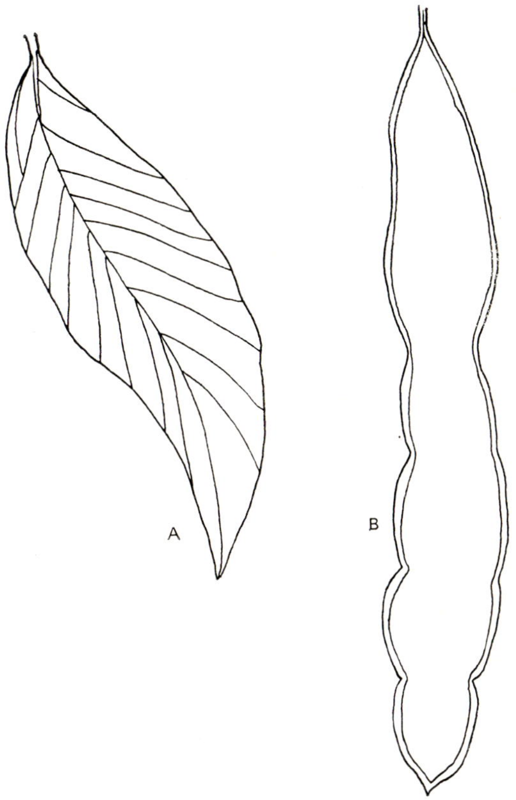


FIG. 95 *Acacia falciformis*: A, phyllode; B, pod. (Both actual size.)

A widely dispersed species outside this habitat also. At lower elevations sometimes reaching a considerable height but here more commonly a tall shrub with long, falcate, often grey-green phyllodes, prominently veined, narrowed at the base and with blunt tips, 10–20 cm long. Flowers in dense balls. Pod flat and thin, 15–20 cm long.

*A. floribunda.* Light yellow. Montane. N.S.W., Vic. Fig. 96.

An erect, smooth, tall shrub or small tree. Phyllodes narrow, pointed, 1-veined, 7.5–10 cm long. Flowerheads rod-like, in long clusters. Pod narrow, 4–10 cm long.

*A. frigescens*. Bright yellow. Subalpine, montane. Vic. Fig. 97.

This species is closely related to *A. melanoxylon*. Height 3–9 metres. Bark smooth, branches dense. Phyllodes greyish-green, rather narrow, pointed, stalked. Flowers in globular heads. Pod 4–10 cm long, hardly constricted between the seeds but often coiled and twisted; seeds with a small white aril.

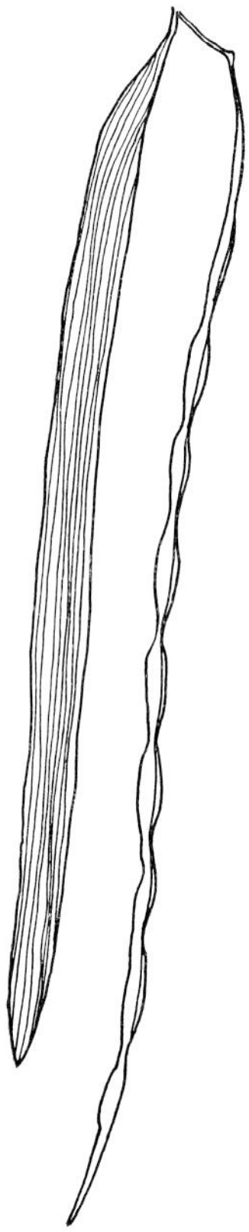

FIG. 96 *Acacia floribunda*: phyllode and pod (actual size).

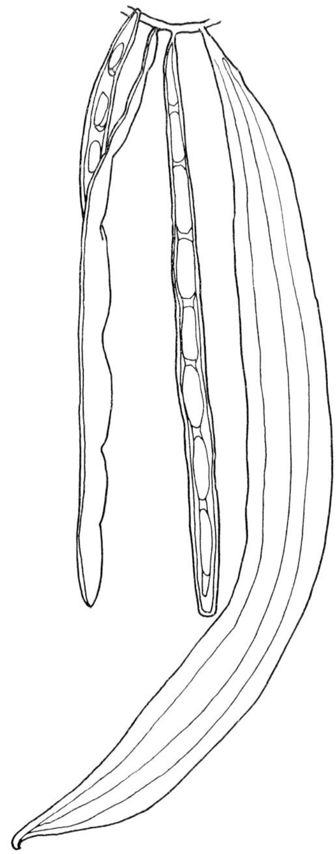

FIG. 98 *Acacia glaucescens*: phyllode and pods (actual size).

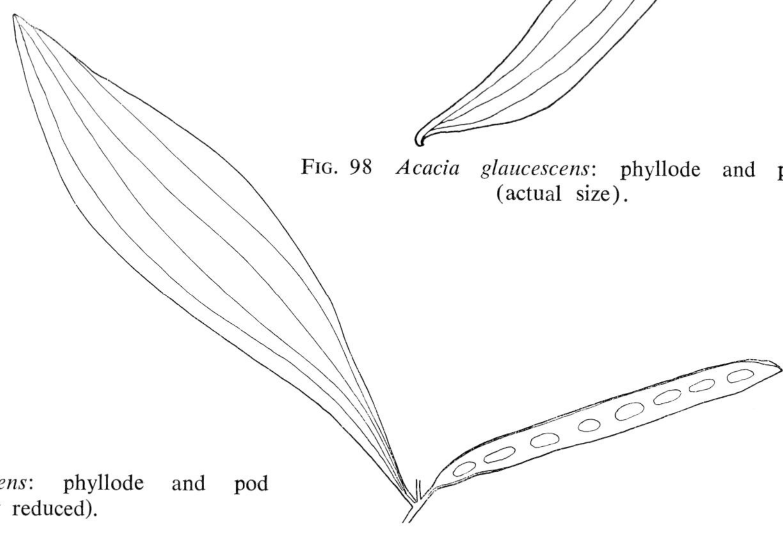

FIG. 97 *Acacia frigescens*: phyllode and pod (slightly reduced).

*A. glaucescens*. Light yellow. Montane. N.S.W. Fig. 98.

A small, spreading tree. Phyllodes grey-green, finely hairy, fairly broad, curved, narrowed at both ends, many-veined, 10–15 cm long. Flowers in rod-shaped heads in dense clusters. Pod about 3 mm broad.

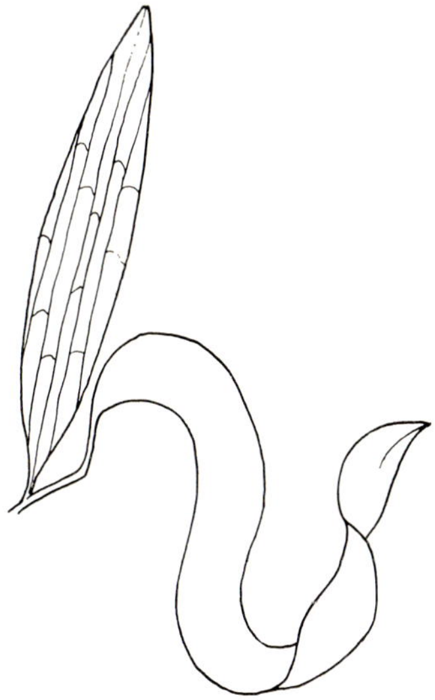

FIG. 99 *Acacia implexa*: pod and phyllode (actual size).

*A. implexa*. Lightwood. Yellow. Montane. N.S.W., Vic. Fig. 99.

A smooth, somewhat grey-green small tree. Phyllodes fairly broad, curved, many-veined, 5–15 cm long. Flowerheads globular, in dense clusters. Pod narrow, much curved and twisted, about 6 mm broad, contracted between the seeds.

*A. kettlewelliae*. Buffalo Wattle. Golden yellow. Montane. Vic. Fig. 100.

A showy wattle growing to the height of a small tree or tall shrub with angular branches. Phyllodes silvery, rather broad and curved, 5–15 cm long, with a prominent marginal vein and a prominent midvein. Flowers in dense balls on stalks shorter than the phyllodes and with about 10 flowers in each head. Pod flat, fairly broad, constricted between seeds, 6–10 cm long.

*A. oreades* (syn. *A. kybeanensis*). Yellow. Montane. N.S.W., Vic. Fig. 101.

A dense shrub 2–3 metres high with narrow, hoary branches. Phyllodes slightly hairy when young, fairly narrow, flat, thin, slightly wavy with a short point, many-veined, 2–2.5 cm long. Pod oblong, very flat, slightly grey-green, 3.5–6 cm long, 1.5–2 cm broad.

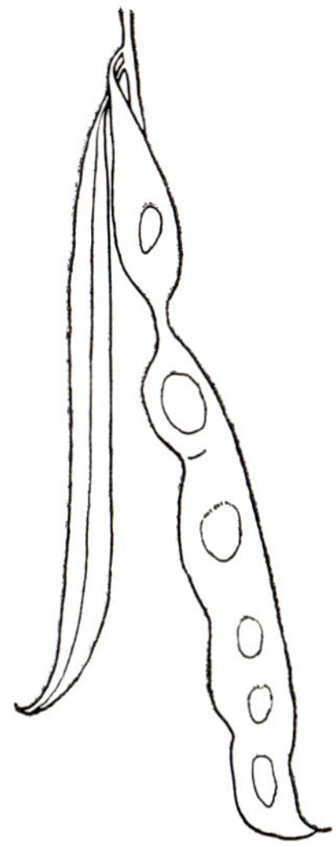

FIG. 100 *Acacia kettlewelliae*: phyllode and pod (actual size).

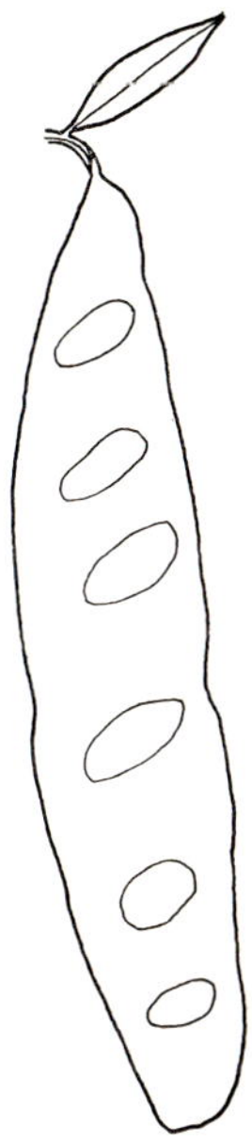

FIG. 101 *Acacia oreades*: phyllode and pod (actual size).

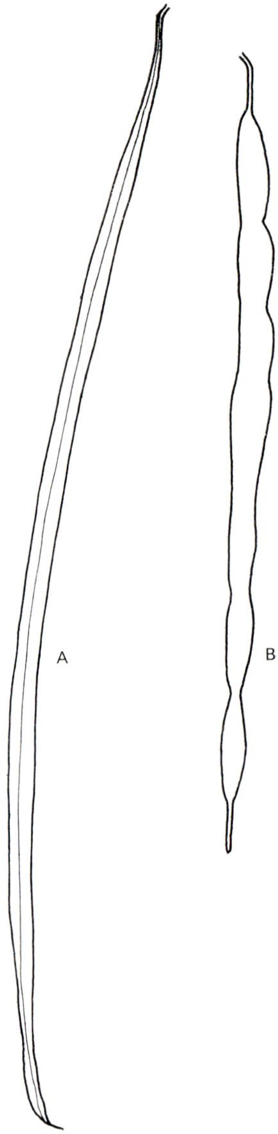


FIG. 102 *Acacia longissima*: A, phyllode; B, pod. (Both actual size.)

*A. longissima.* Pale yellow. Montane. N.S.W., Vic. Fig. 102.

A small tree or tall shrub. Phyllodes narrow, green, 10–15 cm long. Flowers in loose, slender interrupted rods about 2.5–5 cm long. Pod narrow, fairly long.

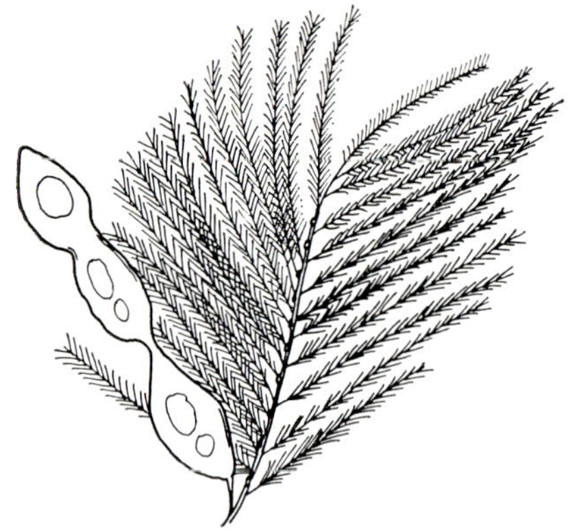

FIG. 103 *Acacia mearnsii*: pod and leaf (actual size).

*A. mearnsii* (syn. *A. mollissima*). Pale yellow. Montane. N.S.W., Vic. Fig. 103.

A medium to fairly tall tree, rather like silver wattle but the leaves bright-green, feathery, divided, only slightly hairy. Flowers in dense balls. Pod elongated, 4–10 cm long.

*A. melanoxylon.* Blackwood. Pale yellow. Montane. N.S.W., Vic., Tas. Fig. 104.

A well-known timber tree used in furniture-making, of rather sombre but very attractive appearance in the field, bushy and smooth, 8–30 metres high, with angular branches. Phyllodes fairly broad, blunt, leathery, often grey-green, pendulous, 4–10 cm long. Flowers in dense balls, 30–40 flowers in each head. Pod flat, much curved, 5–12 cm long, a bright-orange aril attached to each shining black seed making an attractive picture as the pods open.

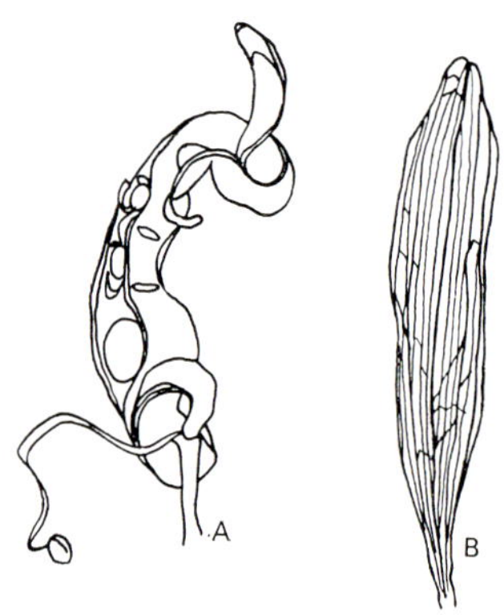


FIG. 104 *Acacia melanoxylon*: A, pod; B, phyllode. (Both actual size.)

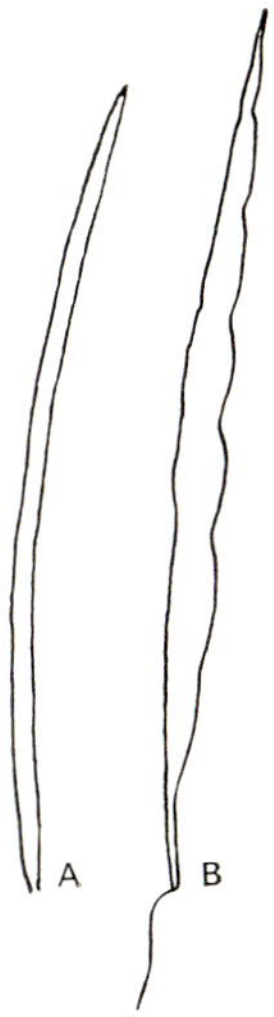


FIG. 105 *Acacia mucronata*: A, phyllode; B, pod. (Both actual size.)

*A. mucronata.* Yellow. Montane. N.S.W., Vic. Fig. 105.

A shrub 1–3 metres high. Phyllodes fairly narrow, rather stiff, leathery and blunt, with a short point, short with 3 prominent veins. Flowerheads globular, in clusters. Pod fairly narrow.

*A. penninervis.* Hickory Wattle. Golden yellow. Montane. N.S.W., Vic. Fig. 106.

A very widely distributed small tree or tall shrub with angular branches. Phyllodes rather thin, narrowed at each end, slightly grey-green. Flowers in loose heads, about 20 flowers in each. Pod very flat and thin, 10–20 cm long.

*A. pravissima.* Ovens Wattle. Golden yellow. Montane. N.S.W., Vic. Fig. 107.

A small tree or tall shrub with smooth, angular branches. Phyllodes very distinctive in shape, trapezoid or somewhat egg-shaped, 6–12 mm long. Flowers in small balls of 10–20 arranged in spikes much longer than the phyllodes. Pod. flat and smooth.

*A. rubida.* Red Stem Wattle. Golden yellow. Montane. N.S.W., Vic. Fig. 108.

A very handsome tall shrub frequently with reddish stems and branches. Phyllodes broad, sharp-pointed, rather thick, 7–9 cm long, often

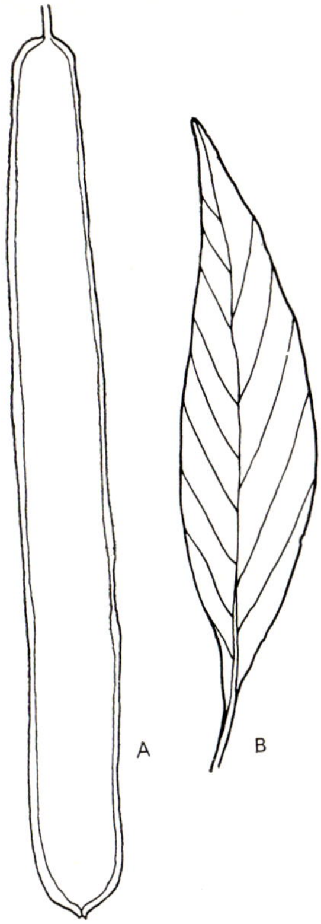


FIG. 106 *Acacia penninervis*: A, pod; B, phyllode. (Both × ½.)

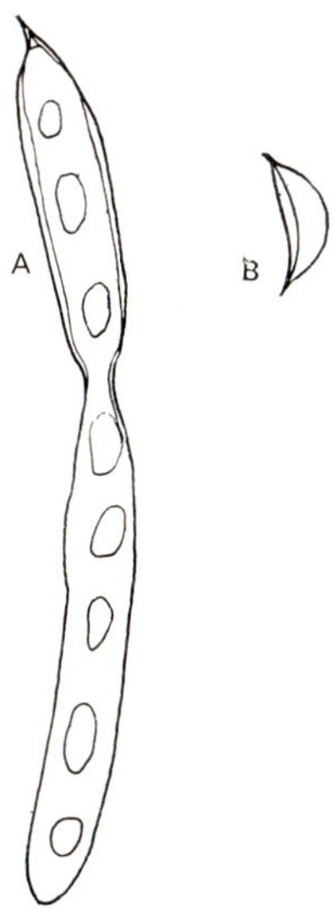


FIG. 107 *Acacia pravissima*: A, pod; B, phyllode. (Both actual size.)

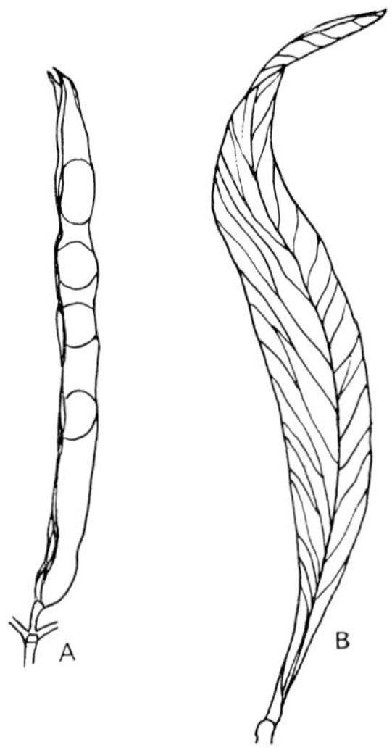


FIG. 108 *Acacia rubida*: A, pod; B, phyllode. (Both actual size.)

retaining a few true feathery leaves at the ends of the phyllodes. Flowers in clustered balls on stalks shorter than the phyllodes, with 10–15 flowers in each head. Pod nearly straight, 5–8 cm long.

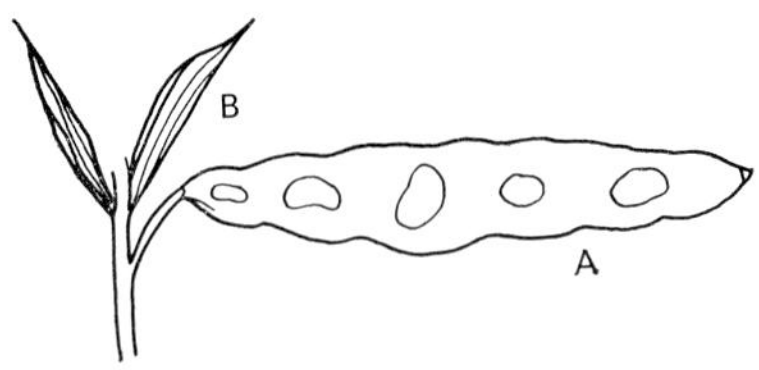


FIG. 109 *Acacia siculiformis*: A, pod; B, phyllode. (Both actual size.)

*A. siculiformis*. Yellow. Montane. N.S.W., Vic., Tas. Fig. 109.
A stiff shrub about 1 metre high restricted almost entirely to montane habitats. Branches narrow, brown, rather scaly. Phyllodes stiff, tapering to a sharp point, 1.5–2.5 cm long. Flowers numerous, in dense balls. Pod flat, narrow, about 2–4 cm long.

*A. silvestris*. Yellow. Montane. Vic., N.S.W. Fig. 110.
Descending to lower elevations also, a tree 7.5–30 metres high with a smooth grey trunk and silvery-hairy branches. Leaves divided with 5–12 divisions and numerous leaflets 4–8 mm long with 2–3 glands between the branches. Flowers in globular heads. Pod narrow-oblong, 6–12 cm long, grey-green, constricted between the seeds.

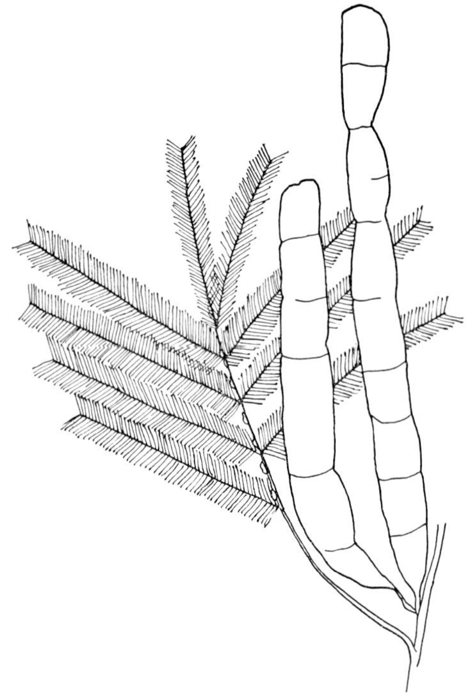

FIG. 110 *Acacia silvestris*: leaf and pods (actual size).

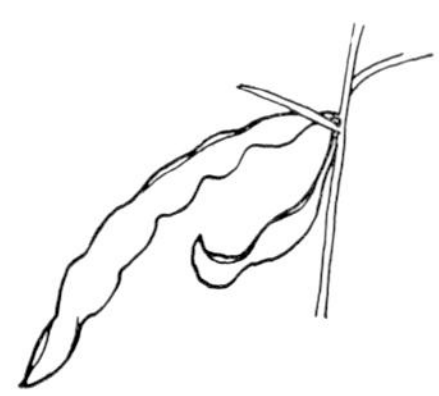

FIG. 111 *Acacia ulicifolia*: pod and phyllode (actual size).

*A. ulicifolia*. Yellow. Montane. N.S.W., Vic., Tas. Fig. 111.
A small, stiff shrub 30–90 cm high. Phyllodes narrow, stiff, sharply pointed, with a prominent vein, about 1 cm long. Flowers in globular heads, solitary on slender stalks. Pod straight or slightly curved, 2.4–4.5 cm long, constricted between the seeds.

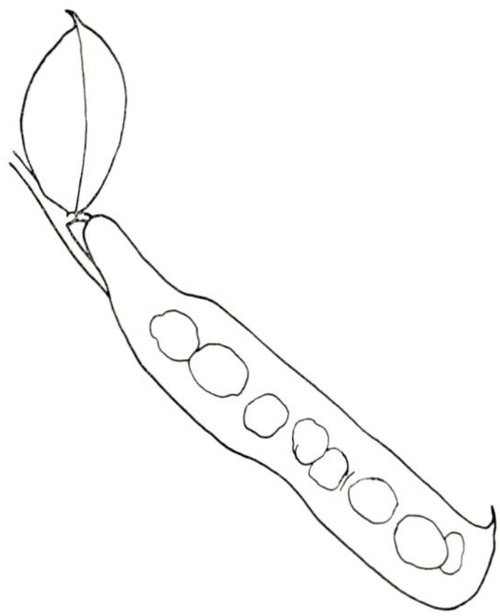

FIG. 112 *Acacia undulifolia*: pod and phyllode (actual size).

*A. undulifolia.* Bright yellow. Montane. N.S.W. Fig. 112.
A handsome spreading shrub, 1 metre or less in height. Phyllodes almost round, grey-green, oblique, wavy, usually pointed. Flowers in globular heads clustered on rather long stalks. Pod 1–2 cm wide, broad and curved.

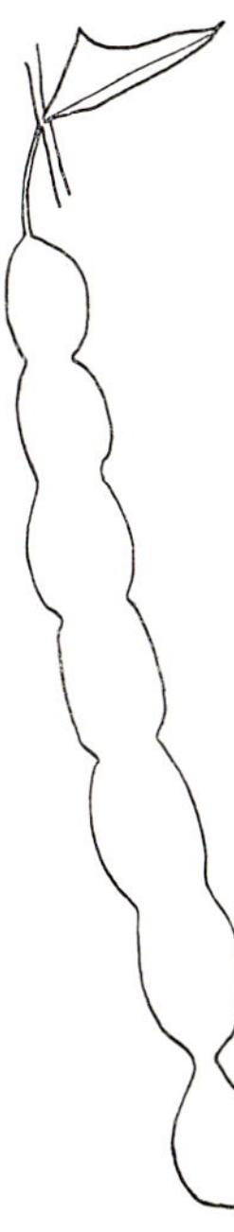

FIG. 113 *Acacia vomeriformis* pod and phyllode (× 2).

*A. vomeriformis.* Ploughshare Wattle. Golden yellow. Montane. N.S.W., Vic., Tas. Fig. 113.
An interesting little wattle, varying in habit from completely prostrate to a height of 20–90 cm. Phyllodes very distinctive in form, broadly triangular, stiff, pointed, 5–10 mm long. Flowers in balls with about 30 flowers in each ball. Pod narrow, flat, 6–8 cm long.

**Family PAPILIONACEAE**

Pea-flowers are widely distributed throughout the world and are found abundantly in every ecological situation in Australia, including the alpine areas. Plants of this family are easily recognized by their characteristically shaped flowers (see Fig. 114). These have 5 sepals and

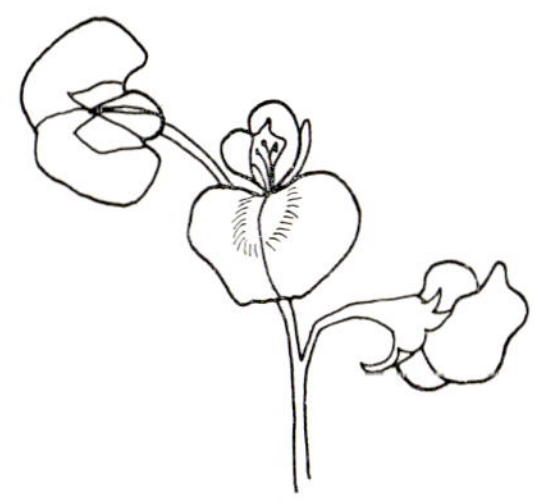

FIG. 114 A typical pea-flower (actual size).

5 petals arranged with one at the back, often much larger than the others, called a standard, two at the side called wings, and two joined in the centre to form a keel. The keel petals cover the stamens, of which there are 10 inserted slightly above the other flower parts. The fruit is a pod or legume.

A number of members of this family common at lower elevations have found their way into the montane and even higher areas, but there are also a number of species which are restricted to the higher altitudes.

*BOSSIAEA* (after Bossieu de la Martinière, botanist on the expedition of La Pérouse, 1785-8). Fig. 115. A genus of pea-flowers in which the stamens are united into a tube free at the top and the leaves are undivided. The stems are

FIG. 115 *Bossiaea*: pod (actual size).

usually more or less flattened, but in a few cases they are quite flat and without leaves. The pods are long and flat.

*B. buxifolia.* Orange-yellow. Montane. N.S.W., Vic.
A shrub with somewhat spiny branches. Leaves alternate, round or heart-shaped, paler on the undersurface, arranged in opposite rows, 3–10 mm in length. Bracts at base of flower fused.

*B. cordigera.* Orange-yellow. Subalpine, montane. Vic., Tas.
A scrambling plant 30–40 cm high, with slender branches. Leaves opposite, rounded, heart-shaped at the base, smooth, rather thin, bright-green, 3–6 mm long. Flowers solitary in the leaf axils; sepals smooth, about 6 mm long, the standard twice as long. Pod about 12–18 mm long.

*B. foliosa.* Bright yellow. Subalpine, montane. N.S.W., Vic.
An early-flowering plant covering considerable areas, very floriferous and spectacular, 2–3 metres high, with rather prickly, robust branches and somewhat flattened stems. Leaves small, stiff, rounded, in opposite rows. Flowers shortly stalked. Pod rounded, compressed, rich bronzy-brown when young, and covered with dense hairs.

*B. riparia.* Yellow and crimson. Montane. N.S.W., Vic., Tas.
A straggling undershrub with flattened branches spreading 15–60 cm. Older leaves reduced to small brown scales arranged alternately on the stem. Flowers in the upper leaf axils, shortly stalked. Sepals smooth, about 4 mm long, the standard about twice as long. Pod smooth, 12–18 mm long, on a fairly long stalk.

*DAVIESIA* (after Rev. Hugh Davies, Welsh botanist, 1739-1821). Fig. 116. A genus of shrubs in which the stamens are all free, the leaves are undivided and the pod is triangular.

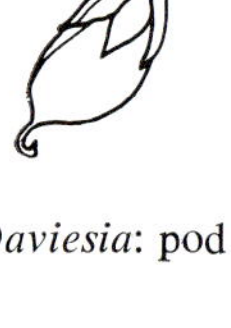

FIG. 116 *Daviesia*: pod (actual size).

*D. corymbosa.* Yellow and red. Montane. N.S.W.
A very widely distributed species reaching a height of 60–90 cm. Leaves rather variable in shape though always fairly narrow, stiff or, when broader, strongly veined. Flowers small, in short heads, flowering from above the middle or from the end only and produced in great profusion; sepals about 2 mm long, the standard about twice as long.

*D. latifolia.* Orange-yellow. Montane. N.S.W., Vic., Tas.
A smooth plant 60 cm–2 metres high with angular branches. Leaves very variable in shape though always fairly broad, 2.5–8 cm long. Flowers in short clusters in leaf axils surrounded by conspicuous brown bracts; sepals about 3 mm long, standard more than twice this length, flushed with purple at the base. In Tasmania, found at lower elevations also.

*D. mimosoides.* Orange-yellow. Montane. N.S.W., Vic.
A shrub 1–2 metres high. Leaves variable, usually rather falcate, 3–8 cm long though sometimes longer, thick. Flowers in short dense clusters on slender stalks; sepals 2–3 mm long, the standard twice that length.

*D. ulicifolia.* Yellow, orange-red. Montane. N.S.W., Vic.
A prickly plant 30 cm–1 metre high, common also at lower elevations, smooth, much branched, sometimes scrambling, the branches ending in spines. Leaves stiff, pointed, rather broad but variable, 0.6–3.6 cm long. Flowers in short heads in the leaf axils; sepals about 3 mm long; standard very broad, more than twice the length of the sepals. In Tasmania, found at lower elevations only.

*D. virgata.* Yellow. Subalpine, montane. N.S.W., Vic.
A shrub 1–5 metres high. Leaves stiff, rather narrow, up to 9 cm long, thick and somewhat wrinkled, with conspicuous veins. Flowers 0.4–0.6 cm long in bunches 1–2 cm long on slender stalks. Pod 0.6–1 cm in diameter.

*DILLWYNIA* (after L. W. Dillwyn, English botanist, 1788-1855). Fig. 117. A genus of shrubs in which the stamens are all free, the

FIG. 117 *Dillwynia*: leaf (× 2).

leaves are usually small and undivided and the pods are spherical and small.

*D. floribunda.* Orange-red. Montane. N.S.W., Vic., Tas.
A hairy shrub 30–60 cm high, very widely distributed in coastal situations also. Leaves stiff, narrow, blunt, crowded, 4–10 mm long. Flowers almost stalkless, 1–2 together in the leaf axils but the leaves so crowded that the flowers appear to be leafy spikes; calyx hairy, 5–7 mm long; standard kidney-shaped. Pod silky-hairy, not much longer than the calyx.

*D. prostrata.* Yellow. Montane. N.S.W.
A prostrate plant. Leaves blunt, narrow, smooth, about 3–5 mm long. Flowers stalkless, solitary or a few together in the axils of the upper leaves; sepals smooth, 4–5 mm long, standard about twice as long.

*D. sericea.* Orange-red. Montane. N.S.W., Vic., Tas.
Common in other habitats also. A hairy shrub about 1 metre high closely related to D. *floribunda.* Leaves narrow, blunt, often covered with tubercles, 5–20 mm long. Flowers stalkless in pairs in the leaf axils; sepals 4–7 mm long, standard twice that length.

*GLYCINE* (from Gk *glykys*, sweet). Fig. 118.
*G. clandestina.* Blue. Montane. N.S.W., Vic., Tas.
A slender, dainty, trailing plant with stems 15–90 cm long. Leaves soft, stalked, trifoliate; leaflets narrow, 1–4 cm long. Flowers in slender heads of 3–6 in the leaf axils; sepals about 4 mm long, the standard scarcely twice that length; stamens united in a tube.

*GOMPHOLOBIUM* (from Gk *gomphos*, a club, and *lobos*, a segment, referring to the

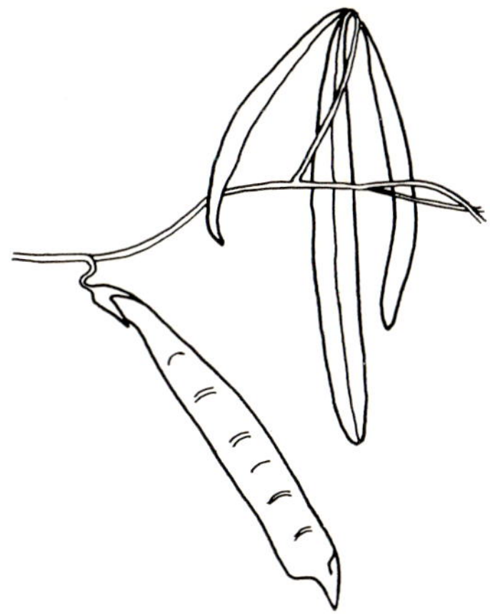

FIG. 118 *Glycine*: pod (× 2).

FIG. 119 *Gompholobium*: fruit (× 2).

fruit). Fig. 119. A genus of plants in which the stamens are all free, the leaves may be divided or undivided, and the pod is ovoid or spherical and inflated.

*G. huegelii.* Creamy-yellow, dark on back. Montane. N.S.W., Vic., Tas.
A delightful, slender little undershrub with spreading branches 15–30 cm high. Leaves stalked, trifoliate, with narrow, blunt leaflets, smooth, grey-green, 6–16 mm long. Flowers solitary or 2–3 together on short stalks; sepals 8–10 mm long standard twice or more than twice that length, rather broad.

*GOODIA* (after Peter Good, d. 1803, botanical collector under Robert Brown). Fig. 120.

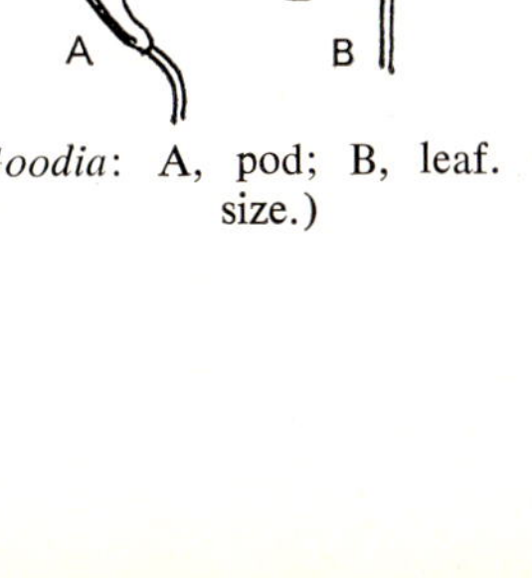


FIG. 120 *Goodia*: A, pod; B, leaf. (Both actual size.)

*G. lotifolia.* Clear yellow. Montane. N.S.W., Vic., Tas.
A soft plant with many slender spreading brittle branches. It reaches a height of 1–4 metres. Leaves divided, on slender stems; leaflets stalkless, broad and rounded, smooth, 1–2.5 cm long. Flowers in loose heads 4–12 cm long; petals often marked with purple; calyx 5–6 mm long, the standard twice that length; stamens united in a tube open on one side.

*HARDENBERGIA* (after the Countess von Hardenberg, sister of Baron von Hugel, who collected plants in W.A. in 1833). Fig. 121.

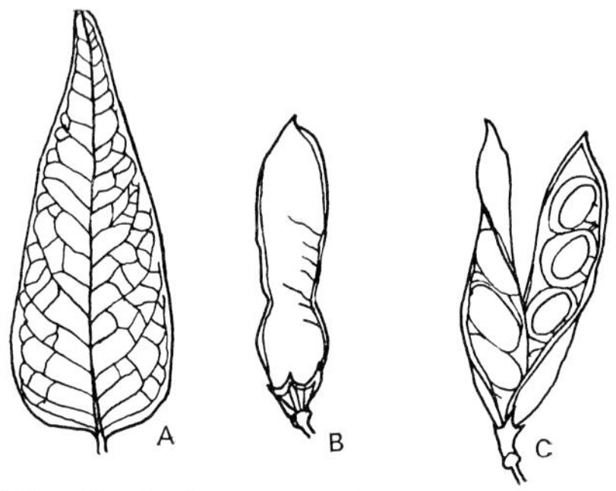


FIG. 121 *Hardenbergia*: A, leaf; B, pod; C, opened pod. (All actual size.)

*H. violacea.* Deep purple. Montane. N.S.W., Vic., Tas.
A widely spread plant, smooth with wiry branches, usually a creeper but sometimes erect. Leaves simple, hard, strongly veined, shortly stalked. Flowers in dense heads, each about 1 cm long; stamens united except along one side. Pod flat and stalkless, with a pithy substance between the seeds.

*HOVEA* (after A. P. Hove, Polish botanist). Fig. 122. Purple-flowering peas in which the stamens are all united and the leaves are simple. The pods are turgid, each containing 2 seeds.

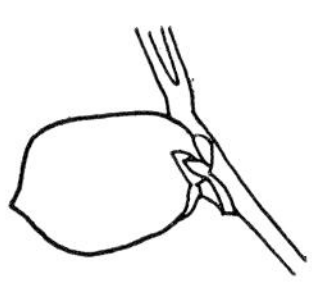

FIG. 122 *Hovea*: fruit (× 2).

*H. linearis.* Bluish-purple, occasionally white. Montane. N.S.W., Vic., Tas.
A slender, trailing plant. Leaves undivided, stalked, variable in shape, smooth, strongly veined, 1–3 cm long. Flowers on short stalks, solitary or 2–3 together, the keel deeply coloured at the tip; sepals brown and woolly, 5–8 mm long.

*H. longifolia.* Purple. Alpine, subalpine. N.S.W., Vic., Tas.
A variable shrub of wide distribution ranging from a rock-clinging, scrambling, almost prostrate plant to a shrub 1–2 metres high. It is likely that a revision of this species will separate the alpine form, which differs considerably from those of lower altitudes. Branches woolly; leaves stalked, oblong, blunt, often with a sharp point, thick, the upper surface smooth, the lower surface covered with dense brown hair, 1–7 mm long. Flowers on short stalks in the leaf axils, 2–3 together or sometimes in short heads; calyx brown, woolly, 5–6 mm long. Pod about 1 cm long.

*INDIGOFERA* (neo-Lat., indigo-bearing; some species provide commercial indigo). Fig. 123.

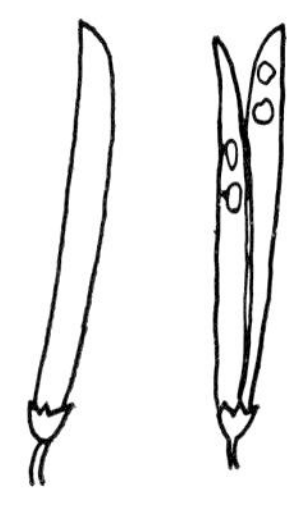

FIG. 123 *Indigofera*: pod (actual size).

*I. australis.* Native Indigo. Pinkish-purple. Montane. N.S.W., Vic., Tas.
An elegant, spreading shrub 50 cm–1.5 metres high, distributed throughout Australia in a variety of habitats. Leaves with soft, blunt, oblong leaflets 1–3 cm long. Flowers stalked, in loose drooping heads of many flowers; sepals brown, hairy, about 2 mm long; stamens united except down one side, standard about 8 mm long. Pod stalkless, brown, fairly long, cylindrical, many-seeded.

*LESPEDEZA* (after M. Lespedez, one-time governor of Florida). Fig. 124.

FIG. 124 *Lespedeza*: fruit (× 3).

*L. juncea* ssp. *sericea*. Purplish-pink. Montane. N.S.W., Vic.
A slender, hairy shrub 60–80 cm high with trifoliate leaves; leaflets 10–20 mm long. Flowers about 6 mm long, in clusters of 2–4; stamens joined, the upper one free. Pod almost round and stalkless, not splitting, 2–3 mm in diameter.

*LOTUS* (Greco-Lat. name for several different plants). Fig. 125. A genus of shrubs in which

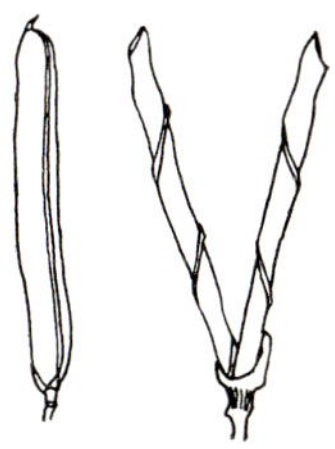

FIG. 125 *Lotus*: pod (× 1½).

the leaves are divided and the stamens are joined. The pods are long and narrow, several-seeded, with partitions between the seeds.

*L. australis*. Pale pink, cream, red. Montane. N.S.W., Vic., Tas.
A smooth, usually compact plant growing to a height of about 60 cm, common in grasslands. Leaflets egg-shaped, soft, light-green, blunt, 1–3 cm long. Flowers in short clusters in the leaf axils; standard 15–18 mm long, the keel shorter and darker. Pod straight, about 3.5 cm long.

*L. corniculatus*. Orange and yellow. Montane. N.S.W., Vic., Tas. (Introduced.)
Usually a smaller plant than *L. australis*. Leaflets broadly egg-shaped, 5–10 mm long. Flowers in short heads on rather stout stalks, 2–8 together; sepals 6–8 mm long, the petals about twice that length. Pod straight, 2–3 cm long.

*MIRBELIA* (after C. F. B. Mirbel, French botanist). Fig. 126.

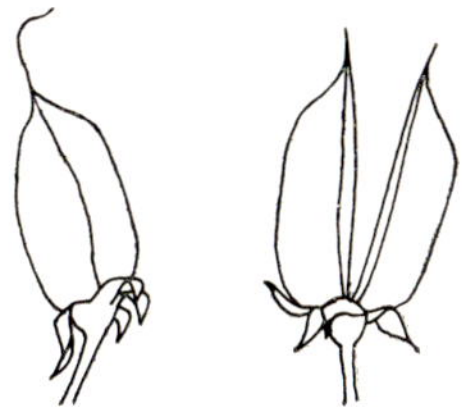

FIG. 126 *Mirbelia*: pod (× 2).

*M. oxylobioides*. Orange and yellow. Montane. N.S.W., Vic.
A fairly tall shrub 1–2.5 metres high with simple, rather broad, slightly pointed leaves in rings of 3, each 3–6 mm long. Flowers small in terminal spikes on very short stalks, the keel dark-purple; stamens all free. Pod oval, turgid, divided into 2 cells. Rather resembling *Bossiaea foliosa* in general appearance.

*OXYLOBIUM* (from Gk *oxys*, sharp, and *lobos*, a segment, referring to the pointed leaf segment in some species). Fig. 127. A genus of

FIG. 127 *Oxylobium*: pod (× 2).

pea-flowers in which the stamens are free, the leaves simple and the pod ovoid or oblong.

*O. alpestre*. Orange. Alpine, subalpine. N.S.W., Vic.
A spectacular little prostrate plant displaying its brilliant blossoms to great advantage as it spreads like a carpet over the grasslands or scrambles over rocks or banks. Leaves opposite or in rings of 3, oblong, sharp, leathery, 2.5–3.5 cm long. Flowers in long heads in the leaf axils or at the ends of the branches; sepals silky. Pod about 1.5 cm long.

*O. ellipticum*. Orange. Subalpine, montane. N.S.W., Vic., Tas.
An erect shrub up to 2.5 metres high, though often shorter. Leaves rather hairy, shortly

stalked, fairly narrow, leathery, dark-green, strongly veined, the lower surface woolly, 1–4 cm long. Flowers in loose heads, usually terminal, on short stalks; sepals woolly, about 7 mm long, the standard longer—about 10 mm. Pod about 7 mm long.

*O. ilicifolium.* Native Holly. Orange. Montane. N.S.W., Vic.
A widely distributed variable shrub common in coastal habitats also, reaching a height of 2–5 metres. Leaves opposite, sharply pointed, very variable but often incised like holly leaves, strongly veined and shining, woolly underneath, about 3.5 cm long, sometimes longer. Flowers in loose clusters at the ends of branches and in the leaf axils; calyx slightly hairy, about 4 mm long, the standard twice as long. Pod short and narrow.

*O. procumbens.* Orange. Montane. N.S.W., Vic.
A prostrate plant, often woolly. Leaves irregularly opposite, leathery, pointed, shining, strongly veined, about 2–2.5 cm long. Flowers fairly large, about 1 cm long in few-flowered heads at the ends of the branches and in the leaf axils. Pod woolly, about 1.5 cm long.

*PLATYLOBIUM* (from Gk *platys*, short, and *lobos*, a pod). Fig. 128.

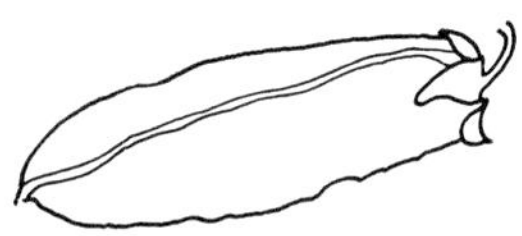

FIG. 128 *Platylobium*: pod (actual size).

*P. formosum.* Deep yellow and red. Montane. N.S.W., Vic.
A widely distributed species outside this habitat also, usually rather straggling and up to 60 cm high. Leaves broadly triangular or heart-shaped, 2–5 cm long Flowers solitary or a few together with brown bracts on the stalks; standard (back) petal about 18 mm long, stamens united into a tube. Pod flat, oblong, 2–4 cm long, opening and rolling back.

*PSORALEA* (from Gk *psoraleas*, scabby, in reference to the glands). Fig. 129. A genus comprising slight plants with divided leaves and

FIG. 129 *Psoralea*: pod (× 2).

conspicuous stipules. The stamens are united into a tube, the upper one free.

*P. adscendens.* Purple. Montane. N.S.W., Vic., Tas.
A small plant with straggling stems 30–60 cm long. Leaves trifoliate, the leaflets broad with a sharp point, soft, strongly veined, 2–5 cm long. Flowers small, in close spikes longer than the leaves; sepals covered with white or black silky hairs and about 4.5 mm long. Pod egg-shaped, short, blackish, not splitting open.

*P. tenax.* Bluish. Montane. N.S.W., Vic.
A straggling plant about 30–60 cm high with 5–7-foliate leaves and shortly stalked flowers in heads 2.5–10 cm long on very long slender stalks.

*PULTENAEA* (after Richard Pulteney, 1730-1801, English botanist). Fig. 130. A genus of shrubs with yellow or, more commonly, red and

FIG. 130 *Pultenaea*: pod (× ½).

yellow flowers. The stamens are all free from one another. The leaves have stipules, though these are small in some species. There are small bracts at the base of the calyx. The pods are small and oval in shape. Most of the pultenaeas found in this area do not ascend beyond subalpine habitats.

*P. altissima.* Yellow. Montane. N.S.W., Vic.
A tall, smooth shrub sometimes attaining the height of a small tree. Leaves narrow, blunt, 12–15 mm long. Flowers several together in close clusters.

*P. cunninghamii.* Yellow. Montane. N.S.W., Vic.
An erect, grey-green, smooth shrub with attractive, broadly rhomboidal, stalkless leaves 6–20 mm long in rings of 3 round the stem. Flowers

solitary or 1–2 together in the upper leaf axils on short stalks.

*P. fasciculata.* Yellow. Montane. N.S.W., Vic., Tas.
A much-branched, creeping plant with hairy limbs, 10–20 cm long. Leaves long and narrow, the edges rolled in leaving only a channel along the upper surface, 4–6 mm long. Flowers in leaf axils on very short stalks, each flower with a broad, lobed bract; sepals hairy, 4–5 mm long; corolla with a standard twice as long as the other petals. Flowers abundantly produced, forming a bright carpet on the ground.

*P. gunnii.* Orange-yellow. Montane. N.S.W., Vic., Tas.
A shrub about 1.5 metres high but sometimes prostrate, with hairy, slender branches. Leaves variable, 6–10 mm long. Flowers on short stalks in dense heads at the ends of the branches, with short bracts at the base; sepals hairy, 4–5 mm long; standard 8–10 mm long; keel brownish.

*P. hibbertioides.* Orange-yellow. Montane. N.S.W., Vic., Tas.
A woody undershrub extending to lower elevations, with erect, hairy branches 30–50 cm long. Leaves crowded, spreading, narrow, very concave, 10–12 mm long. Flowers small on short stalks in dense heads at the ends of branches, with numerous stiff, overlapping bracts at the base; sepals about 4–5 mm long; standard petal 8–10 mm long.

*P. juniperina.* Orange-yellow. Montane. N.S.W., Vic.
A much-branched shrub 40 cm–2 metres high. Leaves crowded, on short stalks set at right angles to the main stem, stiff, pointed, long, variable in shape, 8–12 mm long. Flowers solitary or 2–3 together with short bracts; sepals hairy, 4–5 mm long; standard 8–10 mm long; keel dark.

*P. microphylla.* Orange-yellow. Montane. N.S.W.
A small, spreading shrub extending to the coast, with hoary branches. Leaves narrow, woolly on the underside, 12–15 mm long. Flowers in short terminal heads; sepals 4–6 mm long, the standard twice that length.

*P. polifolia.* Yellow. Montane. N.S.W., Vic.
A soft woolly shrub common in other habitats also. Leaves fairly broad, silky-hairy beneath, pointed, 1–4 cm long.

*P. procumbens* (syn. *P. styphelioides*). Yellow. Montane. N.S.W., Vic.
A hairy shrub varying from 3 metres in height to an almost prostrate form. Leaves somewhat stem-clasping, stiff, round, pointed, very small, with conspicuous stipules. Flowers almost stalkless in the upper leaf axils, the bracteoles similar to the leaves; petals a little longer than sepals. Pod turgid.

*P. subspicata.* Orange-yellow. Montane. N.S.W.
A diffuse shrub. Leaves narrow, small, sometimes with sharp points; stipules on floral leaves conspicuous. Flowers small, usually in leafy spikes; calyx 4–6 mm long, the standard twice that length; bracteoles at base of flower 2-lobed.

*P. subumbellata.* Orange-yellow. Subalpine, montane. N.S.W., Vic., Tas.
A slender shrub 30–90 cm high. Leaves crowded, rather narrow, blunt, smooth, 5–8 mm long. Flowers on very short stalks in terminal heads with few bracts; calyx hairy, 4–5 mm long; standard 8–10 mm long with purple markings which appear also on the wings.

*P. tenella.* Orange-yellow. Subalpine, montane, Vic.
A lovely little prostrate plant with silky branches. Leaves narrow, oblong, blunt, in rings of 3, each 4–6 mm long; stipules very small. Flowers in leaf axils stalked, often longer than the leaves; bracts narrow; sepals about 6 mm long.

*SWAINSONA* (after Isaac Swainson, d. 1806, owner of a private botanical garden near

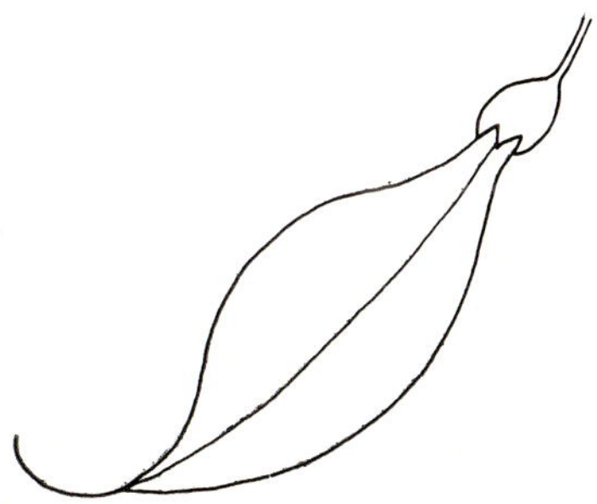

FIG. 131 *Swainsona*: pod (actual size).

London). Fig. 131. A genus of plants not commonly associated with high altitudes, but 2 species, *S. oroboides* and *S. monticola*, occur at montane levels. The stamens are united in a tube with the upper stamen free, the leaves are divided, the style is bearded and the pod is inflated. *S. oroboides* occurs in New South Wales and Victoria as a small, straggling plant. Leaves 2–4 cm long, leaflets 5–7. Flowers bluish, purple or red, 7–9 mm long, in loose heads up to 15 cm long with 3–5 flowers in each head. Pod downy, 8–20 mm long.

**Family ROSACEAE**

A few members of the Rose family occur in alpine regions, all small plants. The leaves have stipules at the base. The flowers are regular with 4–5 sepals, sometimes missing, 4–5 petals, also sometimes missing, and a large number of stamens, 2–4 times as many as the petals, as a rule. The ovary is below the insertion of the other flower parts.

*ACAENA* (from Gk *akaina*, a thorn, referring to the prickles on the fruit). Fig. 132. A genus

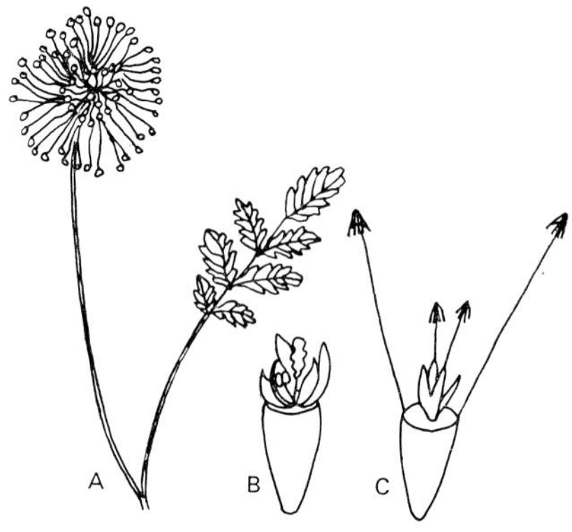


FIG. 132 *Acaena*: A, flowerhead (actual size); B, flower (× 1½); C, fruit (× 1½).

of peculiar little plants familiar by their conspicuous and extremely adhesive burred fruits. They are soft plants with divided leaves and stipules joined to the leaf-stalk. Flowers in an inconspicuous spherical head or elongated interrupted spike on long stems; sepals 3–7; no petals; stamens 2–10. Fruit in reddish, globular or elongated heads.

*A. anserinifolia.* Green, purplish. Alpine, subalpine, montane. N.S.W., Vic., Tas.
A widespread small plant extending sometimes to the coast. Leaves divided into 3–4 leaflets, the upper leaves 5–10 mm long, the others shorter, all rounded, smooth and serrate. Flowers very small in dense rounded heads, 6–12 mm in diameter on long stems. The spines of the fruits about 1 cm in length, barbed at the tip and very adhesive, making the plant most unpopular.

*A. montana.* Purplish. Alpine, subalpine. Tas.
A smaller plant than *A. anserinifolia*, with leaflets sometimes silky-hairy. Fruiting head ovoid, about 2 mm long; spines 1–2 mm long.

*A. ovina.* Purplish. Alpine, subalpine, montane. N.S.W., Vic., Tas.
A very widespread species extending to sea-level in some places. Stems erect, leafy, 20–60 cm high. Additional leaves present as a basal rosette with pairs of distant leaflets, rather rounded, lobed, the upper surface smooth, the lower surface hairy, 0.5–1.5 cm long. Flowers small in a long, interrupted spike, denser at the tip; stamens purple. Fruiting head ovoid, 4–6 mm long with numerous small and unequal prickles.

*ALCHEMILLA* (apparently named for its supposed use in alchemy). Fig. 133.

FIG. 133 *Alchemilla*: A, branch with leaves and fruiting head (actual size); B, fruit (× 3).

*A. xanthochlora.* Greenish. Subalpine, montane. N.S.W.
A swamp-lover with large palmately shaped leaves on long stalks, rounded, 7–9 lobed and

toothed. Flowering stems with conspicuous, leafy, toothed stipules, and clusters of very small flowers; each flower has 4 small sepals, no petals and 4 or fewer stamens. Fruit dry, enclosed in the calyx tube.

*GEUM* (old Lat. name for herb bennet, from Gk *geuo*, an agreeable flavour, the roots of some species being aromatic). Fig. 134. A genus of small, soft plants with irregularly divided leaves with few segments. The flowers are open and have an epicalyx, 5 sepals and 5 petals and about 20 stamens.

FIG. 134 *Geum*: A, flower; B, fruits. (Both actual size.)

*G. renifolium*. White, bluish-pink. Alpine. Tas. (South).

A small, hairy plant. Leaves basal with only 1 leaflet, stalked, kidney-shaped and rather obscurely lobed, the upper surface very wrinkled, the lower surface showing the veins clearly, 5–10 cm broad. Flowering stems up to 60 cm high, with 3 stalkless, toothed, bract-like leaves and a single flower; sepals 5, small, with a conspicuous epicalyx; petals 5, open, about 25 mm long; stamens numerous, bright-yellow. Bushwalkers delight in finding patches of this lovely little plant on their alpine walks.

*G. urbanum*. Bright yellow. Subalpine, montane. N.S.W., Vic.

A slender plant 20-50 cm high usually growing in wet or boggy patches. Leaves toothed with a large terminal segment and 2–4 smaller basal ones, 8–10 cm long and with large toothed or lobed stipules. Flowers few, in loose heads on long stalks, about 20 cm in diameter; petals rounded, spreading. Fruiting heads globular, reddish, with protruding, hooked awns.

*RUBUS* (old Roman name, probably derived from Lat. *ruber*, red, the colour of the fruits of many species). Fig. 135. An interesting little group of plants with 2 alpine species. The leaves are divided and the flowers rose-like with 5 or more petals and a fruit consisting of a number of small berry-like structures.

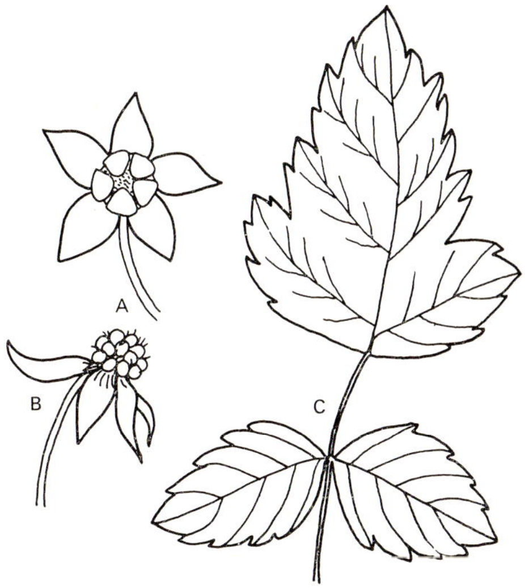


FIG. 135 *Rubus*: A, flower; B, fruit; C, leaves. (All actual size.)

*R. gunnianus*. Creamy white. Alpine, subalpine. Tas.

A dwarf plant sometimes covering extensive patches of ground with slender, wiry, woody stems and leaves in a rosette at the end of each branch. Leaves stalked; leaflets 1–2 cm long, egg-shaped, serrate. Flowers solitary on stalks longer than the leaves; petals about 10 mm long. Fruit a cluster of 3–5 small oval berry-like structures, shining red, very attractive and edible.

*R. parvifolius*. Pink, reddish. Montane, subalpine. N.S.W., Vic., Tas.

A much more widely distributed plant than *R. gunnianus*, extending sometimes to the coast. A slender, prickly, rather straggling bush with stems 1–2 metres long. Leaves stalked, with rounded leaflets, pointed, the terminal one 2–4 cm long, the others shorter. Flowers solitary or in heads shorter than the leaves. Fruit a head of small berry-like structures, edible.

### Family CUNONIACEAE

*ANODOPETALUM* (from Gk *an*, without, *odon*, *odonotos*, a tooth, and *petalon*, a petal, the petals being entire). Fig. 136.

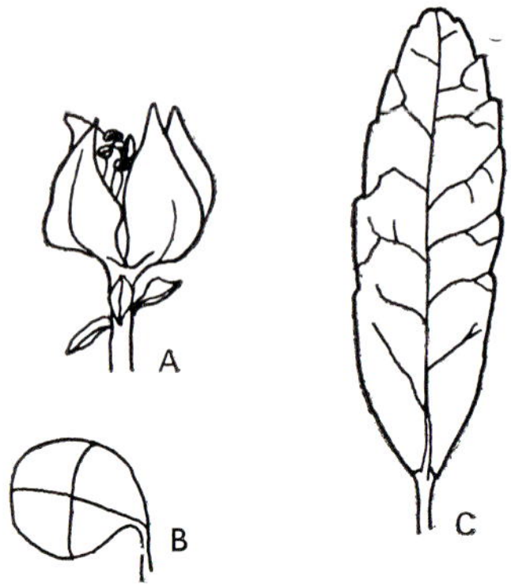


FIG. 136 *Anodopetalum*: A, flower; B, fruit; C, leaf. (All × 3.)

*A. biglandulosum*. Horizontal. Greenish. Temp. rainforest. Tas.

A slender tree up to 15 metres high, typically bending over to a horizontal position, with its intertwining branches forming a platform above and level with the ground. Leaves opposite, on short stalks, narrow, blunt and bluntly toothed, 2–6 cm long. Flowers solitary or 2–3 together on short stalks in the leaf axils; sepals 4–5, triangular; petals 4–5, narrow, shorter than sepals; stamens 8, inserted on a disc. Fruit a fleshy nut about 6 mm long.

### Family BAUERACEAE

*BAUERA* (after Ferdinand Bauer, 1760-1826, Austrian botanical artist who accompanied Robert Brown to Australia). Fig. 137.

*B. rubioides*. Native Dogrose. Pink, white. Alpine, subalpine, montane. Tas.

Essentially a plant of the coast and dividing range on the mainland, this extraordinary little wiry shrub finds its way from the sea coast to over 4,000 feet altitude in Tasmania. It ranges from an erect, spreading shrub 6 metres or more

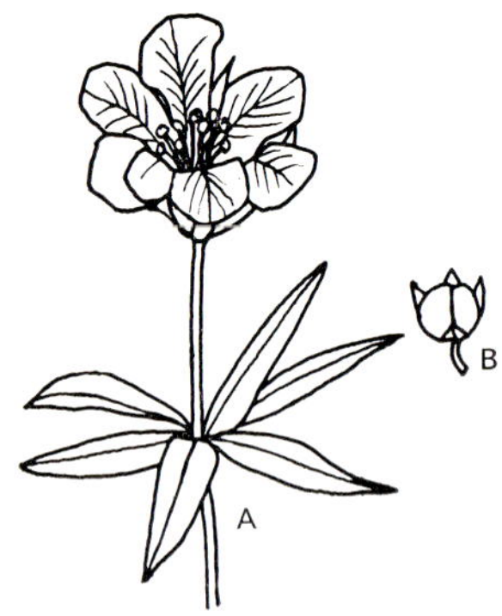


FIG. 137 *Bauera*: A, flower (× 3); B, fruit (actual size).

in height with intertwining branches in the rainforests to an entirely prostrate plant in the alpine regions. Leaves opposite, trifoliate, almost stalkless, looking like a ring of 6 simple leaves; leaflets fairly broad, smooth, often shining, 4–14 mm long. Flowers solitary, very dainty and attractive, 12–18 mm in diameter, sometimes crowded at the ends of the branches on long, slender stalks; sepals and petals commonly 6, the petals narrow, blunt, spreading widely and surrounding the numerous bright-yellow stamens. Fruit a short capsule opening in 2 valves.

### Family ESCALLONIACEAE

A family of woody plants with simple, alternate leaves, usually toothed. There are 4–6 sepals, petals and stamens, the sepals joined at the base and persistent, the stamens inserted on the margin of a nectary. There are 1–6 carpels with joined styles. The fruit is a follicle, capsule or berry.

*ANOPTERUS* (from Gk *ana*, upwards, and *pteron*, a wing, referring to the ascending wings on the fruit). Fig. 138.

*A. glandulosus*. Tasmanian Laurel. White, blush pink. Temp. rainforest, wet sclerophyll. Tas.

A tall shrub or small tree. Leaves clustered at ends of branches, spreading, broad, pointed, shortly stalked, thick, smooth, shining, toothed, 7–17 cm long. Flowers in clusters; individual flowers on slender stems, often recurved; sepals 6, joined at the base, lobes spreading; petals 6, egg-shaped, free, 10–12 mm long; stamens 6, opposite the sepals, filaments shorter than the petals. Fruit a capsule opening in 2 valves.

FIG. 138 *Anopterus*: A, flowerhead; B, flower; C, fruit. (All actual size.)

*TETRACARPAEA* (from Gk *tetra*, four, and *karpheion*, ripe fruit). Fig. 139.

FIG. 139 *Tetracarpaea*: A, flower (× 2); B, flowerhead (actual size); C, fruit (× 2).

*T. tasmanica.* White. Subalpine. Tas.
A small shrub 30–100 cm high. Leaves egg-shaped, blunt, shortly stalked, thick, shining, unequally and bluntly toothed, 1.5–3 cm long. Flowers clustered, small; sepals 4, joined at the base, spreading; petals 4, egg-shaped, clawed, falling soon after opening, about 6 mm long; stamens 8. Fruit a group of follicles, each with 4 parts.

## Family EUCRYPHIACEAE

*EUCRYPHIA* (from Gk *eu*, well, and *kryptos*, covered, the petals being concealed by the calyx in early life). Fig. 140.

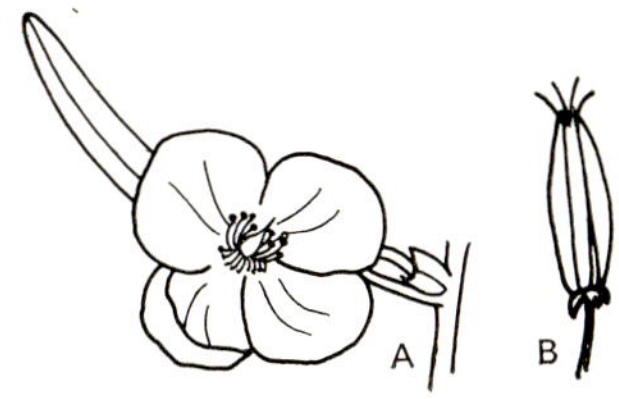


FIG. 140 *Eucryphia*: A, flower; B, fruit. (Both actual size.)

*E. lucida.* Leatherwood. White. Temp. rain-forests. Tas.
An attractive tall shrub or small tree with opposite, simple, shortly stalked, fairly broad, blunt leaves rounded at the ends, leathery, the lower surface white, 2.5–4.5 cm long; stipules small, 2 fused at the base of each pair of opposite leaves; young leaves and stipules covered with a gummy substance. Flowers solitary, showy, in axils of upper leaves on stems 1 cm long with cup-shaped bracts at the base; sepals 4, cohering at the apex, forming an operculum which is pushed off on opening; petals 4, free, spreading, orb-shaped, 18 mm in diameter; stamens numerous in several rows; filaments inserted on a nectar-secreting disc. Fruit a capsule.

A smaller plant, var. *milliganii*, with smaller leaves and flowers, occurs in areas of high rainfall and higher altitudes.

## Family DROSERACEAE

Members of the Sundew family are all insectivorous. The leaves are commonly in the form of a basal rosette, occasionally alternate and clustered or erect, the upper surface of the leaf being covered with stalked, irritable hairs. The leaf surface is frequently coloured and glistens with an exudation from the glandular hairs. This serves to attract small creatures which, crawling over or alighting on the leaf, get caught in the sticky exudation. Their struggles to free themselves irritate the sensitive hairs, which move inwards over the struggling victim. Once the creature is securely bound by these, the digestive exudation serves to extract the nitrogenous material from the body of the prisoner. When this feat is accomplished the hairs uncoil to release the now shrivelled remains, which roll off, blow away or may remain adhering to the leaf surface. The leaf is now ready for another victim. Because of this special adaptation such plants are able to live in soils deficient in nitrogen and so are common inhabitants of badly drained, swampy areas.

Sundews produce flowers on long, slender stalks, sometimes solitary, sometimes in slender heads; sepals, petals and stamens 4–5; petals papery or transparent, free. Fruit a capsule with fine seeds.

*DROSERA* (from Gk *droseros*, dewy, referring to the glandular exudation). Fig. 141.

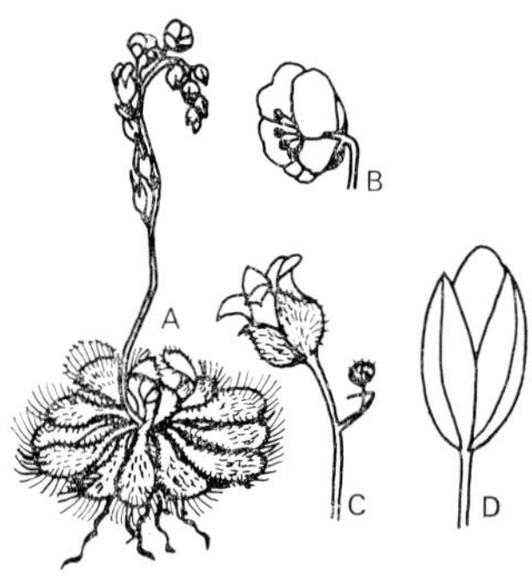


FIG. 141 *Drosera*: A, flowerhead (× ½); B, flower side view; C, flower (both actual size); D, fruit (× 2).

*D. arcturi.* Papery white. Alpine, subalpine, montane. N.S.W., Vic., Tas.

A dainty species common in fens and bogs. It sometimes covers large areas of bog lands and is able to survive long periods completely submerged. Leaves in a basal rosette, glistening, long and narrow, blunt, 3–6 cm long, the upper part covered with masses of coloured, glandular hairs. Flowers solitary; sepals 5–6 mm long; petals egg-shaped with a rounded apex, about 10 mm long; styles 3–4, short, thick.

*D. auriculata.* White, pink. Montane. N.S.W., Vic.

A plant with stems up to 30 cm high and stem leaves with 2 angular lobes on each side of the stalk, the basal leaves in a small rosette or reduced to scales; the styles are divided into tufts.

*D. gracilis.* White. Subalpine. Tas.

A dainty little species of the wet, peaty heaths, closely related to *D. peltata* and for a long time regarded as a variety of it. Resembles *D. peltata* in all its parts but is smaller, the leaves are red and the petals are narrow.

*D. peltata.* White. Subalpine, montane. N.S.W., Vic., Tas.

Extending to other habitats also. Leaves in a basal rosette in addition to others alternate on an erect, thin leafy stem about 6–30 cm high; basal leaves round or heart-shaped on short, broad, flattened stalks, the outer ones markedly longer than the inner; stem leaves round or crescent-shaped, 4–6 mm in diameter, on fairly long stems; both types of leaves beset with reddish or yellowish-green glandular hairs. Flowers few in a terminal head; sepals round, fringed with fine hairs and 3–4 mm long; petals translucent, about twice as long as sepals; stamens much branched.

*D. pygmaea.* Pale pink. Alpine, subalpine, montane. N.S.W., Vic., Tas.

A tiny species, as its name implies, occurring from sea-level to 4,000 feet or higher. Leaves forming a rosette 15–30 mm in diameter, the individual leaves round, on slender stalks, red, the upper surface covered with glandular hairs. Flowers solitary, with 4 sepals and 4 petals.

## Family HALORAGACEAE

*HALORAGIS* (from Gk *hals*, *halos*, the sea, and *rhax*, *ragos*, a grape-berry, alluding to the habitat and appearance of the fruit of a New Zealand species). The raspworts are rather insignificant plants but are so commonly found in alpine parts that they cannot be overlooked. They are scabrous, moisture-loving plants with variable, usually opposite leaves along the stem, and small inconspicuous drooping flowers in terminal heads with 4 very small sepals, 4 somewhat larger boat-shaped petals and usually twice as many stamens. There are 4 common species of raspwort in these parts.

*H. micrantha*, creeping raspwort, has simple leaves and tiny pendulous flowers in a terminal head and is restricted to subalpine habitats on the mainland and in Tasmania, usually in damp places.

*H. tetragyna*, poverty raspwort, has solitary flowers about 4 mm long in the leaf axils and grows in dry stony places on the mainland and in Tasmania.

*H. heterophylla*, rough raspwort, from New South Wales, Victoria and Tasmania, has divided leaves and 1–4 flowers together in a terminal head.

*H. depressa* from New South Wales is prostrate, with simple, toothed leaves, and flowers in short, interrupted heads.

*GUNNERA* (after J. E. Gunnerus, 1718-73, Norwegian naturalist). *G. cordifolia* is restricted to subalpine habitats in Tasmania near streams and in swampy places. It is a small plant with a basal rosette of broad leaves on long stalks and small flowers, male and female in separate crowded heads. The fruits are succulent, conspicuous, bright-red, 2–3 mm long. The plant is fleshy and often forms extensive mats.

### Family CRASSULACEAE

*CRASSULA* (Lat., fem. dimin. of *crassus*, thick, referring to the leaves and branches). A genus with 4 species in alpine habitats. They are herbs with opposite, fleshy, cylindrical or near-cylindrical leaves. The flowers have 4–5 sepals and petals and 4–5 stamens inserted below or around the other flower parts. The ovary has the same number of parts as the petals, usually separate from each other and forming separate, splitting follicles.

*C. sieberana* is a small fleshy plant about 5 cm high of alpine, subalpine and montane areas of the mainland and Tasmania. The reddish small flowers are in spikes and have 4 sepals, 4 petals and 4 stamens.

*C. helmsii* and *C. macrantha* are restricted to montane habitats, the former in New South Wales, the latter in Victoria and Tasmania as well. *C. helmsii* has solitary reddish or white flowers 2–4 mm in diameter; the leaves are narrow, 2–6 mm long, and the stems weak; it grows in water or on swampy land. *C. macrantha* grows 2–6 cm high and is often reddish; the leaves are opposite, narrow, fleshy, 3–5 mm long; the pink flowers have 4 sepals, 4 petals and 4 stamens and form clusters on fairly long stems; it likes wet situations.

### Family MYRTACEAE

This family is represented above the treeline by three genera, *Baeckea*, *Callistemon* and *Kunzea*. In subalpine and montane areas there is a much wider range of both genera and species, including many eucalypts (*Eucalyptus*) which have been described earlier in this book. Members of the family have oil dots on their leaves and a characteristic aromatic odour. The flowers are conspicuous with 5 (occasionally 4) free sepals and petals, 5 or (more commonly) numerous stamens in a ring round an ovary of one or many parts sunk in a pit formed by the bases of the petals which are fused to it. In some plants of this family, such as *Baeckea* and *Leptospermum*, the petals are rose-like and conspicuous and the stamens are short; in others, like *Kunzea*, *Callistemon* and *Eucalyptus*, the petals are small or absent, sometimes deciduous, and the stamens are long and often coloured.

*BAECKEA* (after Dr Abraham Baeck, 1713-95, Swedish naturalist and physician). Fig. 142. A genus of shrubs with opposite leaves, the

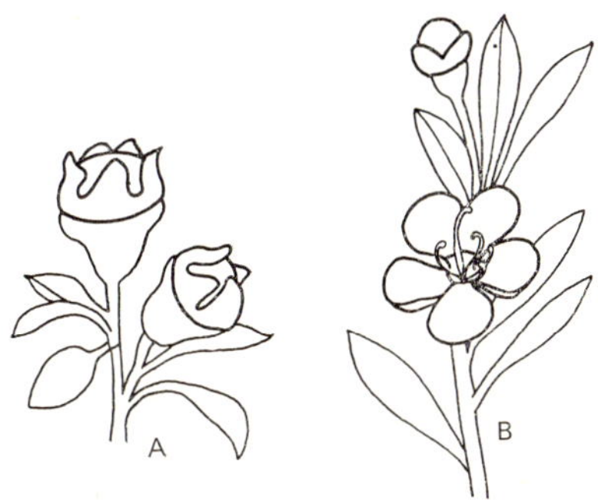


FIG. 142 *Baeckea*: A, fruits; B, flowerhead. (Both × 2.)

flowers tea-tree-like with 5 free rose-like petals and inconspicuous stamens in a single row, usually 5, 10 or 15 in one ring. The fruits are capsules with 2–3 valves, and surrounded by the persistent sepals.

*B. crenatifolia.* White. Alpine. Vic.
A tall shrub up to 30 metres high fringing creeks and springs. Leaves round, 5–7 mm long, 2–3 together on slender stems.

*B. denticulata.* White, pale-pink. Montane. N.S.W.
A prostrate or diffuse smooth shrub with narrow branches. Leaves spreading, shortly stalked, narrow, 4–6 mm long. Flowers in short clusters (very rare in baeckeas) of 2–8 flowers on slender stems terminating in 2 bracts; petals rounded, about 3 mm long; stamens 20–25 in regular rows.

*B. gunniana.* White, pink. Alpine, subalpine, montane. N.S.W., Vic., Tas.
An interesting and attractive little shrub with a wide range of habitats from 6,000 feet to montane habitats and even lower in Tasmania. At lower altitudes a tall, dense shrub to 2 metres in height; in the highest areas a scrambling plant, often rock-clasping. Leaves fine, narrow, close together, with conspicuous oil dots and 2–10 mm long. Flowers solitary on short stalks, produced in great abundance; petals rounded, broad, 3–4 mm long; stamens 5–7, not opposite the petals.

J. H. Willis considers that a subalpine and montane mainland form with leaves 4–10 mm long, flowers on longer stalks and 7–10 stamens should be regarded as *B. utilis*.

*CALLISTEMON* (from Gk *kallos*, beauty and *stemon*, a stamen). Fig. 143. A genus of shrubs rather similar to *Kunzea*, but the stamens, which are very numerous, are much longer and the sepals are deciduous. The fruit is a woody capsule, closely set on the branch and remaining unopened on the plant, often for several years.

*C. pallidus.* Lemon-yellow. Montane. N.S.W., Vic., Tas.
A rather stiff, erect shrub 2–4 metres high from well-drained swamps and river banks with slender spreading branches. Leaves narrow, 3–6 cm long, smooth when mature, densely silky-hairy when young and sometimes delicately greyish-pink at this stage. Brushes about 4–7 cm long, the stamens numerous, about 1.5 cm long.

*C. sieberi.* Yellow, rarely pink. Subalpine, montane. N.S.W., Vic.
A lovely shrub common in wet depressions, often in great stands. A tall compact shrub or small tree. Leaves narrow, sharply pointed, 1–3 cm long. Individual flowers small, stalkless, compact, forming dense, bright brushes; sepals short and blunt, petals round, stamens very numerous, in 2 rows, less than twice the length of the petals.

*C. viridiflorus.* Greenish-yellow. Montane. Tas.
An attractive spreading shrub 1–2 metres high from wet areas. Leaves stiff, narrow, pointed, crowded to overlapping, about 2 cm long. Flower-spikes 4–7 cm long; sepals rounded, about 2.5 mm long, lightly fringed with hairs on the edges; petals broad, twice as long as the sepals; stamens very numerous in several rows, about 1.5 cm long.

*CALYTRIX* (from Gk *kalyx*, cup, and *thrix*, *trikos*, a hair, the sepals ending in long hairs). Fig. 144.

FIG. 143 *Callistemon*: A, flowerhead (× ½); B, flower (× 2); C, fruits (× 2).

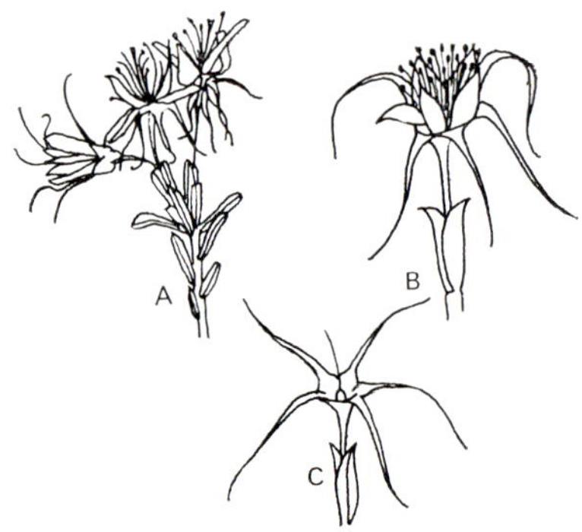


FIG. 144 *Calytrix*: A, flowerhead (actual size); B, flower (× 2); C, young fruit (× 2).

*C. tetragona*. White, pink. Montane. N.S.W., Vic., Tas.

The only species of the genus in this area, this is recognized by the very long hairs extending from the sepals and by the ovary which has only one division. A very common bushy shrub of coastal habitats, up to 2 metres high. Leaves narrow, blunt, spirally arranged, 4–8 mm long. Flowers up to 12 in terminal heads; calyx enlarges after flowers fall and turns bronze-red, particularly at higher elevations, and is most attractive; petals 5, spreading; stamens numerous in several rings. Fruit non-splitting, surrounded by the elongated, persistent sepals.

*EUCALYPTUS*. See note on page 44.

*KUNZEA* (after Gustav Kunze, 1793-1851, German botanist, entomologist and physician). Fig. 145. An easily recognized genus with very

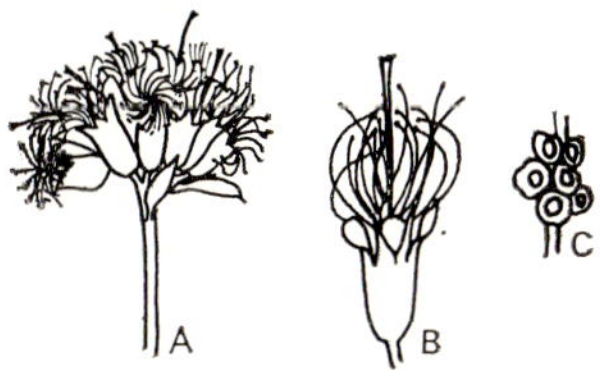


FIG. 145 *Kunzea*: A, flowerhead (actual size); B, flower (× 2); C, fruit (actual size).

small and inconspicuous petals, and numerous stamens, conspicuous and longer than the petals; the flowers are in heads at the ends of the branches or in the upper leaf axils. The fruit is a capsule opening at the top and surrounded by the persistent sepals.

*K. muelleri*. Yellow. Alpine, subalpine, montane. N.S.W., Vic.

A low, bushy shrub, often completely prostrate and rock-clambering, more common at the higher altitudes. Leaves small, rounded, about 10–12 mm long. Flowers stalkless in small, dense heads; petals about 3 mm in diameter; stamens very numerous, less than twice as long as the petals. Very floriferous, often forming stands along the ground, over rocks or in wet depressions.

*K. parvifolia*. Crimson. Montane. N.S.W., Vic.

A small, erect shrub. Leaves oblong, concave, 1–2 mm long. Flowers small, few in heads; stamens numerous.

*LEPTOSPERMUM* (from Gk *leptos*, slender, and *sperma*, a seed). Fig. 146. A well-known

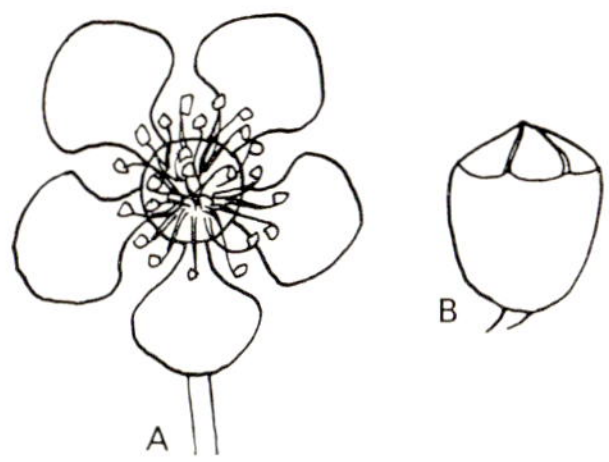


FIG. 146 *Leptospermum*: A, flower; B, fruit. (Both × 1½.)

and widely distributed genus with species ranging in habitat from alpine to sea-level. They are shrubs, often tall with small, alternate leaves (opposite in *Baeckea*), the flowers occurring singly in the leaf axils; the stamens numerous, in a single row, shorter than the rose-shaped, spreading petals and often clustered in groups opposite the petals. The style is short with a rounded stigma. The fruit is a capsule opening at the top.

*L. arachnoides*. White. Montane. N.S.W.

Also extending to coastal habitats, this is a rigid shrub up to 1 metre high. Leaves narrow, about 1 mm wide and 1–2 cm long, pointed. Flowers fairly large, woolly on the back at the base. Summit of ovary not hairy. Fruit fairly large, protruding.

*L. attenuatum*. White. Subalpine, montane. N.S.W., Vic.

Also extending to coastal situations in New South Wales, this is a tall shrub 1–2 metres high. Leaves 3–5-nerved, flat, rather narrow, 12–15 mm long. Flowers 5–12 mm in diameter, on very short stalks; summit of ovary hairy.

*L. flavescens*. Tantoon Tea-tree. Creamy-yellow. Montane. N.S.W., Vic.

A shrub of very wide distribution, 2–4 metres high with drooping branches. Leaves almost smooth, pointed, 1.5 mm wide, 5–20 mm long. Flowers 5–12 mm in diameter, with a sweet perfume; summit of ovary not hairy. Fruit small.

*L. humifusum.* White. Alpine, subalpine. Tas.
A lovely little prostrate plant, often scrambling over boulders at higher elevations but erect and reaching 1.5 metres at lower elevations. Leaves crowded, blunt, rounded, rather thick and shining, 5–8 mm long. Flowers solitary, produced in great abundance, 12–16 mm in diameter; summit of ovary not hairy. Fruit about 5 mm in diameter, woody and scaly.

*L. lanigerum.* Woolly Tea-tree. White, cream. Montane. N.S.W., Vic., Tas.
A slender tree or bushy shrub up to 18 metres high often fringing river banks, the young shoots woolly and often pinkish-purple or bronze in colour. Leaves flat, spreading, narrow to egg-shaped, woolly, 8.18 mm long. Flowers solitary, 18 mm in diameter, hairy on the back; summit of ovary not hairy. Fruit 7–9 mm in diameter, sepals persistent until fruit is mature, top protruding. Var. *macrocarpa* has yellow or deep cream flowers and beautiful bronze foliage.

*L. myrsinoides.* White. Subalpine, montane. N.S.W., Vic.
A shrub 30–50 cm high, often growing near water and extending to other habitats. Leaves 10–12 mm long, fairly broad. Flowers stalkless, solitary in leaf axils; calyx hairy. Capsule flat-topped with 5 divisions.

*L. myrtifolium.* White. Montane. N.S.W., Vic.
A widely distributed species extending to the coast, this is an erect shrub 3–4 metres high. Leaves small, rounded, blunt, about 1–1.5 mm long. Flowers solitary, stalked; summit of ovary not hairy. Capsule hemispherical, about 6 mm in diameter.

*L. nitidum.* White. Montane, subalpine. Tas.
A rigid, much-branched shrub, widely distributed, extending to the coast and commonly on wet peaty heaths, 2–4 metres high but at higher altitudes often stunted. Leaves smooth and shining, pointed and narrow, 8–20 mm long. Flowers solitary, 18–24 mm in diameter. Fruit nearly 1 cm in diameter, almost flat. Closely related to *L. lanigerum* and at one time included in it.

*L. phylicoides* (syn. *Kunzea peduncularis*). White, rarely pink. Montane. N.S.W., Vic.
Extends to the tablelands and shows a wide variation in habit from a tall shrub 3–4 metres high at lower elevations to a graceful, willowy, small plant as it ascends. Common on sandy flats near watercourses. Leaves rather narrow, 2–3 cm long. Flowers in graceful clusters, crowded, on short stalks, 2–3 together; petals rounded, 5–6 mm long; stamens 20–30, fairly long but on slender stalks so that the petals are well revealed.

*MELALEUCA* (from Gk *melas*, black, and *leucos*, white, referring to the black trunk and white branches of an Asiatic species). Fig. 147.

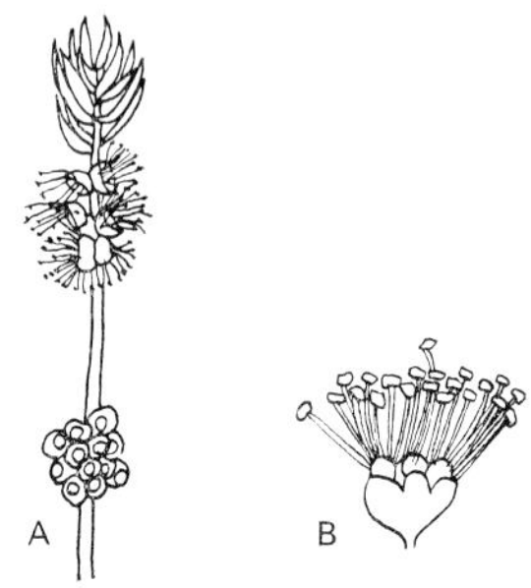


FIG. 147 *Melaleuca:* A, spike with flowers and fruit. (× ½); B, flower (× 2).

A large Australian genus conspicuously absent from alpine habitats and represented by only one species. Petals small and inconspicuous; stamens conspicuous, often brightly coloured, arranged in 5 groups opposite the petals.

*M. squamea.* Purple, white, yellowish. Alpine, subalpine, montane. N.S.W., Vic., Tas.
An erect shrub of wide distribution extending to the coast, about 6 metres high with woolly young shoots. Leaves crowded, alternate, rather broad, pointed, 4–8 mm long and usually with 3 distinct longitudinal veins. Flowers crowded, in stalkless, terminal heads about 15 mm in diameter. Fruit a capsule set into the stem, opening at the top.

**Family LYTHRACEAE**

*LYTHRUM* (from Gk *lythron*, gore, an old name given by Dioscorides). Fig. 148.

*L. salicaria.* Purple Loosestrife. Purple. Montane. N.S.W., Vic., Tas.
A marsh or water dweller growing to about 60–100 cm with angular branches, the upper leaves opposite but the lower ones sometimes in rings

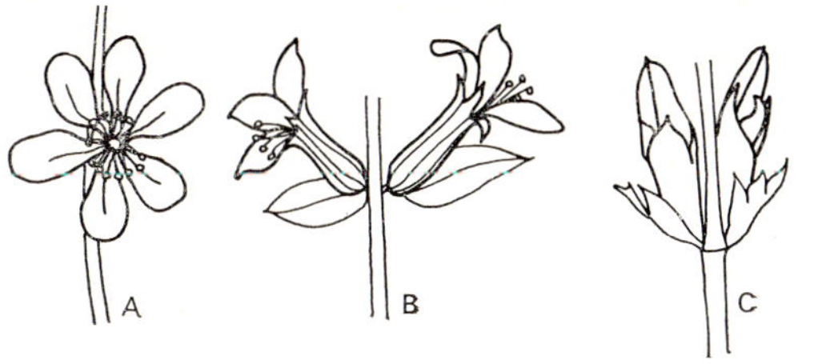


FIG. 148 *Lythrum*: A, flower (actual size); B, flowerhead (actual size); C, fruit (× 2).

of 3 without stalks, slightly stem-clasping, fairly broad, pointed, 2–6 cm long. Flowers in a loose fluffy spike, each flower with 5–6 petals often 12–15 mm long; stamens usually 12, 6 of them longer and protruding; ovary inserted above other flower parts. Fruit a capsule.

## Family ONAGRACEAE

*EPILOBIUM* (from Gk *epi*, upon, and *lobos*, a pod, referring to the flower resting on the pod-like fruit). Fig. 149. A genus of soft, slender

FIG. 149 *Epilobium*: A, flowerhead; B, unopened fruit; C, opened fruit. (All actual size.)

plants belonging to the same small family to which the well-known garden fuchsia belongs. There are 4 sepals, 4 free pink or white petals and 8 stamens in 2 rings of 4. The seedbox is conspicuous below the petals and stamens. The fruit is an elongated capsule with plumed seeds.

*E. billardierum*. Pale pink. Subalpine, montane. N.S.W., Vic., Tas.

Extending from sea-level to about 3,000 feet, mostly in damp places. About 20–30 cm high, at first running along the ground and then erect, rather reddish. Leaves egg-shaped or heart-shaped, blunt with a toothed margin, 2–4 cm long. Flowers solitary in leaf axils, 10–15 mm in diameter, on short stalks; sepals rather broad, blunt, about 5–7 mm long; petals heart-shaped at tip, 6–8 mm long.

*E. curtisii* (syn. *Boisduvalia tasmanica*). Pale purplish-pink. Montane. Tas.

A slender, trailing plant 2–20 cm long with almost stalkless, usually opposite leaves, blunt with crisped edges, 8–15 mm long. Flowers few, solitary, almost or quite stalkless; petals rounded with a notch at the top, 6–7 mm long. Seeds not plumed.

*E. erosum*. White. Montane. Tas.

Very similar to *E. sarmentaceum*, but stems without the downy covering on the upper part, and broader, finely and irregularly toothed leaves.

*E. glabellum*. Pink. Alpine, subalpine, montane. N.S.W., Vic.

A very variable plant extending to the tablelands in wet places, up to 50 cm high but usually smaller. Leaves narrow to broadly egg-shaped, often purplish, opposite, finely toothed, 12–18 mm long. Flowers solitary on short stalks; sepals 3–5 mm long, petals slightly longer.

*E. gunnianum*. Purple-pink. Alpine, subalpine. Vic., Tas.

A hairy plant reaching a height of 5–30 cm. Leaves opposite or in rings of 3, rather broad with a blunt tip, without stalks, often stem clasping, 1.5–2.5 cm long. Flowers solitary in the leaf axils, 18–20 mm in diameter; sepals narrow, tipped with a sharp point, covered with scattered hairs and often crimson or purplish, about 7 mm long; petals egg-shaped, with a distinct apical notch, 12–14 mm in length.

*E. sarmentaceum*. Pale lilac, pink. Alpine. Tas.

A plant 5–20 cm high with rather flaccid stems and with a greyish downy covering of hairs. Leaves mostly opposite, rather narrow, blunt, covered with short hairs and on short pale-green to grey stems and about 1.5–2 cm long. Sepals

5–6 mm long, narrow and pointed, greyish-green; petals 7–8 mm in length, egg-shaped with a narrow tube.

*E. tasmanicum*. Pink. Alpine, subalpine. N.S.W., Vic., Tas.

A small plant from wet places on plateaux with almost smooth, creeping stems. Leaves opposite, meeting across the stem, rather crowded, sometimes with short stalks, sometimes stalkless, egg-shaped, thick and shining, with a blunt tip 8–10 mm long. Flowers usually 2 together in the axils of the upper leaves, appearing terminal, on stout, woolly stems; sepals narrow, 3–4 mm long; petals slightly longer, oval to egg-shaped, notched at tip.

**Family UMBELLIFERAE**

This family, to which the carrot and parsley belong, and also the native flannel flower, contains a number of plants which are apparently very dissimilar. This is because the flowers, which are produced in a cluster on stems all radiating from one point (hence the family name) are frequently subtended by bracts of differing kinds. In flannel flowers, for instance, the flowers are crowded into a compact head surrounded by radiating bracts.

In this family the flowers are always small with 5 sepals (or none), 5 petals and 5 stamens. The ovary is inserted below the other flower parts. The fruit is made up of 2 separate non-splitting parts which are called mericarps.

The family is well represented in alpine habitats.

*ACIPHYLLA* (from Gk *akros*, sharp, and *phyllon*, a leaf). Fig. 150. This genus comprises one of the more conspicuous and attractive members of the family, with divided leaves and rather large heads of radiating flowers. The individual flowers are small without sepals and with 6 white or cream incurved petals. The fruit is dry, flattened, with narrow, winged ribs.

*A. glacialis*. Native Celery. Cream. Alpine. N.S.W., Vic.

A charming plant of the alpine herbfields, reaching a height of 30–40 cm. Leaves once or twice divided, with narrow leaflets, rather stiff. Flowers in dense heads on fleshy stems, often red, longer than the leaves.

*A. procumbens*. Cream. Alpine. Tas.

This species commonly forms dense cushion-like tufts up to 2 metres in diameter. Branches spread along the ground, bare except at the ends where they are leafy. Leaves stalked, twice divided with narrow leaflets, 2–5 cm long. Flowers in compound heads on stalks radiating from a single point with numerous small flowers.

*A. simplicifolia*. Brownish. Alpine, subalpine. N.S.W., Vic.

A more slender and less conspicuous plant than *A. glacialis*, but dainty, usually occurring in grasslands and bogs. Leaves narrow, 15–40 cm long. Flower heads open, fairly small.

*ACTINOTUS* (Lat., furnished with rays). Fig. 151. Although the mainland flannel flowers do not extend to alpine areas, 3 small species from Tasmania are found in alpine, subalpine and montane areas. These plants are distinguished by the presence of a ring of spreading bracts in a fashion superficially resembling the ray

FIG. 150 *Aciphylla*: A, flowerhead (actual size); B, leaves (actual size); C, fruits (× 2).

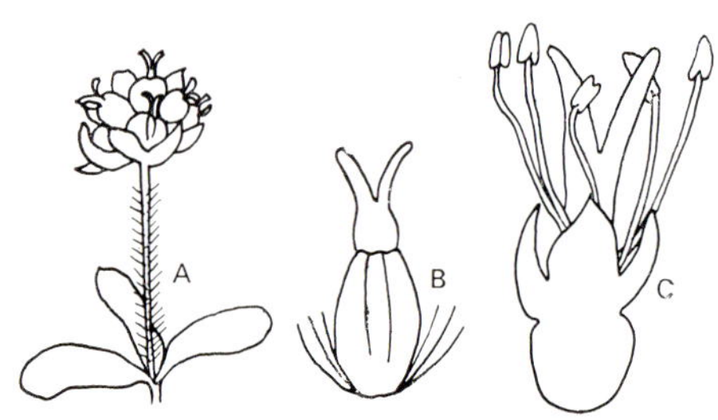


FIG. 151 *Actinotus*: A, flowerhead (× 2); B, fruit (× 4); C, flower (× 4).

florets of the daisies. The bracts are covered with closely set short hairs giving a flannel-like appearance to them, hence the common name. They surround a central portion of small white flowers.

*A. bellidioides.* White. Alpine, subalpine, montane. Vic., Tas.
A small plant with a rosette of spreading leaves, spoon-shaped or rounded, blunt, about 5 mm long, thick, sometimes 3-lobed and on stalks reaching 3 cm in length. Flowering head at first short but elongating to 3 cm, bearing 6–18 flowers surrounded by 6–10 bracts, joined at the base but spreading and forming a shallow cup about 6 mm in diameter. Flowers with 5 sepals and 5 small petals or none.

*A. moorei* is confined to wet places in the west and south-west mountains of the central plateau. A trailing leafy plant, woolly when young with crowded, hairy trifoliate or deeply 3-cleft leaves, often again divided, round in outline. Flowers in small heads of 5–6 with about 6 bracts forming a cup about 4 mm in diameter; sepals 4; petals absent.

*A. suffocatus* is a rosette-forming species restricted to wet places, with thick, blunt, rounded leaves 4–5 mm long. Flowers with neither sepals nor petals, in heads of 6–10 surrounded by 6–10 bracts joined at the base but spreading to form a shallow cup about 6 mm in diameter, on stalks reaching 3 cm in length.

*DICHOSCHIADIUM* (from Gk *dickos*, in two ways, and *schiadeion*, affording shade, as an umbel, probably referring to the flower arrangement). Fig. 152.

*D. ranunculaceum.* Gleaming white. Alpine. N.S.W., Vic., Tas.
If ever there was need for a common name it is for this plant. But alas, there is none and so one of the loveliest of our alpines must be known by this imposing title. It is a hairy, creeping plant of the wet patches, sometimes completely submerged and revelling in the cold waters on the margins of the melting snow. Leaves in a basal rosette, rather fleshy, bright-green, resembling the leaves of some of the buttercups—hence the specific name—rounded with 5–7 lobes, bluntly 3-toothed, 10–35 mm in diameter. Flowers in

FIG. 152 *Dichoschiadium*: A, flowerhead; B, flower; C, fruits; D, leaf. (All × 1½ except A, actual size.)

heads of 3–6, radiating from a common point on an umbel which is often branched, produced in great profusion, frequently in a complete circle; sepals and petals similar, small, deciduous, about 2–5 mm long. Fruit flattened on top.

*DIPLASPIS* (from Gk *diplos*, double, referring to the fruit). A small genus of creeping plants with simple leaves in rosettes and numerous small flowers in heads. The fruit is very flat, dry, consisting of 2 non-splitting 1-seeded parts.

*D. cordifolia.* Yellowish-green. Alpine. Tas.
A common species in peaty pockets. Leaves often purplish, rounded, heart-shaped, 7–15 mm in diameter. Flowers more than 20 in a radiating head on stalks 4–5 mm long.

*D. hydrocotyle.* Yellowish-green. Alpine, subalpine. N.S.W., Vic., Tas.
Leaves 5–10 mm long, heart-shaped or kidney-shaped, rather fleshy. Flowers shorter than the leaves, all radiating from a common point, usually less than 20 in a head; sepals very small; petals 5, small, rounded.

*HYDROCOTYLE* (from Gk *hydor*, water and *kotyle*, a dish or plate, referring to the leaves being slightly depressed in the centre). Fig. 153.

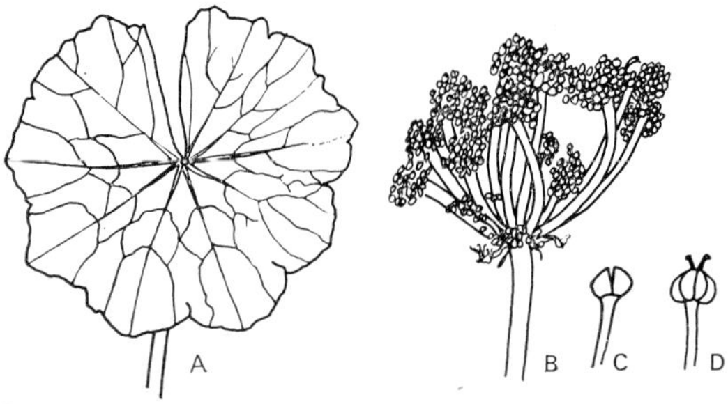


FIG. 153 *Hydrocotyle*: A, leaf (actual size); B, flowerhead (× 2); C, fruit (× 2); D, flower (× 4).

A genus of water-inhabiting plants with attractive round leaves, common in alpine pools or along the edges of waterways. The flowers are small and produced in inconspicuous heads.

*H. muscosa.* Brownish. Montane. N.S.W., Vic., Tas.
A smooth, creeping plant, sometimes forming dense mats. Leaves rounded, stalked, 3–6 mm in diameter, divided to the base into 3–5 segments. Flowers minute, in terminal heads each with 3–4 flowers.

*H. sibthorpioides.* Brownish. Montane. N.S.W., Vic., Tas.
Occurs also in rainforest in Tasmania. A slender plant sometimes forming dense mats and often colonizing banks of streams. Leaves stalked, crowded, with rounded blades 3–6 mm in diameter with broad lobes, upper surface hairy. Flowers in small heads. Fruit in rounded heads.

*OREOMYRRHIS* (from Gk *oros*, *orios*, mountain, and *myrrhis*, a fragrant herb). Fig. 154. Essentially an alpine genus, with 2 species on the mainland and one in Tasmania. They are small tufted herbs, sometimes forming cushions with divided leaves, with heads of white or reddish flowers arranged like the spokes in a wheel and surrounded by rather broad bracts.

*O. argentea* (syn. *O. andicola*). White. Alpine, subalpine, montane. N.S.W., Vic., Tas.
A tufted, silvery-white plant of the meadows. Leaves in a basal rosette, stalked, finely divided, 2.5–8 cm long. Flowers small, 5–15 in a head, surrounded by 6–11 rounded bracts; sepals tiny; petals 5.

*O. ciliata.* White. Alpine. Tas.
A lover of damp spots, often forming cushions. Leaves rather similar to those of *O. sessiliflora.* Flowers 5–25 in an umbrella-like head, almost stalkless; petals 5; stamens 5, short with brownish-crimson tips.

*O. eriopoda.* White with red centres. Montane. N.S.W., Vic., Tas.
Leaves in a rosette, twice divided, stalked, 2.5–20 cm long. Flowerheads on long, fairly stout stalks of 12–25 small flowers surrounded by 8–12 bracts, joined at the base.

*O. pulvinifica.* White. Alpine. N.S.W., Vic.
An attractive alpine plant forming cushions along watercourses. Leaves on long stalks, smooth, soft, with narrow leaflets. Flowers in heads of about 20, on very short stalks and surrounded by about 7 joined bracts, reddish in centre.

FIG. 154 *Oreomyrrhis*: A, flower (× 2); B, leaf (× 1½); C, fruiting head (× 1½).

*O. sessiliflora.* White. Alpine. Tas.
A plant with an odour like that of carrots (which are members of the same family). Leaves basal, stalked, smooth, finely divided, 1–9 cm long. Flowers almost stalkless, 10–20 in an umbrella-like head, small, with 9 or 10 joined bracts at the base.

*OSCHATZIA* (from Gk *oscha*, a sucker). Fig. 155.

*O. cuneifolia.* White. Alpine, subalpine. N.S.W., Vic.
A small, smooth plant of fens and bogs with characteristic wedge-shaped leaves in a tuft at the base of the plant on fairly long stalks and

with 3–7 shallow lobes at the tip. Flowers 1–several in heads, small. Fruit egg-shaped, about 3 mm long.

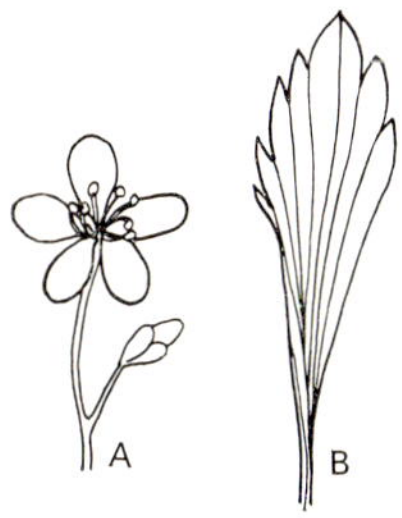


FIG. 155 *Oschatzia*: A, flower with fruit; B, leaf. (All × 2½.)

*PLATYSACE* (from Gk *platys*, flat, and *sakkos*, a sack, referring to the fruit). Fig. 156. This genus was at one time included with *Trachymene*.

FIG. 156 *Platysace*: A, flowerhead (× ½); B, flower (× 6); C, fruit (× 6).

*P. lanceolata*. White. Montane. N.S.W., Vic.
A slender shrub 30–90 cm high with broad leaves 12–25 mm long and without stalks. Flowers small, shortly stalked, with small narrow bracts at the base with several heads rising from a single stalk. Fruit ribbed, about 2 mm long, with a flat disc at the summit.

*SESELI* (from Gk *seselis*, a shrub). Fig. 157. *S. algens*, a small, somewhat creeping plant from alpine and subalpine regions of New South Wales, is the only species of *Seseli* occurring at these levels. The leaves are grey-green and pinnate, the lobes deeply toothed and narrow. The flowers are white, in small heads, each small flower with 5 prominent pointed sepals and 5 egg-shaped petals. The fruit is often turgid.

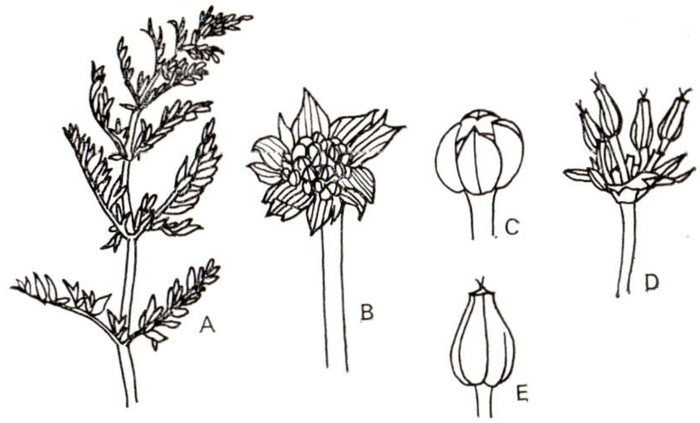


FIG. 157 *Seseli*: A, flowerhead (actual size); B, flower (× 2); C, bud (× 2); D, fruiting head (actual size); E, single fruit (× 2).

*TRACHYMENE* (from Gk *trakhys*, rough, and *mene*, moon, referring to the fruit). Fig. 158. An interesting genus with some alpine species

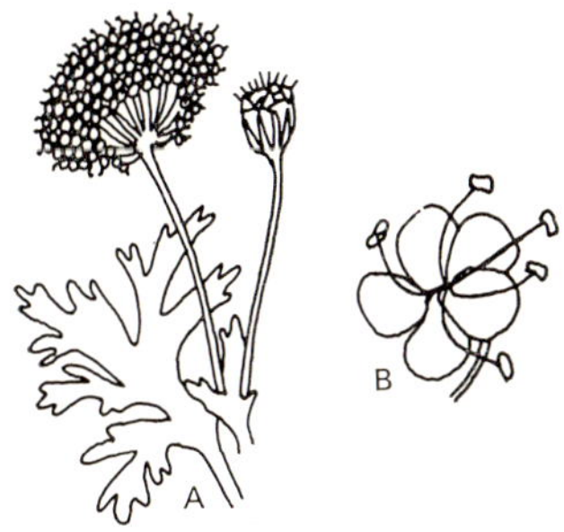


FIG. 158 *Trachymene*: A, flowerhead (actual size); B, single flower (× 4).

still under review. The leaves are divided, with 3–5 lobes usually again divided, and without stipules. The flowers are in heads surrounded by narrow, short bracts. The fruit is usually flat and notched.

*T. anisocarpa*. White. Subalpine, montane. N.S.W., Vic., Tas.
An attractive, short-lived, tall herb 40–120 cm high. Lower leaves stalked, 2.5–10 cm in diameter, deeply divided into wedge-shaped, toothed segments, the undersurface hairy; upper stem leaves smaller with shorter stalks and 3 narrow lobes. Flowers several in a head 2–3 cm in diameter with many heads on a single, long, somewhat succulent stalk. Fruit often with only one part, about 25 mm, broad and flat.

*T. humilis*. White, pink. Alpine, subalpine, montane. N.S.W., Vic., Tas.

A charming little rosette plant preferring wet situations. Leaves numerous, stalked, flat on the ground, often concealed by surrounding growth, oval or egg-shaped, entire or bluntly lobed, smooth or with a few scattered hairs, 1–2.5 cm long. Stalks of flowers leafless and often very short so that the flowerheads sit on top of the little plant, but sometimes erect and longer than the leaves; each stalk bears a small umbrella-like head of flowers 1–2 cm in diameter and surrounded by bracts; individual flowers numerous, the petals scarcely 2 mm long.

## Family ARALIACEAE

*POLYSCIAS* (from Gk *polys*, many, and *skia*, a shadow, possibly in reference to the large leaves and spreading habit). Fig. 159.

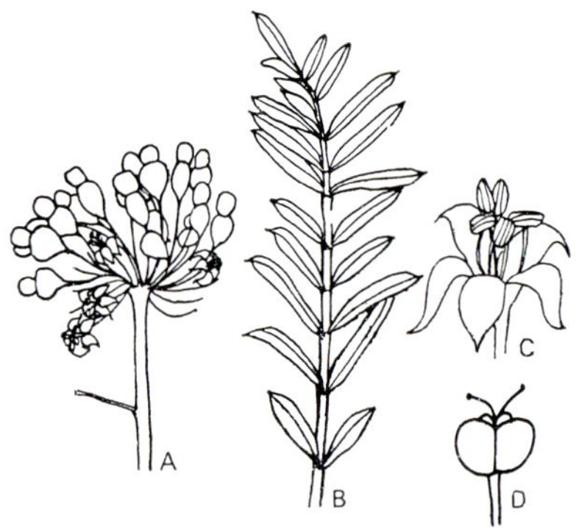


FIG. 159 *Polyscias*: A, flowerhead (actual size); B, leaves (actual size); C, flower (× 3); D, fruit (actual size).

*P. sambucifolia* (syn. *Tieghmanopanax sambucifolia*). Elderberry Panax. White. Montane. N.S.W., Vic.

An elegant shrub 60–100 cm high with smooth stems and leaves divided into broad leaflets, the whole leaf 4–6 cm long. Flowers in loose heads; 5 small sepals; 5 petals, often cohering at the tip; 5 stamens inserted above the other flower parts. Fruit succulent, with 2 compartments, bluish in colour.

## Family CAPRIFOLIACEAE

Two species of Elderberry are found in montane areas, both of them in moist and often dense scrub. They are tall, smooth shrubs with soft, light-green divided leaves, small tubular flowers in dense drooping heads, making an attractive show in full flower. There are usually 4–5 sepals and 4–5 petals, with 4–5 stamens usually attached to the petals. The ovary is inserted below the other flower parts and the fruit is succulent.

*SAMBUCUS* (old Lat. name for elder tree, possibly from *sambuca*, a musical instrument made of elder wood). Fig. 160.

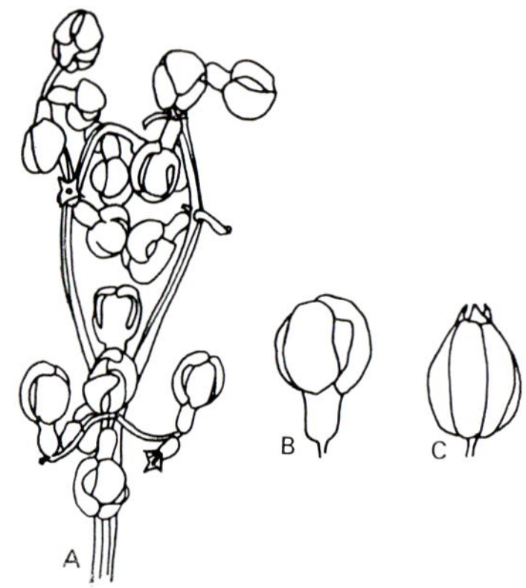


FIG. 160 *Sambucus*: A, flowerhead (× 1½); B, flower (× 2); C, fruit (× 2).

*S. australasica*. Yellow. Montane. N.S.W.

A tall shrub 5–10 metres high. Leaves 5–8 cm long; leaflets 3–5, stalked, often sharply toothed. Flowers tubular with spreading lobes; sepals and petals 3–4, small; stamens 3–5. Fruit yellow with 3–5 stones each containing 1 seed.

*S. gaudichaudiana*. White Elderberry. Cream. Montane. N.S.W., Vic., Tas.

A shrub 1–2 metres high or taller. Leaves with 4–11 leaflets, coarsely toothed, 4–10 cm long. Flowers shortly tubular with spreading lobes; 4 tiny sepals; petals 4; stamens 4. Fruit yellowish-white with 3–5 stones surrounding a single seed.

## Family RUBIACEAE

There are a number of genera in this family in alpine areas, the most conspicuous and attractive of which is *Coprosma*. The family is not very common in other Australian habitats. Members are usually shrubs with their leaves opposite or arranged in rings, usually with stipules. There are commonly 4–5 free sepals, often enlarging in fruit, 4 or occasionally more joined, with an equal number of stamens joined to them. The ovary is inserted below the other flower parts. The flowers are often unisexual. The fruit is usually succulent, more conspicuous than the flowers.

*ASPERULA* (from Lat. *asper*, rough, referring to the stiff, hairy covering). Fig. 161. A genus with several species in these habitats. It comprises small but abundant plants in heath and woodland situations. They are weak shrubs with quadrangular stems and whorled leaves with stipules. The flowers are small, funnel-shaped and in small heads, often male or female only; the flower parts are commonly in fours. The fruit is a small nut.

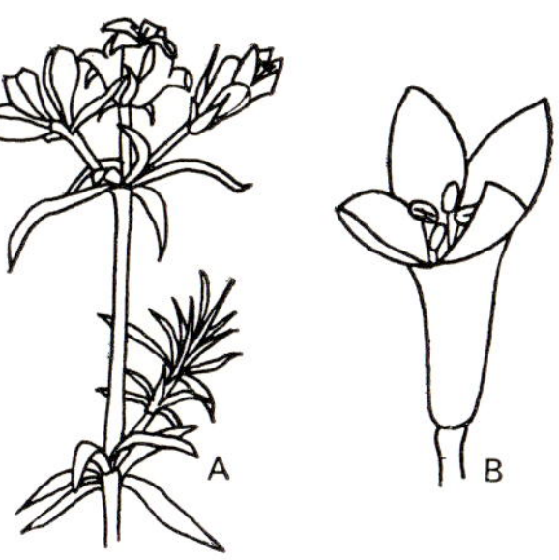


FIG. 161 *Asperula*: A, flowerhead (× 2); B, flower (× 4).

*A. conferta*. White. Alpine, subalpine, montane. N.S.W., Vic., Tas.

A branched, erect plant 3–30 cm high. Leaves mostly 6 in a whorl, narrow, without stalks, 4–11 mm long. Flowers solitary or a few together, usually 3, often either male or female.

*A. gunnii*. Creamish-white. Alpine, subalpine, montane. N.S.W., Vic., Tas.

A tufted plant, often very small at high altitudes. Leaves 3–6 in a whorl, rather broad and somewhat leathery, 3–10 mm long. Flowers in small clusters of 3–5, sweetly scented.

*A. pusilla*. White. Alpine, subalpine. N.S.W., Vic., Tas.

A small, tufted, rough plant, densely leafy. Leaves 6–8 in a whorl, stiff, spreading, rather narrow, 2–8 mm long. Flowers solitary or in threes in the leaf axils.

*A. scoparia*. White. Alpine, subalpine, montane. N.S.W., Vic., Tas.

A straggling plant with a long creeping stem, grooved and woolly, much branched at the base. Leaves in whorls of 6, smooth, 4–9 mm long. Flowers in small clusters in the leaf axils.

*COPROSMA* (from Gk *kopros*, dung, *osme*, a smell, alluding to unpleasant odour). Fig. 162.

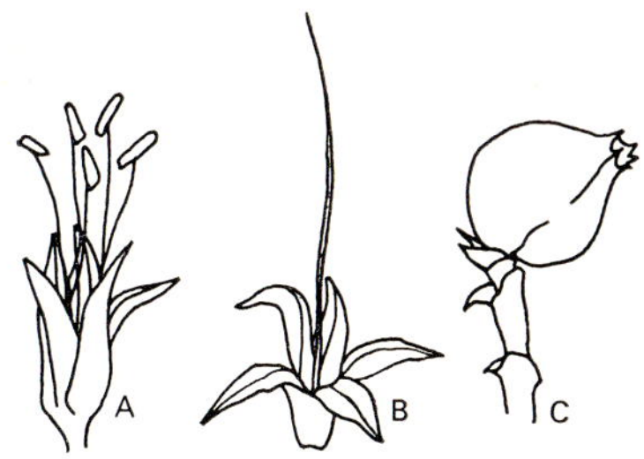


FIG. 162 *Coprosma*: A, male flower (× 3); B, female flower (× 3); C, fruit (× 2).

A genus of shrubby or creeping plants with small flowers, usually either staminate or pistillate on each plant, and attractive, large, shining, often brightly coloured succulent fruits.

*C. hirtella*. Coffee Berry. White; berries red or black. Subalpine, montane. N.S.W., Vic., Tas.

A widespread, stiff, sometimes spiny, attractive shrub, often found on rocky hillsides. Leaves broad, pointed, slightly stalked, shining on top, 2–6 cm long and with broad, short stipules on the stems between the leaves (not, as in most plants, at the base of the leaf). Flowers small in short heads. Fruit round, shining, 5–7 mm in diameter, edible but not very pleasant to the taste.

*C. moorei*. White; berries blue. Alpine, subalpine, montane. Vic., Tas.

Usually a prostrate shrub. Leaves egg-shaped, thick, smooth, shining, 3–5 mm long, each with a short stalk fused with that of the opposite leaf to form a sheath. Flowers solitary, bell-shaped; petals 2–3 mm long; stamens with stalks joined to the base of the petals and protruding; ovary with 2 protruding stigmas. Fruit round, 5–6 mm in diameter.

*C. nitida.* White; berries pale orange-red. Alpine, subalpine. Vic., Tas.

A small shrub, sometimes creeping, often found in exposed situations. Branches sometimes ending in spines. Leaves numerous, broad, blunt, sharp-pointed, shining, leathery, 4–10 mm long, on short stalks. Flowers solitary, each flower subtended by a 4-lobed cup formed from 2 bracts fused with 2 stipules; petals 4, in staminate flowers 5–6 mm long, smaller in pistillate flowers; stamens, 4 protruding; stigmas 2, protruding, 8–10 mm long. Fruit spherical, 6–10 mm long, shining, almost translucent, edible.

*C. pumila.* White; berries orange or red. Alpine. N.S.W., Vic., Tas.

A prostrate plant with densely matted branches. Leaves numerous, broad, thick, smooth, shining, 5–10 mm long on short stalks joined to the stipules to form sheaths. Flowers solitary, stalkless, funnel-shaped; staminate flowers about 8 mm long; stamens on long protruding stalks; pistillate flowers small with 4 stigmas protruding. Fruit round, 6–8 mm in diameter.

*C. quadrifida.* Native Currant. White; berries orange or red. Montane. Vic., Tas.

Occurring also in Tasmania at lower elevations. A shrub 2–4 metres high with slender, hairy branches terminating in fine spines. Leaves numerous, usually broad but variable, 4–25 mm long. Flowers solitary, very similar to those of *C. nitida.* Fruit also similar to that of *C. nitida*, but slightly smaller.

*GALIUM* (Greco-Lat. *galion*, a plant name, perhaps from Gk *gala*, milk, because the leaves of the plant were used to curdle milk). Fig. 163.

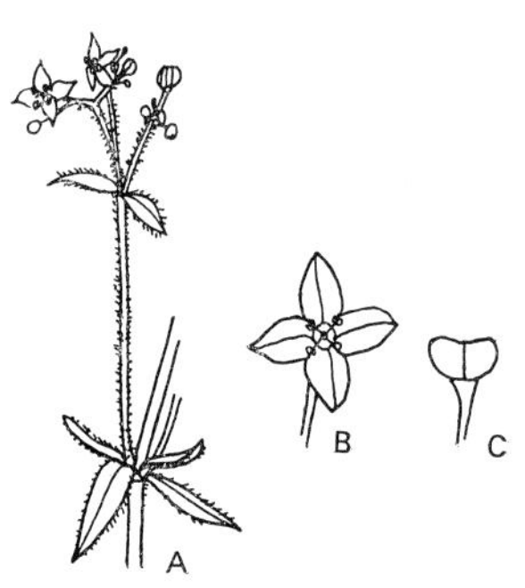


FIG. 163 *Galium*: A, flowerhead (× 2); B, flower (× 4); C, fruit (× 4).

A genus of small, widespread plants with slender, quadrangular stems with leaves and leaf-like stipules in whorls and small, usually unisexual flowers. The flowers, in small heads, are shortly tubular with 4 lobes. The fruit is dry, separating into 2 one-seeded parts when ripe. *G. gaudichaudii* is a small, straggling, hairy plant 10–30 cm long from montane areas on the mainland and Tasmania. There are 4 leaves in each whorl. The tiny white flowers are set in the leaf axils, 1–2 together on short stalks. The fruit is kidney-shaped.

*NERTERA* (from Gk *nerteros*, low down, referring to its prostrate habit). A genus of slender, creeping plants with thin, bright-green leaves sometimes spreading for considerable distances. *N. granadensis* is an alpine and subalpine species from the mainland and Tasmania with broadly triangular, opposite, fused stipules and opposite, round or rhomboid, blunt leaves. The flowers are tiny, with petals about 2 mm long. The fruit is red and succulent, about 4 mm in diameter.

*OPERCULARIA* (from Lat. *operculum*, a lid, referring to the caps on the fruits). A genus of slender, stiff plants with simple, opposite leaves and small stipules fused with the leaf-stalks to form a short sheath round the stem. The flowers are small, bell-shaped, 3–5 lobed, with 2–5 stamens. The fruit is a small capsule of 2 parts, several joined to form a compound fruit. *O. ovata* is common in montane habitats on the mainland and in Tasmania, with egg-shaped or rhomboid leaves and a strong foetid smell when bruised; the outer face of the seeds is usually smooth. *O. varia,* found in similar habitats, has narrow leaves, and the outer face of the seeds is usually rugose.

## Family COMPOSITAE

The Daisy family is one of the largest in the world and the most widely distributed. It is common in many parts of Australia but in the alpine areas it is, together with the buttercups, one of the outstanding features of the vegetation. One is inclined to think of buttercups and daisies as representatives of more gentle climates than we mostly have to offer in Australia but in our limited alpine parts these are among the most

TOP The slopes of Cradle Mountain, dissected by fens and bogs

BOTTOM *Sprengelia incarnata* *Pimelea alpina*

TOP Subalpine grasslands

BOTTOM *Bauera rubioides* *Hibbertia procumbens*

*Eucalyptus rubida*

*Eucalyptus dalrympleana*

TOP *Ranunculus anemoneus* *R. millanii*

BOTTOM *Billardiera longiflora* *Geum renifolium*

TOP *Eucryphia lucida* *Aristotelia peduncularis*

BOTTOM *Drosera arcturi* *Linum marginale*

TOP *Kunzea muelleri* *Baeckea gunniana*

BOTTOM *Rubus parvifolius* *Sambucus gaudichaudiana*

TOP *Dichoschiadium ranunculaceum* *Trachymene humilis*

BOTTOM *Trachymene anisocarpa* *Callistemon sieberi*

TOP *Helichrysum thyrsoideum* *Ewartia nubigena*

BOTTOM *Ewartia meredithii* *Celmisia longifolia* with *Euphrasia glacialis*

TOP *Helichrysum adenophorum var. waddellii* *Craspedia uniflora*

BOTTOM *Helipterum albicans* *H. anthemoides*

TOP *Prionotes cerinthoides* *Wahlenbergia gloriosa*

BOTTOM *Gentianella diemensis* *Richea scoparia*

TOP *Plantago gunnii* *Isophysis tasmanica*

BOTTOM *Herpolirion novae-zelandiae* *Pimelea ligustrina*

common, the loveliest and the most varied of our plants.

As might be expected in such a large group there is a very wide range of form among the genera. The alpine representatives are shrubs or herbs; some are soft and fragile plants, some are hard and woody; some have leaves in basal rosettes, entire or divided, some have stem-clasping leaves; in some the flowers are large and brightly coloured, in others they are pale and inconspicuous.

Although they present great differences in general appearance, close examination readily identifies them. The daisy flower, so-called, consists of a group of small flowers closely arranged on the expanded end of the stalk, rather resembling a single, large flower. Sometimes these heads are solitary but at other times they are arranged in groups at the end of a single stem. In a typical daisy flower (Fig. 164) such

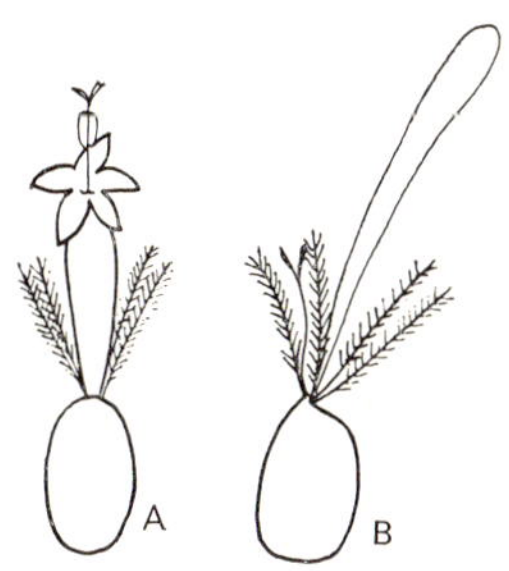


FIG. 164 A typical daisy flower: A, flower; B, fruit. (Both × 2.)

as cosmos, Michaelmas daisy or marigold, the centre of the daisy consists of a number of small tubular flowers, usually called disc florets. There are no sepals or they are modified to form a group of hairs or spines on the top of the seed-box, called a pappus and used to help disperse the single-seeded fruits as in the well-known "Father Christmas". There are 5 petals joined together to form a tube, 5 stamens with the pollen boxes adhering to one another, and a divided stigma. The outer parts of the daisy, the so-called "petals", are also really flowers, though usually incomplete ones, called ray florets. Here again there are no sepals; the petals are joined to form what appears to be a strap-like single petal. The stamens are frequently missing and sometimes the pistil also, but if it is present it has a divided stigma at the tip.

Each head of flowers, whether single or one of a cluster, is surrounded by a ring of one or more rows of bracts (modified leaves) which may be scaly, stiff or green and leaf-like or otherwise coloured, persistent or deciduous. In the paper daisies these bracts are often large, coloured and numerous and the outer florets are missing. Sometimes there are additional leaf-like bracts forming a calyx-like structure; in some cases a compound head is surrounded by a set of common bracts.

In some members of this family there are no ray or outer florets at all, the flowerhead consisting of tubular florets only with small bracts or with large white or coloured bracts which—in the case of the paper daisies, for instance—serve to attract pollinating agents as do the ray florets in other kinds. In *Craspedia*, billy buttons, the flowerheads consist entirely of closely packed tubular florets.

*BRACHYCOME* (from Gk *brachys*, short, and *comus*, hair, referring to the short pappus). Fig. 165. Some authors prefer the name *Brachyscome* for this genus and it was used by Cassini

FIG. 165 *Brachycome*: A, flowerhead (× ½); B, group of flowers (actual size); C, unopened flowerhead (× ½); D, fruit (× 2).

in 1817, but in 1825 he used the name *Brachycome* with the footnote: "It is thus that the generic name should be written instead of *Brachyscome*." It seems clear, therefore, that the first spelling was an orthographic error and that the well-established name *Brachycome* should be retained.

This genus has a considerable number of species in these habitats, making it difficult to distinguish them from one another. They are herbs with the typical daisy flower with a large number of tubular disc florets, usually pistillate in the centre and surrounded by white, lilac, blue or purple, or rarely yellow, strap-shaped ray florets. The leaves are either tufted in a basal rosette or placed along an erect stem or both. There are usually 2 rows of leafy dry bracts, sometimes with purplish margins, at the back of the flowerhead. The fruit is usually flattened, sometimes winged, with a pappus of short bristles or none at all and is important in distinguishing one species from another.

*B. aculeata* (syn. *B. scapiformis*). Purple, lilac. Alpine, subalpine, montane. N.S.W., Vic., Tas.

A widely distributed species. Leaves egg-shaped or narrower, in a basal rosette, toothed or wavy on the margin, dark-green, rather thick, almost smooth, narrowed into a stalk 3–6 cm long. Flowering stems usually leafless, up to 40 cm long. Flowerheads solitary, 8–15 mm in diameter; ray florets numerous. Fruit with flattened wings.

*B. alpina*. Bluish. Subalpine. Vic.

Restricted to Bogong High Plains, Victoria. Leaves basal, spoon-shaped, 2.5 cm or less in length, rough, toothed or wavy on the margins, narrowed into a broad stalk. Flowering stalk 5–10 cm long with a few scattered leaves. Flowerheads 8–10 mm in diameter, solitary; ray florets numerous. Fruit egg-shaped, flattened, with a thickened margin.

*B. angustifolia*. Lilac, pale pink. Subalpine, montane. N.S.W., Vic., Tas.

A widespread small species from sea-level to 3,000 feet in Tasmania and common in montane habitats on the mainland. Leaves basal, clustered and also arranged alternately on the erect stems, entire, pointed and narrow, 3–6 cm long. Flowering stalks 3–10 mm long. Flowerheads about 5 mm in diameter, solitary; ray florets numerous, 6–10 mm long. Fruit narrow, flattened, with winged margin.

*B. decipiens*. White tinged with blue. Montane. N.S.W., Vic., Tas.

A rather large-flowered, widely distributed species of grassy places, not unlike the commonly cultivated *Bellis perennis* but stouter and taller. Leaves in a basal rosette, elliptical or oblong, entire or sometimes lobed, blunt, with a sheathing base. Flowerheads solitary on stalks 8–15 cm long, the margins of the bracts often deep purple; ray florets numerous, about 8 mm long. Fruit flattened, narrow, tuberculate, margin winged.

*B. diversifolia*. White. Subalpine, montane. N.S.W., Vic., Tas.

Sometimes fairly tall, up to 50 cm high but more commonly shorter. Leaves in a basal rosette and also on the erect stems, the lower leaves up to 14 cm long, those on the stems shorter; broad at the base but not sheathing, narrowing and coarsely toothed or finely or sometimes deeply lobed, hairy. Flowering stems stout. Flowerheads solitary, large, 1.2–1.8 cm in diameter; ray florets up to 15 mm long. Achenes narrow, somewhat flattened but thick, the faces with 2 longitudinal folds; the pappus fairly small and spreading.

*B. graminea*. Pale pink, mauve. Montane. N.S.W., Vic., Tas.

A small plant usually confined to wet habitats. Leaves alternate, on rather weak ascending stems and often also clustered at the base, broad, sharp-pointed, tapering at the base, the lower ones 14.5 cm long, the upper ones smaller. Flowerheads solitary on stalks 15 cm or more in length; ray florets about 9 cm long. Fruit egg-shaped, flattened but with a thick, rounded margin.

*B. marginata*. Yellow. Montane. N.S.W., Vic.

A small plant. Leaves fairly narrow, pointed, sometimes with a few teeth or notched. Fruit flattened, rough with a broad, lobed wing; pappus conspicuous.

*B. nivalis*. White, rarely blue. Alpine, subalpine, montane. N.S.W., Vic., Tas.

One of the loveliest of the alpine daisies. Found in wet montane habitats and grasslands in Tasmania, but at higher elevations on the mainland, growing in attractive little tufted clumps, often in rock crevices where there is constant running water. Leaves in basal rosettes and also along the erect stems; basal leaves usually several times divided but sometimes entire, spoon-shaped, narrow and stem-clasping, 2–9 cm long. Flowerheads solitary, large, on stalks about twice as long as the leaves; ray florets about 10 mm long, numerous, strap-shaped. Fruit rounded, winged, with a conspicuous pappus.

*B. obovata*. White, bluish. Subalpine. Vic.
A plant of the sphagnum bogs at about 4,700 feet. Leaves in a basal rosette, narrow, egg-shaped, entire. Flowerheads stalked, 1–2 on each plant; ray florets fairly numerous, about 5 mm long. Fruit with a pappus of short, straw-like bristles.

*B. radicans*. Mauve, white. Alpine, subalpine, montane. N.S.W., Vic., Tas.
A common inhabitant of swampy ground. Leaves in a basal rosette, narrow, blunt or with a few lobes in the upper part but very variable, 2.5–5 cm long. Flowerheads solitary, on stalks 5–20 cm long with a few alternate leaves; ray florets about 6 mm long, numerous. Fruit broad, flattened, striate with a thickened margin and a short pappus.

*B. rigidula*. Pink, mauve. Subalpine, montane. N.S.W., Vic., Tas.
A rather weak little plant with wiry stems, spreading 10–40 cm, found also at lower elevations in Tasmania. Stem leaves very much divided, pointed and smooth. Flowerheads solitary, on slender stalks 0.8–3 cm long; ray florets 5–8 mm long. Fruit broad, flattened, light brown to dark brown with lighter coloured wings and a fairly long pappus.

*B. scapigera*. Mauve, white. Alpine, subalpine, montane. N.S.W., Vic.
A widely distributed species. Leaves in a basal rosette, broad, entire, stalked, up to 19 cm long. Flowerheads solitary on long, slender stalks 7–20 cm high sometimes bearing a few leaves; ray florets numerous. Fruit compressed, rounded, without wings.

*B. stolonifera*. White. Alpine, subalpine. N.S.W.
A very dainty but sturdy little plant. Leaves in a basal rosette fairly broad, entire, blunt, with a sheathing base. Flowerheads solitary on stout stalks considerably longer than the leaves; ray florets numerous, about 6 mm long.

*B. tenuiscapa*. White, mauve. Alpine, subalpine. N.S.W., Vic., Tas.
A sturdy plant with several varieties. Leaves in a basal rosette, the fibrous remains of the old leaves round the base broad, often thick and fleshy with a narrow base sheathing the stem, blunt with a wavy or toothed margin. Flowerheads solitary on erect stems up to 15 cm long; ray florets numerous, about 7 mm long. Fruit wedge-shaped, flattened, with a thick margin but not winged, smooth, dark brown to black. In the most common mainland form the leaves are rigid, hairy and up to 10 cm long, and the flowering stalk about twice that length, the plant closely resembling *B. decipiens*.

*CASSINIA* (after Alexandre Henri Gabriel Cassini, 1784-1832, French botanist). Fig. 166.
A genus of tall shrubs with alternate leaves and small flowerheads in clusters, rather similar to *Helichrysum*, but differing in the possession of scales between the florets. The latter are few, all tubular. The fruit is short, angular, usually hairy, with a pappus of several hairs in a single row slightly cohering at the base.

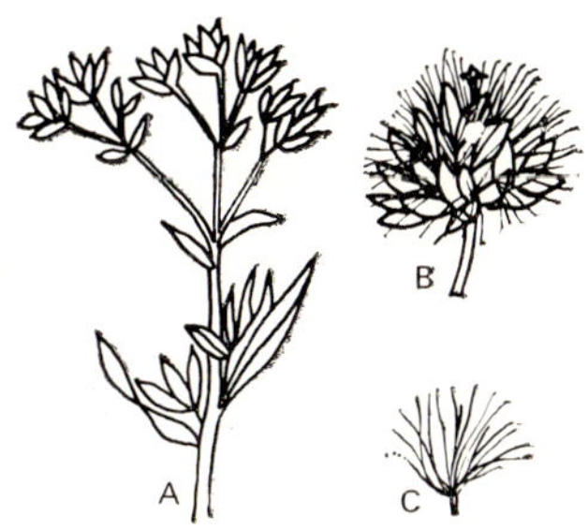


FIG. 166 *Cassinia*: A, flowerhead (× ½); B, fruiting head (actual size); C, single fruit (× 2).

*C. aculeata*. White, pale brown, pink. Alpine, subalpine, montane. N.S.W., Vic., Tas.
A very showy plant widely distributed from the sea-coast to almost 6,000 feet, often occurring in large stands and making a spectacular display in late summer. A sturdy, compact shrub 1–4 feet high, rather woolly. Leaves narrow, pointed, the upper surface smooth, the lower surface woolly, 1–3 cm long. Flowerheads numerous in loose clusters, cylindrical, small but conspicuous because of their large number, not spreading

until the fruits are ripe; bracts papery; florets 6–12, about as long as the bracts. Fruit hairy, with slender pappus bristles.

*C. cunninghamii.* Straw-coloured. Montane. N.S.W.

At one time known as *Helichrysum cunninghamii*, this is a heath-like plant with woolly branches. Leaves narrow, stiff, pointed, with the margins rolled under, smooth above, woolly below, 1–1.5 cm long. Flowerheads small and numerous in close clusters; florets usually 3. Fruit smooth; pappus bristles slender.

*C. longifolia.* Cream. Montane. N.S.W., Vic., Tas.

A shrub or small tree with woolly branches. Leaves narrow, sharp, shortly stalked, upper surface dark green and smooth, lower brownish or white, woolly, 2–7 cm long. Flowerheads numerous in broad, dense clusters, not spreading until the fruits are ripe. Fruit hairy, with slender pappus bristles.

*CELMISIA* (after Celmisius, a son of the nymph Alciope). Fig. 167. Everyone who visits alpine country knows and loves the snow or silver daisies. Occurring as mat plants at the very summits of the mountains and extending down to something under 6,000 feet on the mainland, these silver-leafed plants make wonderful ground cover and help to retain the soil against the severe natural erosion that occurs here. They make a pleasing picture, even without their attractive flowers, as the soft silver-grey leaves dance in the frequent winds.

These tufted plants, of which there are three common species, have their leaves clustered into a close rosette, and the silvery appearance is due to the many scattered hairs covering them. The flowerheads are large, solitary, on tall, straight, woolly stems; each head is surrounded by a number of rows of soft, usually woolly bracts with dry crisp margins; there are many disc florets in the centre and a single row of strap-like pistillate ray florets, white, sometimes pink or purplish on the tip and back, especially in bud. The single-seeded fruits have a number of fine, unequal pappus bristles.

*C. longifolia.* Snow or Silver Daisy. White. Alpine, subalpine. N.S.W., Vic., Tas.

This much-loved daisy is the only species occurring in New South Wales, but it makes up for this by its abundance, covering considerable areas of the highest slopes. Leaves silvery-grey in a close rosette, 5–18 cm in length, pointed, 2–8 mm broad; as they grow older the silvery covering wears off, leaving the upper surface blue and shining. Flowering stalks stout and woolly, varying greatly in height, in rare cases reaching 30 cm but usually much less. Flowerheads 3–4.5 cm in diameter, often a large number on one plant; buds often rosy-pink before opening.

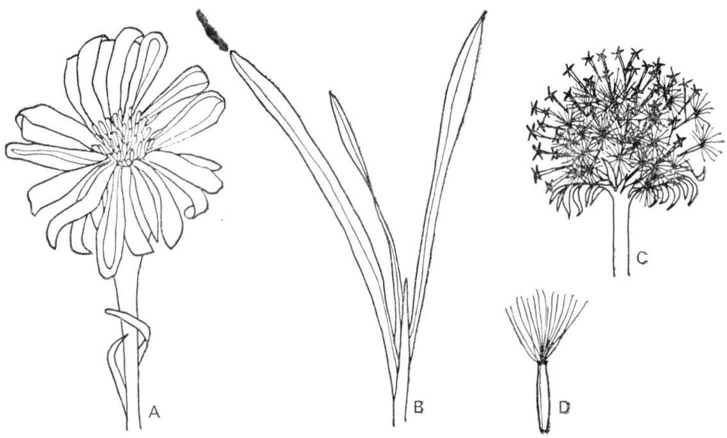


FIG. 167 *Celmisia*: A, flower (× 1½); B, leaves (actual size); C, fruiting head (× 1½); D, single fruit (× 3).

*C. saxifraga.* White. Alpine, subalpine. Tas.

Found on mountains or slopes or in depressions where snow has been lying. Leaves crowded, 1.5–3 cm long, 3–4 mm broad, the upper surface yellowish or grey-green, the undersurface covered in white or yellowish woolly hairs.

Flowering stems up to 15 cm long, woolly, bearing flowerheads rather smaller than those of the two other species.

*C. sericophylla*. Snow Daisy. White. Alpine, subalpine. Vic.

Abundant on the Bogong High Plains at an elevation of 5,000 to 5,500 feet, following the streams of swiftly flowing melt waters down the slopes. A lovely plant, very similar to the more common *C. longifolia* but larger in all its parts and with a more spreading habit. Leaves silvery, about twice as broad as those of *C. longifolia* and longer. Flowers well displayed on long stalks.

*CRASPEDIA* (from Gk *kraspedon*, a border or hem, referring to the fringe on the leaves of some New Zealand forms of *C. uniflora*). Fig. 168. Commonly known as billy buttons and a feature

FIG. 168 *Craspedia*: A, flowerhead (× ½); B, fruiting head (actual size); C, single fruit (× 2).

of alpine slopes, sometimes completely covering a hillside or a slope as far as the eye can see, and conspicuous because of their brilliant button-like orange-coloured heads. In a summer snowstorm they make a wonderful sight, for the snowflakes tend to gather in a frill round the edges of the yellow button-like faces of these quaint little daisies. The florets are all tubular, surrounded by short bracts.

*C. alpina*. Yellow. Alpine, subalpine. Tas.

A species differing only from *C. glauca* in the fact that the stems and leaves are covered with a dense mat of white, woolly hairs.

*C. glauca*. Yellow, cream, white. Montane. Tas.

A very variable plant extending to sea-level. Leaves in a basal rosette, broad or narrow, smooth or woolly, stalked or stalkless, 3–25 cm long. Flowerheads 1–3 cm in diameter, clustered. There are several recognized varieties.

*C. uniflora*. Orange, yellow, cream. Alpine, subalpine, montane. N.S.W., Vic.

An extremely variable plant which will undoubtedly be divided up into several species or subspecies upon further investigation. At its best it forms considerable stands and is a fine spectacle. A rather hairy plant with a basal rosette of stalked leaves often quite long but variable, and narrow stem-clasping leaves as well. Flowerheads globular, up to 2–2.5 cm in diameter on woolly stalks, variable in size and number of flowers. Fruit silky, with 10–15 plumose bristles.

*ERIGERON* (Greco-Lat. name of groundsel, from Gk *eri*, early, and *geron*, an old man, perhaps referring to the bald receptacle after

FIG. 169 *Erigeron*: fruit (× 3).

fruiting). Fig. 169. A fairly large genus with a few alpine species. There are 2 distinct types of florets. In many species the ray florets are small; in others large and then the flower has the appearance of a *Brachycome*. The ray florets are narrow, in several rows. The pappus on the fruit is long and brittle and the hairs separate readily from one another, distinguishing the genus from *Brachycome* in which the hairs are very short or absent.

*E. pappocromus*. White, violet. Alpine, subalpine. N.S.W., Vic., Tas.

A delightful little daisy plant with leaves in a basal rosette. Leaves rather broad, blunt, smooth, often stalked. Flowerheads solitary on stalks 12–15 cm high, sometimes with a few leaves; ray florets numerous, strap-shaped, in about 2 rows. Fruit smooth, with a long, whitish pappus. A species readily confused with *Brachycome*.

Other less conspicuous species of *Erigeron* are *E. conyzoides*, a rather rare, taller species from montane habitats in New South Wales and Victoria with a rosette of stalked leaves in clusters, and flowerheads with numerous ray florets; and *E. stellatus*, a subalpine species from New South Wales and Tasmania with narrow, almost stalkless leaves and solitary flowerheads on long stalks with numerous ray florets.

*EWARTIA* (after A. J. Ewart, 1872-1934, professor of botany, Melbourne University). Fig. 170. Sometimes known as Australian edelweiss. A genus of dwarf plants forming hummocks or cushions, easily recognized by their attractive silver-grey foliage. Leaves small, densely hairy, soft. Flowerheads small and solitary but often produced in such numbers as to give the whole surface of the cushion a starry white or pink appearance. When the fruits are ripe the pappus-bearing heads make the little grey mats particularly attractive.

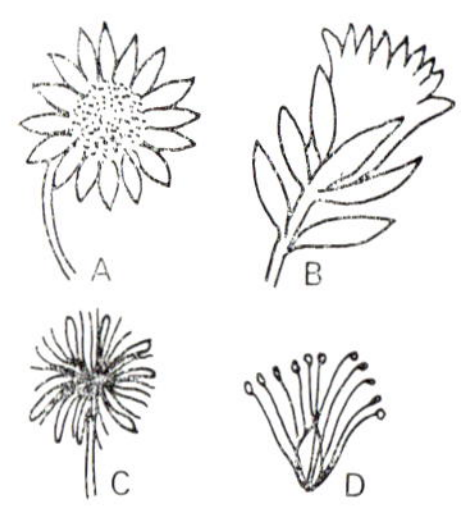


FIG. 170 *Ewartia*: A, flowerhead; B, flowerhead side view; C, head of fruits; D, single flower. (All × 2 except D × 6.)

*E. catipes*. White, crimson or rusty-brown. Alpine. Tas.
Found in wet, exposed places. Leaves broad, crowded, 2.5–4.5 mm long, blunt, woolly, with silvery-white hairs. Flowers in terminal leaf clusters. Pappus bristles slightly cohering towards the tip or free or cohering and falling off in clusters.

*E. meridithii*. White. Alpine. Tas.
Found in wet, exposed places. Leaves crowded, broad or spoon-shaped, blunt, covered with yellowish, woolly hairs 5–7 mm long. Flowers longer than the brown bracts. Pappus bristles slightly thickened towards the tip.

*E. nubigena*. Brown Edelweiss. Brownish. Alpine. N.S.W., Vic.
Common at 6,000 feet and over, often scrambling over rocks. Leaves silvery-brown, rather broad and soft. Pollen boxes of disc florets dark-coloured, giving a brownish appearance to the flowers. Pappus with a single row of bristles, barbellate at the tip.

*E. planchonii*. Brownish. Alpine. Tas.
Found on mountain tops at 4,000 feet and over. Leaves short and thick, both surfaces covered with woolly-white hairs, often rust-coloured when young. Flowers small, solitary, with stalks shorter than the leaves and consequently hidden but the stems lengthening as the fruits ripen; floral bracts longer than the white and brownish flowers. Pappus bristles free or cohering and falling in small clusters.

*HELICHRYSUM* and *HELIPTERUM*. These are the paper daisies, so called because of the papery nature of the conspicuous part of the flowerhead—usually the bracts, which are so dry that they will remain on the cut flowerhead for many years although the centre of softer florets will fall unless they are unopened when cut. These plants occur in many parts of the alpine areas and are usually small and attractive and cover extensive areas of open country, often as far as the eye can see.

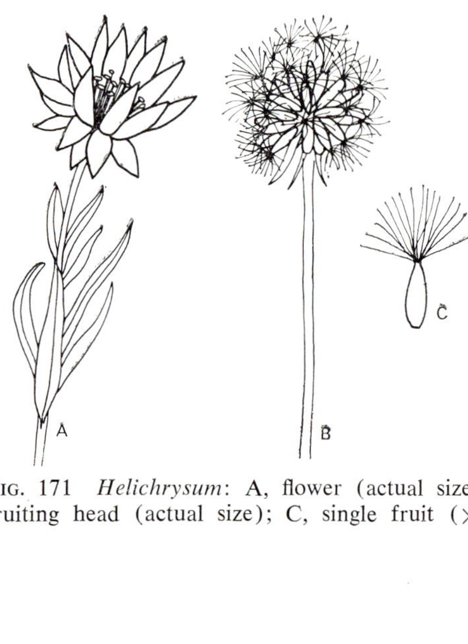


FIG. 171 *Helichrysum*: A, flower (actual size); B, fruiting head (actual size); C, single fruit (× 2).

*HELICHRYSUM* (Greco-Lat. name for a yellow everlasting, from Gk *helios*, the sun, and *khrysos*, gold). Fig. 171. A genus well represented among the paper daisies. It consists of herbs or shrubs, usually woolly, with alternate entire leaves. The florets are all tubular and surrounded by radiating papery bracts, often large and coloured. In some the flowerheads are solitary on long stalks; in others they are small and borne in clusters on a single stalk. The pappus on the fruit consists of simple or barbed bristles.

*H. acuminatum.* Golden yellow. Alpine, subalpine, montane. N.S.W., Vic., Tas.
A slender plant of wide distribution. Leaves crowded at the base and borne also on the erect stems, fairly broad with a bluntly toothed end, the base rounded and slightly stem-clasping, rather thin with a conspicuous midrib, both surfaces hairy and with cobweb-like hairs on the margin, 2.5–5 cm long. Flowers solitary on stems 3–15 cm high; bracts conspicuous, shining, the outer ones egg-shaped, often brown or reddish, the inner ones narrower; florets much shorter than the bracts. Pappus bristles often pale-yellow with tiny barbs.

*H. alpinum.* Pink or light brown. Alpine, subalpine. N.S.W., Tas.
An erect shrub 1.5 metres high, the branches slightly hairy but becoming smooth with age. Leaves narrow, egg-shaped or oblong, spreading, blunt, leathery, smooth above, pale or white beneath, 5–7 cm long. Flowerheads small, narrow, somewhat globular, 4.5–5 mm long, in dense clusters; bracts woolly, not expanding; florets 4–6 in each head. Fruit smooth or finely hairy, pappus bristles slightly thickened and finely toothed.

*H. antennarium.* White, backed brown or straw colour. Alpine, subalpine. Tas.
An angular, much-branched shrub 1–1.5 metres high. Leaves broad, blunt, leathery, 1–3 cm long, on short stalks, the upper surface usually sticky and bright-green, the lower surface paler and covered with resinous scales. Flowerheads in loose clusters on slender stalks; bracts numerous, scaly, not expanding; florets 20–28, each 4–5 mm long. Fruit hairy at the top; pappus bristles slender but slightly thickened at the tip.

*H. apiculatum.* Golden yellow. Montane. N.S.W., Vic., Tas.
A widespread and rather variable small plant with woolly-white flowering stems 10–15 cm high. Leaves alternate, broad, 2.5–6 cm long, the upper ones shorter and narrower. Flowerheads in loose clusters, the individual heads small; outer bracts narrow and pointed; florets only slightly shorter than the bracts. Pappus bristles yellow with tiny barbs.

*H. backhousii.* Brown or reddish. Alpine, subalpine. Tas.
A slightly resinous shrub about 1 metre high with yellowish or grey branches. Leaves broad, very blunt, smooth or very finely hairy above, pale below with a fine, downy covering obscuring the midvein, 7–10 mm long. Flowers in globose clusters, turban or bell-shaped, 5–6 mm long; bracts rather wrinkled with a few long, tangled hairs, the outer ones brown or reddish, the inner ones white and radiating; florets 12–18. Fruit smooth or finely hairy, the bristles slightly thickened at the tip.

*H. bracteatum.* Golden yellow, shining. Subalpine, montane. N.S.W., Vic.
The well-known yellow everlasting or paper daisy extending in distribution from the coastal mountains to subalpine situations at 5,000 feet. It varies in size but at higher elevations is a compact small plant. It can, however, reach 25–30 cm or higher. The golden glow of the large, brilliant yellow flowers makes fascinating patterns among the boulders and the other plants of the subalpine meadows. Sometimes the bracts are bronze or straw-coloured but more frequently brilliant orange-gold. Leaves vary from narrow to fairly broad, usually smooth on both surfaces, up to 10 cm long. Flowerheads solitary, 2–4 cm in diameter, with numerous rigid, crisp, papery bracts surrounding a mass of tubular flowers. Pappus hairs simple, 6–8 mm long.

*H. conditum.* Cream or yellow. Montane. N.S.W., Vic., Tas.
A rather sticky shrub about 1 metre in height, aromatic, with densely white, woolly branches. Leaves fairly narrow, 1.5–2.5 cm long, pointed, rather flat, smooth or cobwebby above, rusty

woolly underneath but the margins and midrib conspicuous. Flowerheads large in loose heads, each head somewhat turban-shaped, with about 16 florets; the outer bracts with whitish tips, the inner ones with broad, rather spreading, white tips. Fruit woolly, the pappus bristles fine but thickened towards the tip. Rather like *H. thyrsoideum* but distinguished by the woolly undersurface of the leaves and the pointed bracts.

*H. dealbatum.* Snow-white, backed red or bronze. Subalpine, montane. Vic., Tas.

A short, tufted, woolly plant 5–30 cm high. Leaves in a basal rosette and also along the erect stem, broad and smooth on top, cottony white or silvery below, 2–3 cm long. Flowerheads solitary and terminal, 2–5 cm in diameter; outer bracts red or bronze, the inner ones white and shorter so that the buds are coloured and the open flowers snow-white, making a delightful contrast. Pappus bristles barbed.

*H. ericeteum.* White, backed pale pink or brown. Subalpine. Tas.

A spreading shrub with a sweet aromatic odour, about 3 metres high. Leaves sticky, fairly narrow, blunt and leathery, the under-surface yellowy-woolly, the upper with scattered hairs. Flowerheads profuse in loose, dense clusters on long, slender stalks, each head 4–5 mm long; ray florets 5–6 in a head. Fruit hairy; pappus bristles slender, minutely barbellate.

*H. expansifolium.* White, backed straw colour or red. Alpine, subalpine, montane. Tas.

A tall shrub about 1–2 metres high, rather resmbling *H. hookeri* and at one time included with it, but with stouter branches, larger leaves and more numerous flowerheads in each cluster. Leaves rather narrow but wider at the base, grey-green, sticky, pointed, the undersurface woolly, overlapping, 2–6 mm long. Flowerheads 3–6 together in profusely branched laterals, rather stiff, 3.5–4.5 mm long; outer bracts with woolly hairs at the base; inner bracts with white tips; florets 5–7 in each head. Fruit usually smooth, the pappus bristles slightly thickened above.

*H. hookeri.* Kerosene Bush. Straw colour. Alpine, subalpine. N.S.W., Vic., Tas.

A widely distributed sticky shrub 1–2 metres high. Leaves rather scale-like, overlapping and pressed against the stem, triangular or egg-shaped, the upper surface smooth and sticky, the lower woolly; the margins flanged, 1–2 mm long. Flowerheads without stalks, 5–8 together at the ends of short branches, forming dense compound clusters, rather stiff but very showy, 4 mm long; outer bracts papery, the inner ones white-tipped; florets 2–4 in each head. Fruit finely hairy, with slender pappus bristles slightly thickened at the tips.

An endemic Tasmanian species, very common, often in dense stands at altitudes of 3,000 to 4,000 feet. The vernacular name, kerosene bush, is used for several fairly closely related species, apparently in reference to the odour of the leaves or possibly to its inflammable properties.

*H. ledifolium.* Red, yellow or red-brown. Alpine, subalpine. Tas.

A stout shrub up to 1 metre high, sweet-smelling, and forming a rounded bush. Older stems marked by persistent leaf bases. Leaves alternate, crowded, narrow, thick and leathery, edges rolled under, 7–15 mm long, woolly below, smooth above, sticky. Flowers in terminal clusters; florets 8–15. Fruit hairy, with slender pappus bristles. A sticky plant, sometimes also called kersosene bush.

*H. leucopsideum.* White, backed pink. Subalpine, montane. N.S.W., Vic., Tas.

This is perhaps the best known of all the paper daisies, being widely distributed from the coastal mountains to an elevation of about 4,000 feet on the mainland, and almost entirely restricted to the coast in Tasmania. It is able to tolerate the hot parching sun and dry soils but does not extend to the highest elevations. It grows in compact bushes 50–60 cm high, and, in full flower, the bushes are a mass of starry pink and white daisies which open in the sun to display their shining satin bracts surrounding the bright-yellow tubular flowers in the centre. In the evenings and in dull weather the bracts close over the flowers. The outer bracts are rich pink or bronzy-brown, making an attractive variation amid the open white flowers. In the sunshine this is a lovely picture. Leaves usually pointed, thin, narrow, or oblong, smooth above, woolly white below, up to 4 mm long. Flowerheads solitary,

2.5–3.5 cm in diameter, closely surrounded at the base by 3 small woolly foliage leaves; florets numerous, less than half as long as the bracts. Pappus bristles slender with tiny barbs.

A plant of rather similar appearance, *H. adenophorum* var. *waddellii*, bearing flowerheads 3–4 cm across, is common among rocky outcrops in subalpine and montane situations.

*H. milliganii*. White, backed straw colour or crimson. Alpine. Tas.

A dainty little everlasting found in exposed situations. Leaves crowded at the base, oblong or broad, pointed with a sheathing base, thick, apparently smooth, 1.2–3.5 cm long. Rootstock branched to form a thick cushion from which flower stems arise. Flowerheads solitary, 3–5 cm in diameter; outer bracts egg-shaped, white, straw-coloured or crimson, the inner ones narrower, white and spreading, the florets about half as long. Pappus bristles barbed.

*H. pumilum*. White, backed pink or red. Alpine. Tas.

Common in peaty regions, rather resembling a miniature *Celmisia longifolia* in general appearance. Leaves mostly basal, narrow and sheathing, almost flat, leathery, the upper surface smooth and shining, the undersurface silvery-white, 1.2–4 cm long. Flowerheads 1.2–2.2 cm in diameter, solitary, bracts stiff, rounded and narrow, the outer ones coloured and short, the inner ones spreading, dark at the base; florets less than half as long as the bracts. Pappus bristles papillose.

*H. rogersianum*. Yellow, white. Montane. Vic.

A recently described species. A stout, rigid shrub, 1.5–2.5 metres high with a spicy aroma, covered with a cottony down and with prominent scars from fallen leaves. Leaves spreading, narrow, 1–2 cm long, upper surface finely wrinkled, undersurface cottony-white, often stained yellow with exuded resin. Leaves in dense, compound heads 2–4 cm wide with up to 50 heads together. Pappus bristles 20–30, 2–3 cm long, with fine barbs.

*H. rosmarinifolium*. Light-brown or crimson, centre white. Montane. N.S.W., Vic., Tas.

A widely distributed species reaching a height of 2–5 metres with slender branches, the young parts woolly. Leaves alternate, crowded, narrow and pointed, rough on top, woolly beneath, 1.5–4 cm long. Flowerheads numerous in clusters; inner bracts white, outer bracts dry and papery; florets 6–7, 3.5–5 mm in diameter. Fruit smooth, the pappus bristles very slender.

*H. rutidolepis*. Yellow-brown. Alpine. N.S.W., Vic.

Very closely related to *H. scorpioides* but more freely branched, and the flower stalks almost leafless. Leaves somewhat stem-clasping. Flowerheads small, about 8–10 mm in diameter, the bracts often transversely wrinkled. Fruit 1–2 mm long; pappus hairs few.

*H. scorpioides*. Yellow. Alpine, subalpine, montane. N.S.W., Vic., Tas.

A small plant 5–40 cm high, often cottony-hairy. Leaves alternate, crowded at the base, variable in shape, often rather rounded, pointed, soft. Flowerheads solitary, terminal, 1.5–2.5 cm in diameter, on leafy stalks; florets numerous, about half as long as the bracts. Pappus bristles few, slender, finely barbed.

*H. secundiflorum*. Cascade Everlasting. White, pale pink or brown. Subalpine. N.S.W., Vic.

A shrub to 2 metres high with gently drooping, striate, woolly branches. Leaves very narrow, smooth or slightly webby, dark-green on the upper surface, thickly woolly white beneath but with the midrib and margins not conspicuous, 8–10 mm long. Flowerheads papery, turban or bell-shaped, 4–5 mm long, borne on the tips of the numerous short lateral branches, giving the cascade appearance from which this and *H. thyrsoideum* derive their common name and so numerous that the plant is weighed down with a mass of pink or white blooms; outer bracts greenish, the inner ones white and surrounding the tubular florets which are 10–16 in number. Fruit with slender pappus bristles, the tips thickened.

*H. semipapposum*. Golden yellow. Montane. N.S.W., Vic., Tas.

Widespread in a variety of habitats throughout Australia, this attractive little compact plant up to 50 cm high makes brilliant patches in the grasslands and forests. Basal leaves oblong; stem leaves alternate or clustered, usually narrow with a silvery covering of hairs, giving the plant a soft appearance. Flowerheads small,

3–10 mm in diameter, in dense, globose clusters on unbranched stems, several together, woolly, rigid; bracts woolly, the tubular flowers equal to them in length. Fruit 4–5 mm long, with only a few pappus hairs.

*H. stirlingii.* Pale brown. Alpine, subalpine. N.S.W., Vic.

A spreading, rather slender, sticky, aromatic shrub about 3 metres high, the young branches woolly. Leaves lance-shaped, about 4–8 cm long, 3-nerved, the upper surface with a few hairs, the undersurface densely woolly. Flowerheads small, in dense clusters of 5–6; outer bracts round, softly hairy, inner ones narrower; florets over 50, 6–7 mm long. Fruit rough, hairy, the pappus densely plumose at the top.

*H. thyrsoideum.* White, pale pink, brown. Subalpine, montane. N.S.W., Vic., Tas.

A shrub to 3 metres high with angular branches. Closely related to *H. secundiflorum.* Leaves narrow, much longer, 2.5 cm long, recurved, smooth on top, less woolly beneath, midrib much more clearly defined. Flowers in small heads of 10–20 at the tips and along short, lateral branches, much branched and longer; florets 10–16. Fruit hairy, the pappus bristles distinctly thickened above.

*HELIPTERUM* (from Gk *helios*, the sun, and *pteron*, a wing or feather, referring to the winged pappus). Fig. 172. This is the other genus con-

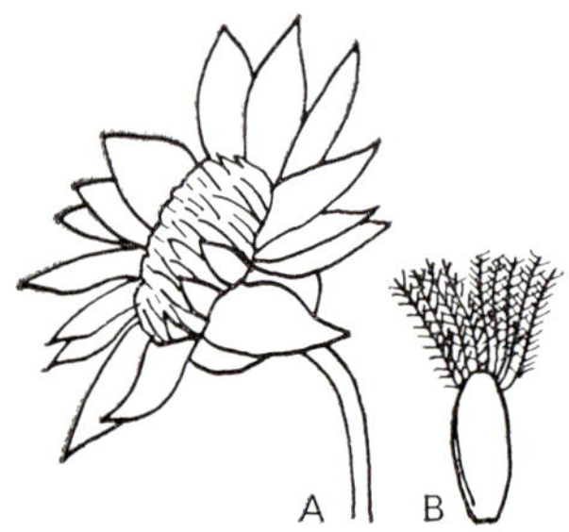


FIG. 172 *Helipterum*: A, flower; B, fruit. (Both actual size.)

taining paper daisies. It differs from *Helichrysum* mainly in having the pappus hairs always feathery (plumose) from the point of insertion at the top of the seedbox, whereas in the latter they are simple or barbed or plumose at the tip only.

*H. albicans.* Alpine Sunray. White, yellow. Subalpine, montane. N.S.W., Vic., Tas.

A species of many forms differing somewhat in its characters in different places. On the mainland at above 6,000 feet one form, sub-species *alpinum*, is a small, almost prostrate plant with a basal rosette of soft, silvery leaves, narrow to broad, spreading itself over the stony surface of the alpine herbfields. Even without its flowers it is a delight to the eye. When the buds appear they dot the silvery cushion with buttons of delicate pink or bronzy-brown, often both colours on the same plant. In full flower the solitary cylindrical heads, 1–2 cm in diameter, are raised on woolly stalks up to 30 cm high but usually less. The inner side of the outer bracts is white; they are pointed, scaly and hyaline, shorter than the inner ones and surrounding a large number of bright-yellow, tubular flowers.

Other forms have bright-yellow or dull-orange bracts and these are also attractive, but the lovely white, yellow-eyed gems are a memory every summer visitor will cherish. The fruit is smooth, the pappus bristles not joined at the base.

*H. anthemoides.* White. Alpine, subalpine, montane. N.S.W., Vic., Tas.

A dainty little paper daisy widely distributed from the drier and higher parts of the coastal plain to 6,000 feet, but more common at 5,000 feet and under. It is a compact plant about 30 cm high in the shape of a little inverted pyramid with many slender stems and terminal flowerheads. Leaves soft green, alternate, narrow, pointed, smooth or with a few scattered hairs, stalkless, the lower ones 6–15 mm long, the upper ones shorter. Flowerheads sitting in the tips of the straight branches, 15–25 mm in diameter with papery-white or rosy-brown bracts, the outer ones short, forming a cup which makes a lovely colour pattern among the widely open flowers; florets pale yellow, few. Pappus hairs scale-like, 2 mm long, sometimes absent altogether.

*H. australe.* Brown. Montane. N.S.W., Vic.

A small plant spreading to lower elevations. Leaves narrow, 2–4 cm long. Flowerheads small, usually terminal, without stalks; outer bracts broad and fringed with long hairs; florets numerous. Fruit smooth, with a pappus of about 10 plumose bristles.

*OLEARIA* (from *olea*, the olive, the leaves of *Olearia dentata* resembling those of the olive). Fig. 173. A genus of daisy bushes with a large number of small flowerheads, sometimes papery

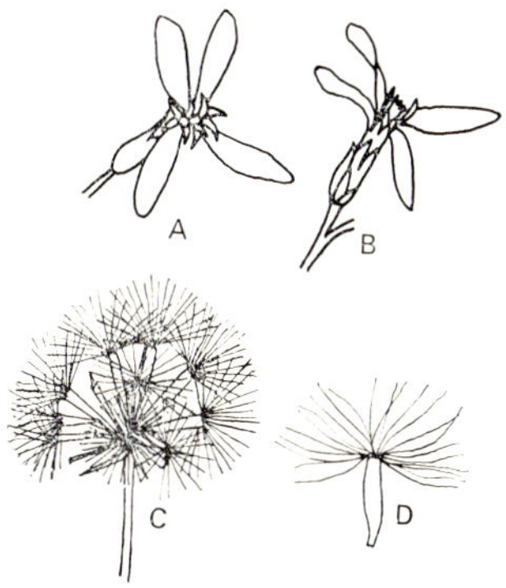


FIG. 173 *Olearia*: A, flower (actual size); B, flower side view (actual size); C, fruits (× 1½); D, single fruit (× 2).

and then referred to as paper daisy bushes. Because the small heads are so abundantly produced, these plants are often very showy and there are so many of them that they are an outstanding feature of these parts. With one marked exception (*O. frostii*) olearias all have flowerheads of this nature.

Most of the daisy bushes are sturdy shrubs, sometimes with a strong, musk-like odour and with white, pink, blue or purple flowerheads. The fruit has a pappus consisting of one ring of numerous, fine, rigid bristles unequal in length or sometimes with a second ring of much shorter, unequal bristles.

*O. adenophora*. White. Montane. N.S.W., Vic.
A sticky, rather hairy tall shrub 9–15 metres high with narrow, often clustered leaves 2.5–3 cm long. Flowerheads small, in leafy clusters on short stalks; ray florets short, about 20 in a head. Fruit short, silky-hairy, with an outer row of numerous short bristles.

*O. algida*. White. Subalpine, montane. N.S.W., Vic., Tas.
A shrub 30–60 cm high, sometimes more, the whole plant covered with short hairs. Leaves alternate or clustered, without stalks, narrow, triangular, blunt, thick, smooth on top, often greyish-green and woolly underneath, 1–3 mm long. Flowerheads terminal or at the ends of the main branches. Fruit smooth.

*O. alpicola*. White. Subalpine. N.S.W., Vic.
A tall shrub with silvery, narrow, leathery leaves 5–8 cm long. Flowerheads small and crowded, making a grand show, each head with 4–6 ray florets. Pappus hairs nearly equal.

*O. alpina*. White. Subalpine. Tas.
A bushy shrub 60 cm—1 metre high with elliptical leaves, shiny on the upper surface, reddish and silky below, 1–2 cm long. Flowerheads stalked, extending beyond the leaves, making the flower clusters more conspicuous; ray florets about 6 in a head. Fruit smooth.

*O. argophylla*. White. Montane. N.S.W., Vic., Tas.
A musk-scented tree 5–15 metres high, common in moist gullies. Leaves broad, stalked, pointed, 6–15 cm long, commonly toothed, silky below, smooth and dark–green above. Flowerheads small but numerous, in dense clusters; florets 9–12. Fruit about 3 mm long, shortly hairy.

*O. asterotricha*. Blue. Montane. N.S.W., Vic.
A shrub about 1 metre high. Leaves narrow, toothed, 2.5 mm or longer, rough on upper surface. Flowerheads solitary or a few together, large; ray florets 20 or more. Fruit hairy.

*O. chrysophylla*. White. Montane. N.S.W.
A tall shrub. Leaves opposite, toothed, thin, hairy below. Flowerheads small, in clusters.

*O. erubescens*. White. Subalpine, montane. N.S.W., Vic., Tas.
A small, bushy plant 1–1.5 cm high, the younger parts hairy. Leaves oblong to lance-shaped, with finely toothed and often wavy margins, shining above, white or yellowish-hairy below, 2.5–4 cm long. Flowerheads solitary or a few together in a leafy head; ray florets 3–5. Fruit smooth, with unequal pappus hairs.

*O. flavescens*. White. Subalpine, montane. N.S.W., Vic.
A medium-sized shrub with long, blunt, wavy or toothed leaves, 2.5–5 cm long. Flowerheads small and numerous on slender, leafless stalks about 2.5 cm long.

*O. floribunda.* White. Subalpine, montane. N.S.W., Vic., Tas.

More widely distributed in Tasmania, where it is common along river banks and drainage channels. A shrub 1–2 metres high with slender branches, the whole plant covered with a cottony down. Flowerheads small, solitary but so close together as to give the appearance of a large cluster; florets 6–10 in a head, half of which are ray florets. Fruit smooth or minutely hairy.

*O. frostii.* White, pink or purple. Alpine. Vic.

A lovely little plant common on the Bogong High Plains and Mount Hotham and restricted to these places. Individual flowerheads so much larger than those of most olearias as to look more like a brachycome. A small shrub 30–60 cm high. Leaves broadish, smooth on the upper surface, softly woolly below, 5–12 cm long. Flowerheads in small clusters or solitary; ray florets 8–10. Fruit hairy.

*O. glandulosa.* White. Montane. N.S.W., Vic., Tas.

A most attractive aromatic shrub about 1 metre high with slender, erect branches covered with glandular tubercles. Leaves alternate, spreading, narrow, without stalks, covered with tubercles and 2–4 cm long. Flowerheads small in clusters; ray florets about 20 in each head. Fruits hairy.

*O. iodochroa.* White with purple centre. Montane. N.S.W., Vic.

One of the early spring bloomers, this lovely daisy-bush is missed by summer visitors. A bushy small plant. Leaves rounded, rather thick, about 1 cm long. Flowerheads solitary or few together; ray florets about 20 in a head. Fruit short and hairy.

*O. ledifolia.* White. Alpine. Tas.

An abundant species, often growing in the shelter of boulders, 30–60 cm high. Leaves alternate, crowded, narrow, blunt, without stalks, rather thick, smooth on the upper surface, rusty or silvery below, 1–3 cm long. Flowerheads solitary on stalks a little longer than the leaves; about 10–12 ray florets in each head. Fruit smooth.

*O. lirata.* White. Montane. N.S.W., Vic., Tas.

A tall shrub 20–50 cm high, found commonly also at lower levels. Leaves broad, pointed, entire, often wrinkled above and hairy below, 5–25 cm long. Flowers in dense clusters partly concealed by the leaves; ray florets 8–12 in each head. Fruit hairy, with a fairly long pappus.

*O. megalophylla.* White. Subalpine. N.S.W., Vic.

A handsome, medium tall plant. Leaves large, dark-green, leathery, rusty below, 5–10 cm long, on short stalks. Flowers abundant in loose, fairly large heads; ray florets 10–12, narrow, with numerous overlapping bracts in each head. Fruit with a pappus of 2 rows, the upper one shorter.

*O. myrsinoides.* White, pinkish-purple. Subalpine, montane. N.S.W., Vic., Tas.

This small shrub about 60 cm high with silky-hairy young branches sometimes descends to lower levels. Leaves alternate, very shortly stalked, broad and blunt with irregularly toothed margins, dark-green, shining and reticulate above, silky hairy below, 5–15 mm long. Flowerheads 1–3 in a cluster; ray florets 2–3 in each head. Fruit smooth, the pappus conspicuous.

*O. obcordata.* White. Subalpine, montane. Tas.

A rather straggling shrub about 60 cm high. Leaves alternate, crowded, stalkless, rather broad, with 3–5 blunt lobes at the tip, almost smooth above and white beneath, 7–10 mm long. Flowerheads solitary on short leafy branches; ray florets 6–7 in each head. Fruit smooth, with a pappus 7 mm long.

*O. persoonioides.* White. Subalpine, montane. Tas.

A shrub 1–3 metres high. Leaves rounded, blunt, shortly stalked, with entire margins, the lower surface white or silky hairy, 1.7–5 cm long. Flowerheads stalked, 3–5 together each with 3–4 ray florets. Fruit hairy, with a pappus 6–7 mm long.

*O. phlogopappa.* White, pink, purple, blue. Subalpine, montane. N.S.W., Vic., Tas.

A common species which has been known under a variety of names, occurring as low as sea level. Very variable at times, reaching a height of 160 cm. Shoots and undersurface of leaves covered with a soft grey or yellowish wool. Leaves usually fairly small, 0.8–3.5 cm long, alternate and oblong or elliptical with a wavy margin, the upper surface with a few scattered hairs. Flowerheads very numerous, stalked, solitary or a few together, each head with 10–13 ray florets. Fruit shortly hairy, with a pappus about 5 mm long in 2 rows, the outer row shorter.

*O. pinifolia.* White. Alpine, subalpine. Tas.
A rigid plant sometimes branched only at the top and then resembling a small tree, 1–3 metres high. Leaves crowded, alternate, stalkless, narrow, pointed, smooth on top except when young, silky-hairy below, 1.5–4.6 cm long. Flowerheads numerous, solitary on fairly long stalks, each head with 8–10 ray florets. Fruit smooth, with a white pappus sometimes drying pinkish-purple.

*O. rosmarinifolia.* White. Montane. N.S.W.
A twiggy, erect shrub covered with fine hairs. Leaves opposite, narrow, smooth above and woolly below, the margins closely rolled under to conceal the lower surface, 3.5–7 cm long. Flowerheads in clusters, each head with 6–8 ray florets. Fruit glandular, with unequal pappus hairs.

*O. subrepanda.* White. Subalpine. N.S.W., Vic.
A medium tall shrub. Leaves egg-shaped, reticulate and smooth above, greenish-yellow and woolly below, 10–12 cm long. Flowerheads in small clusters on slender leafless stalks about 2.5 cm long. Fruit hairy.

### Some Yellow Daisies

Among the yellow flowering daisies are *Bedfordia, Cotula, Craspedia* (already described), *Cymbonotus, Microseris, Podolepis* and *Senecio.*

*BEDFORDIA* (after the Duke of Bedford, 19th-century patron of botany). Fig. 174.

*B. salicina.* Yellow. Montane. N.S.W., Vic., Tas.
A very interesting small woolly tree 2–5 metres high inhabiting the wooded gullies of the Southern Alps and the rainforests of Tasmania, and descending to the coast in some parts. Leaves alternate, broad, blunt, smooth on top, densely woolly-hairy below, 5–15 cm long. Flowerheads narrow in short clusters; bracts in a single row, with dry margins; florets all tubular. Fruit angular or striate with a pappus of numerous toothed bristles.

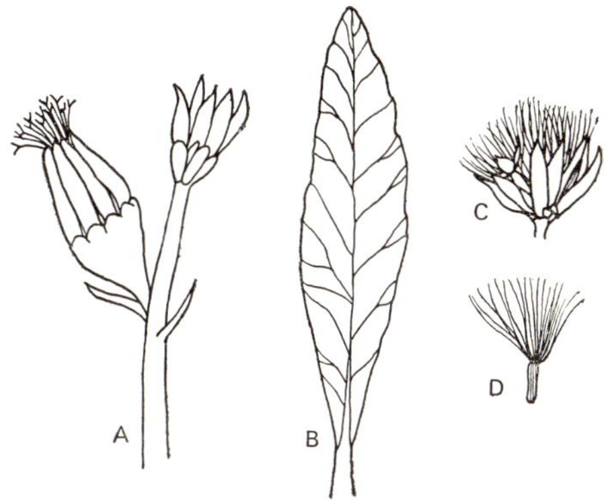


FIG. 174 *Bedfordia*: A, flowerhead; B, leaf; C, fruiting head; D, single fruit. (All × $1\frac{1}{2}$ except B × $\frac{1}{4}$.)

*COTULA* (from Gk *kotyle*, a cup, possibly alluding to the cup-like heads of some species). Fig. 175. A genus of water-loving plants, mostly

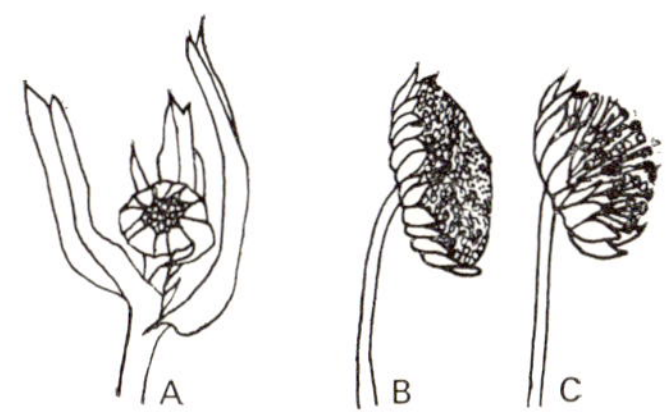


FIG. 175 *Cotula*: A, unopened flowerhead (actual size); B, flowerhead (× 2); C, fruiting head (× 2).

small, often fleshy and widespread. They are creeping herbs with alternate leaves and numerous ray and disc florets. The ray florets are short and tubular, giving a flat, discoid head. The outer florets are without petals, the bracts in 2 rows. The fruit is flattened, sometimes winged, with no pappus.

*C. alpina.* Creamish-yellow. Montane. N.S.W., Vic., Tas.
A small, stout herb with stems 2–8 cm long. Leaves shortly stalked, deeply divided, 0.8–2.2 cm long with a few scattered hairs. Flowerheads about 4 mm long on short, thick, hollow stems. Fruit with thick wings.

*C. filicula.* Yellow. Alpine, subalpine. N.S.W., Vic., Tas.

A small, stout, fleshy herb covered with white or rust-coloured hairs. It occurs also in Tasmanian temperate rainforests. Leaves stalked, divided, the blade egg-shaped, 0.2–2 cm long. Flowerheads about 6 mm in diameter on stout stems. Fruit egg-shaped, dotted with glands and with small, thick wings.

*C. reptans*. Greenish-yellow. Montane. N.S.W., Vic., Tas.

A smooth herb. Leaves variable, divided, 0.6–6 cm long. Flowerheads 3.5 mm in diameter on slender stems. Fruit with thickened margins, about 1 mm long.

*CYMBONOTUS* (from Gk *kymbe*, a boat, and *notos*, back, referring to the convex back of the fruits). Fig. 176.

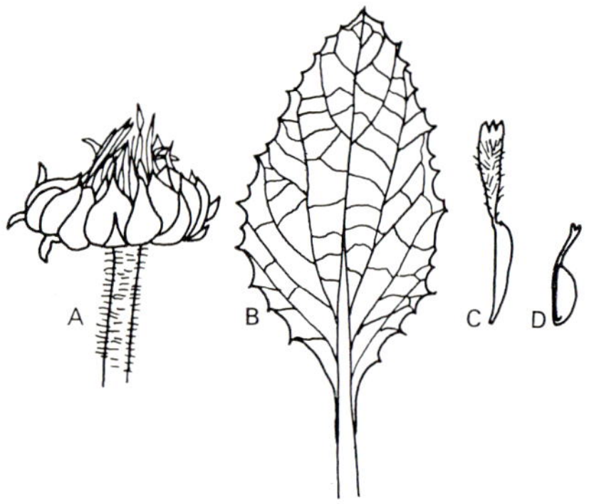


FIG. 176 *Cymbonotus*: A, flowerhead (actual size); B, leaf (× ½); C, flower (× 2); D, fruit (× 2).

*C. lawsonianus*. Yellow. Montane, subalpine. N.S.W., Vic., Tas.

A small herb with a basal rosette of leaves flat on the ground. Newly formed leaves simple or toothed, soft, without stalks, 5–12.5 cm long. Flowerheads solitary in the centre of the leaves, stalkless or with very short stalks, about 1–1.5 cm in diameter, with ray and disc florets; bracts overlapping in several rows. Fruit smooth, with 3–5 ribs on the outer side, without pappus.

*MICROSERIS* (from Gk *mikros*, small, and *seris*, chicory, possibly referring to the fruits). Fig. 177.

*M. scapigera*. Native Dandelion. Yellow. Subalpine, montane. N.S.W., Vic., Tas.

A widely distributed soft herb, this is also known as yam. Leaves all basal, smooth, entire or lobed. Flowerheads semiflocculose, solitary,

FIG. 177 *Microseris*: A, fruit; B, flowerhead. (Both actual size.)

3–5 cm in diameter, on stems 20–60 cm high; all florets strap-shaped, narrow, numerous. Pappus bristles dilated at the base.

*PODOLEPIS* (from Gk *podos*, a foot, and *leptos*, slender, referring to the stalks of the bracts). Fig. 178. A genus of erect herbs,

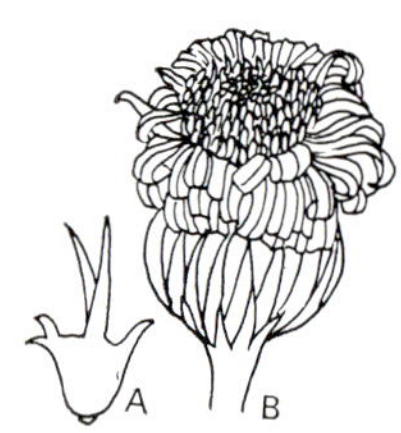


FIG. 178 *Podolepis*: A, fruit (× 2); B, flowerhead (actual size).

slightly woolly when young, with leaves in basal rosette and also alternate on the branches. Flowerheads often solitary, broad and terminal on the ends of straight stems; bracts papery in several rows; florets, both ray and disc, short and inconspicuous, the ray florets lobed or lipped. Pappus bristles slender, slightly cohering at the base, sometimes absent.

*P. hieracioides*. Yellow. Subalpine, montane. N.S.W., Vic.

A robust woolly herb 70 cm high. Basal leaves clustered, elliptical, pointed, stem-clasping, smooth or with a few hairs, 16.5 cm long; stem leaves narrower and shorter. Flowerheads 1.5 cm in diameter; ray florets 15–20. Fruit narrow, contracted at the summit, 3 cm long.

*P. jaceoides*. Yellow. Alpine, subalpine. N.S.W., Vic., Tas.
A soft, hairy herb up to 60 cm high occurring in grasslands at lower elevations also. Basal leaves stalked, fairly broad, up to 12 cm long; stem leaves narrower and stem-clasping. Flowerheads 2–3 cm in diameter; florets about 30–40 and about 2 cm long. Fruit about 1 cm long, with a pappus of fine hairs.

*P. longipedata*. Yellow. Alpine, subalpine. N.S.W., Vic.
A smooth herb 25–90 cm high, loosely woolly. Basal leaves clustered, smooth, or with a few hairs, stem-clasping, 5–20 cm long; stem leaves slightly smaller. Flowerheads stalked, often 3–4 together, 2–3 cm in diameter; bracts yellowish, egg-shaped; ray florets longer than the bracts. Fruit 1.8 mm long, 0.6 mm broad, finely hairy.

*P. robusta*. Yellow. Alpine, subalpine. N.S.W., Vic.
A much stouter and more robust plant than any of the foregoing, this has erect woolly stems up to 60 cm high and sometimes grows in considerable stands, making a spectacular sight in full bloom. Basal leaves up to 20.5 cm long, spoon-shaped, blunt, often with crinkled margins; stem leaves woolly-white, smaller and stem clasping. Flowerheads in a dense, terminal cluster about 2.5 cm in diameter; ray florets 30–40, deeply 4–lobed. Fruit 2.5–4 mm long, narrow, smooth, contracted at the summit.

*SENECIO* (from Lat. *senex*, old (an old man) referring to the white pappus). Fig. 179. A widely distributed genus of variable plants with alternate leaves and flowerheads solitary or in clusters. The disc florets are tubular, the ray florets usually strap-shaped or absent. The fruit has a pappus of several rows of fine hairs.

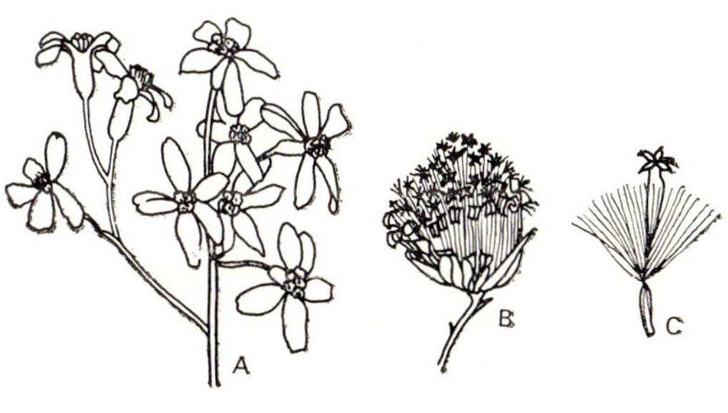


FIG. 179 *Senecio*: A, flowerhead (actual size); B, fruiting head (actual size); C, single fruit (× 2).

*S. biserratus* (syn. *Erechtites prenanthoides*). Greenish-yellow. Subalpine, montane. N.S.W., Vic., Tas.
A small, smooth plant with narrow to broad, toothed, stem-clasping leaves, 2.5–15 cm long. Flowerheads numerous, small and insignificant. Fruit dark-brown, 2.5 mm long with broad ribs, hairy, with a short callus ring at the base of the pappus.

*S. brunonis* (syn. *S. centropappus*). Yellow. Subalpine, montane. Tas.
The tallest of all the senecios and very rare, reaching 2–4 metres as a slender tree, the branches marked with conspicuous leaf scars. Many plants were destroyed in the disastrous Tasmanian bushfires of 1967. Leaves spreading, without stalks, narrow, blunt, 5–9 cm long, aromatic when crushed. Flowerheads in loose clusters; ray florets strap-shaped, spreading. Fruit smooth; pappus bristles rigid, prominently barbed.

*S. cunninghamii*. Yellow. Subalpine, montane. N.S.W., Vic., Tas.
A slender, branching herb, somewhat woody at the base, rarely above 40 cm high and often shorter. Leaves soft, smooth, broad, entire or coarsely toothed, 8–10 cm long, entirely stalkless and lobed at the base. Flowers numerous, in dense heads about 2 cm in diameter, on slender stalks; ray florets 10–12 considerably longer than the bracts. Fruit slightly hairy, the pappus inserted on a ring.

*S. glomeratus* (syn. *Erechtites arguta*). Greenish. Alpine, subalpine. N.S.W., Vic.
A soft, woolly plant about 1 metre high or less with variable toothed, lobed or divided leaves up to 12 cm long. Flowerheads in small, insignificant clusters. Fruit about 2 mm long; pappus hairs 5–6 mm long.

*S. hispidulus* (syn. *Erechtites hispidulus*). Greenish. Montane. Vic., Tas.
A soft plant up to 1 metre high with entire or variable toothed leaves. Flowerheads in a loose, inconspicuous cluster. Fruit slender and striate.

*S. lautus.* Yellow. Alpine, subalpine, montane. N.S.W., Vic., Tas.

A very widely distributed species extending to the coast, smooth, 30–120 cm high. Leaves variable, toothed, entire or deeply lobed, 1.5–10 cm long. Flowerheads in loose clusters, 2–2.5 cm in diameter; ray florets strap-shaped. Fruit rather woolly, 3–4 mm long.

*S. leptocarpus.* Yellow. Subalpine, montane. Tas.

A widely distributed species extending to the coast, closely related to *S. pectinatus*, differing only in having stalked flowerheads, 3–5 together in a loose cluster.

*S. linearifolius* (syn. *S. macrodontus*). Yellow. Subalpine, montane. N.S.W.

A tall, handsome plant. Leaf broad with narrow lobes and reaching 12–15 cm in length, toothed, stalked, but sometimes with auricles at the base. Flowerheads numerous in a loose cluster; ray florets strap-shaped, 4–6 in a head.

*S. papillosus.* Yellow. Alpine, subalpine. Tas.

A smooth, small plant with a basal rosette of leaves, stalked, rounded, blunt, rather thick, hairy on both surfaces but more thickly hairy on the upper surface, 5–20 mm long. Flowerheads solitary, terminal, about 2.5 cm in diameter; ray florets strap-shaped. Fruit smooth, flattened, ribbed.

*S. pectinatus.* Yellow. Alpine, subalpine, montane. N.S.W., Vic., Tas.

A handsome, soft, small plant. Leaves mostly radical, narrow and entire or, more commonly, deeply and regularly lobed, stalked, smooth, 1.5–5 cm long. Flowerheads solitary and terminal, 1.5–2.5 cm in diameter; ray florets strap-like and spreading. Fruit smooth. A Tasmanian form, *S. pectinatus* var. *ochroluca*, has cream flowers.

*S. primulifolius.* Yellow. Alpine, subalpine. Tas.

A small plant with an erect stem. Leaves stalked, broad, blunt, rather thick, coarsely veined, 2–7 cm long, the margins wavy, upper surface slightly hairy, lower surface pale and cobwebby. Flowers solitary or up to 4 in a cluster, about 2.5 cm in diameter; ray florets straplike and spreading. Fruit smooth.

*S. quadridentatus* (syn. *Erechtites quadridentata*). Greenish. Montane. N.S.W., Vic., Tas.

A soft, woolly plant 50–60 cm high with fairly narrow, slightly toothed leaves. Flowerheads in a loose, slender cluster. Fruit striate and contracted at the top.

*S. vagus.* Yellow. Montane, subalpine. N.S.W., Vic.

A smooth plant 70–100 cm high. Leaves stalked, broad, deeply divided, the middle lobe longer, 12–15 cm in length; upper leaves smaller and less divided. Flowerheads large, in loose, stalked clusters; ray florets 10–15; disc florets numerous. Fruit smooth.

*S. velleioides.* Yellow. Montane, subalpine. N.S.W., Vic.

A smooth, erect plant 50 cm–2 metres high. Leaves broad, usually stalkless and stem-clasping, margins coarsely toothed. Flowerheads numerous, in clusters, about 2.5 cm in diameter; ray florets strap-like and spreading, about 10 in a head. Fruit ribbed.

*ERECHTITES* (Gk name for the groundsel). Fig. 180. A small genus closely related to *Senecio*, in which the outer florets are filiform and the central ones tubular. The flowers are usually inconspicuous. There are many introduced species also.

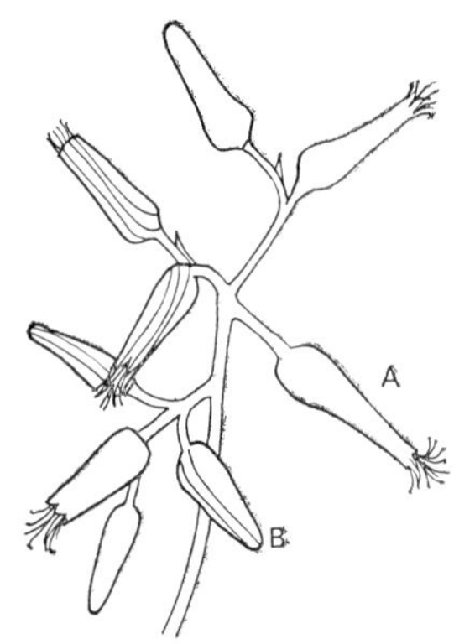


FIG. 180 *Erechtites*: A, flower; B, fruit (× 1½).

*E. mixta.* Greenish. Montane. N.S.W., Vic.

A not very common small plant with finely divided leaves up to 10 cm long, sometimes cottony below. Flowerheads narrow, in long terminal clusters. Fruit with long pappus hairs.

## A Miscellany of Daisies

There are a number of mostly small daisies which do not fit into any of the preceding groups. They are described below.

*CALOTIS* (from Gk *kallos*, beauty, and *otes*, an ear, referring to the pappus scales). Fig. 181. A genus of small herbs restricted to the mainland, rather like miniature brachycomes. The leaves are alternate. The florets are of both types: the ray florets strap-like in a single row, the disc florets numerous, tubular and sterile. The fruiting heads are globular and the fruit is flat and oblong.

FIG. 181 *Calotis*: flowerhead (actual size).

*C. anthemoides*. White. Montane. N.S.W., Vic. A creeping plant. Leaves in a basal rosette, finely lobed, on long stalks and 7–10 mm long. Flowerheads solitary, small, on slender stems. Fruit with 6–8 pappus bristles.

*C. glandulosa*. White, violet, blue. Montane. N.S.W., Vic.

A small plant 15–25 cm high, rather hairy. Leaves shortly stalked, toothed at the ends, oblong, the upper ones narrow and stalkless. Flowerheads rather larger than in the other species. Fruit rough, with a pappus of 2–7 bristles.

*C. lappulacea*. Yellow. Montane. N.S.W., Vic.

A slightly hairy plant 8–14 cm high. Leaves at base and along the erect stems, oblong, toothed or lobed, 8–10 cm long, the lower ones mostly smaller. Flowerheads small, with narrow bracts. Fruit covered with short protuberances, with a pappus of 1–4 ridged barbed bristles and 1 or more very short ones.

*C. scabiosifolia* var. *integrifolia*. White, purple. Alpine, subalpine, montane. N.S.W., Vic.

A densely tufted plant 15–20 cm high. Leaves broad, toothed or lobed, stalked, 5–8 cm long. Flowering stems 15–20 cm high with a few small, stalked leaves. Flowerheads solitary. Fruit softly hairy, with a pappus of about 8 stiff bristles.

*GNAPHALIUM* (from Gk *gnaphalion,* woolly, an allusion to the woolly foliage). Fig. 182. An interesting little genus of rather inconspicuous plants which cannot be overlooked because there are many species fairly widely distributed throughout the alpine areas. The plants are mostly small, sometimes completely prostrate, with grey-green, woolly or cottony leaves. The flowerheads are tiny, rounded, in small clusters, with overlapping scaly or membranous bracts, scarcely opening so that the greenish-yellow flowers are not at all conspicuous, but spreading widely when the fruits ripen. Fruit with one row of bristles.

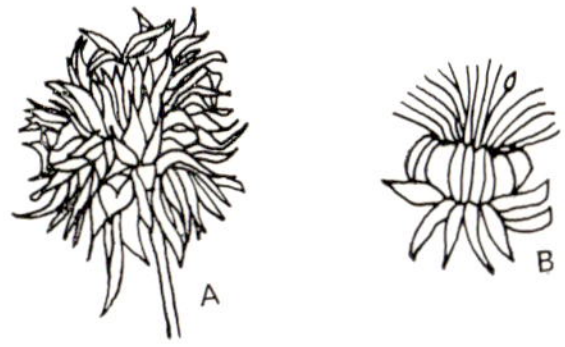


FIG. 182 *Gnaphalium*: A, flowerhead; B, fruiting head. (Both × 3.)

*G. collinum* and *G. argentifolium* occur at all highland levels in New South Wales, Victoria and Tasmania. The former is a small erect plant 6–40 cm high with leaves 2–15 cm long and flowerheads about 4 mm long. The species needs revision. The latter is a tufted plant 3–10 cm high, the leaves 1–3 cm long with comparatively large flowerheads 5–7 mm long and usually 4–7 together.

Two other species are confined to New South Wales and Victoria, *G. luteo-album* and *G. purpureum* in subalpine and montane habitats. *G. luteo-album* is a tall plant with narrow leaves, the upper ones not stalked and flowerheads in dense loose clusters with no subtending leaves. *G. purpureum* is a small plant with stalked spoon-shaped leaves, the upper ones narrower, cottony-white on both sides, and the flowerheads are in leafy clusters.

Two others, *G. umbricola* and *G. traversii*, are confined to Victoria and Tasmania, both extending from alpine to montane habitats. The former is an erect plant 5–15 cm high with flowerheads in spherical clusters, terminal and subtended by 1–3 spreading leaves. The latter is a tufted plant with solitary terminal flowerheads on short stalks bearing narrow leaf-like bracts.

One species, *G. japonicum*, is confined to subalpine and montane habitats in Victoria. It is about 25–30 cm high with broad, stalked leaves and small flowerheads in dense clusters.

*G. supinum* is a Tasmanian montane species common among button-grass. It is a creeping plant with narrow blunt leaves and flowerheads in threes in a terminal cluster.

*LAGENOPHORA* (from Gk *lagenos*, a flask, and *phoros*, bearing, referring to the fruits). Fig. 183.

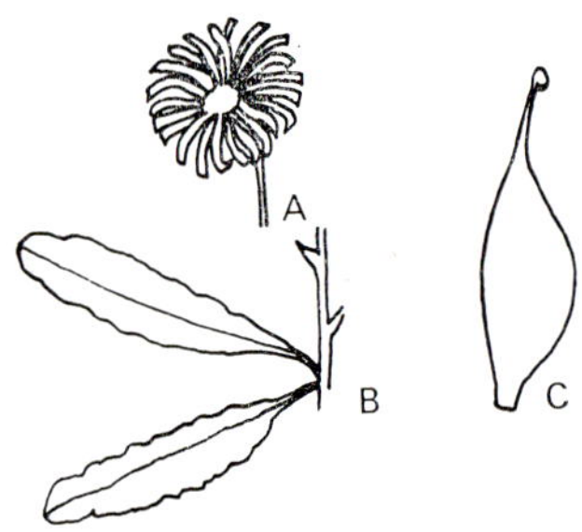


FIG. 183 *Lagenophora*: A, flowerhead (× 2); B, leaves (× 2); C, fruit (× 3).

*L. stipulata*. White, bluish. Alpine, subalpine. Vic., Tas.
A small, inconspicuous but commonly occurring plant with leaves in a basal rosette. Leaves broad and blunt, wavy or toothed, 2.5–5 cm long. Flowerheads solitary, small, 4–8 mm in diameter on long, slender stalks; florets of 2 kinds, the ray florets female, the disc florets male. Fruit compressed, 3–4 mm long, straw-coloured to dark-brown, beaked with no pappus.

*LEPTORRHYNCHOS* (from Gk *leptos*, slender, and *rhynkos*, beak, referring to the fruits). Fig. 184. A small genus of small plants with alternate, entire leaves and solitary, terminal, discoid flowerheads, turban or bell-shaped. The bracts are numerous, overlapping in several rows, thin and papery or leafy with dry tips, not spreading. The florets are all tubular. The fruit is flattened, with a pappus of one row of slender bristles.

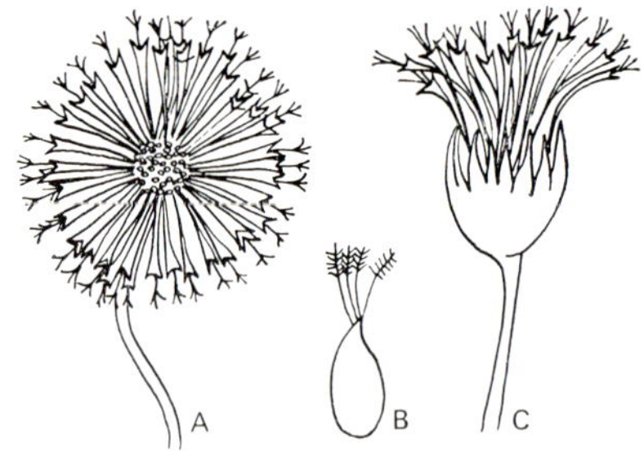


FIG. 184 *Leptorrhynchus*: A, flowerhead (actual size); B, fruit (× 1½); C, fruiting head (actual size).

*L. squamatus*. Yellowish. Alpine, subalpine. N.S.W., Vic., Tas.
A very widely distributed and common little plant, wiry, with leafy stems 10–20 cm long. Leaves without stalks, crowded, rather narrow, pointed, dull-green, smooth on top, woolly-white below, 11–20 mm long. Flowerheads on thin, wiry stalks, small and narrow, surrounded by membranous hyaline bracts with woolly margins. Fruits contracted at the top, 2–3 mm long; pappus hairs 5–6 mm long, connate into a ring at the base.

*L. elongatus* from the mainland montane and subalpine habitats differs from the above species in being rather taller and having beaked fruits with numerous pappus bristles.

*VITTADINIA* (after Carlo Vittadina, 1800-1865, an Italian writer on fungi).
*V. triloba* is a small annual from montane parts of New South Wales and Victoria with stalked, variable leaves 1.2–2.5 cm long, and blue or white solitary flowerheads. There are ray and disc florets, the former narrow. Fruit narrow, usually hairy, with a pappus of numerous, unequal brownish bristles.

**Cushion Bushes**

These are formed by 2 genera of composites that cover wide areas on Tasmanian mountain summits particularly on the wet exposed areas where snow lies for many months. These often grow together in coalescing hummocks with cushion plants of other families and also form persistently wet seed-beds for a number of other plants. With bright-green and grey-green foliage intermixed and dotted with their own white starry flowers and those of their adopted seedlings such as droseras, they make a most attractive and characteristic feature of these parts. Some species occur also on the mainland but do not grow so extensively.

*ABROTANELLA* (from Gk *abrotonum*, an aromatic plant). Fig. 185. A genus with 3 common species. Though only one of the species

FIG. 185 *Abrotanella*: flower (× ¼).

forms a cushion bush, the other two (which are also alpine) are included here for the sake of convenience. These are densely tufted herbs with narrow, rather fleshy leaves. The flowerheads have few bracts in 2–3 rows; the female florets are tubular, 3–4 toothed in 1 row, and the disc florets are 4–toothed and often infertile. The fruit has a minute pappus or none at all.

*A. forsterioides*. White. Alpine, subalpine, montane. Tas.

Commonly associated with *Pterygopappus* in alpine habitats, this is the most widely distributed of the cushion bushes. Leaves bright-green, very thick, pointed, toothed on the margin, with a broad sheathing base and about 2 mm long. Flowerheads very small, solitary, on short stalks, each head with 3–5 erect short bracts surrounding 3–5 florets, longer than the bracts. Fruit 4–angled.

*A. nivigena*. White. Alpine, subalpine. N.S.W., Vic.

A not very common plant forming dense tufts 4.5–7 cm long. Leaves thick, blunt, smooth, narrow. Flowerheads solitary on short stalks. Fruit with roughened angles.

*A. scapigera*. White. Alpine. Tas.

A tufted plant 2–3 cm high. Leaves narrow, thick, sheathing, blunt, 2–8 cm long. Flowerheads few in small, dense, terminal clusters; about 12 florets in a head, inner florets staminate, outer pistillate.

*PTERYGOPAPPUS* (from Gk *pteryx, pterygos,* a wing, and *pappus*, a bristle). Fig. 186.

FIG. 186 *Pterygopappus*: flowerhead (× 1½).

*P. lawrencei*. Pale brown. Alpine. Tas.

A hummocky cushion formed of densely packed stems, the basal part of the stem clothed with the persistent, dead leaves, the cushion 1 metre or more in diameter and so springy and hard that no impression is made by walking on it. Leaves soft sage-green, closely overlapping, broad, sometimes spoon-shaped, sharp, thick, 4 mm long with a band of long, shaggy hairs. Flowerheads solitary on short stalks not much longer than the leaves but elongating in the fruiting stage; bracts 6–10 in each head, papery and pale brown, about 2 mm long and surrounding 6–12 florets. Fruit flattened or 4–angled with 6 or fewer pappus bristles.

## Family STYLIDIACEAE

A small family including 3 alpine genera, *Stylidium, Forstera* and *Phyllacne*. They are all small plants, often with leaves in a basal rosette. The flowers are usually irregular with 5–7 sepals forming a 2–lipped calyx and 5 petals usually joined with a short tube and overlapping lobes, often with 1 petal differing from the others and forming a labellum. There are 2 stamens with the stalks joined to the style, forming a column free from the petals. The ovary is below the other parts of the flower. The fruit is dry, opening by 2 valves.

*FORSTERA* (after J. R. and G. Forster, 18th-century botanists). Fig. 187.

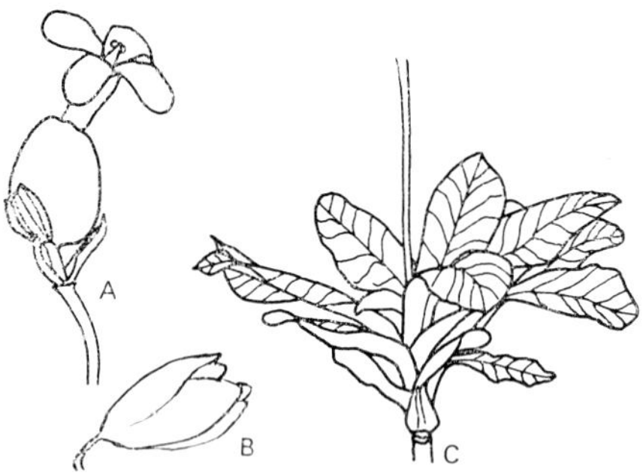


FIG. 187 *Forstera*: A, flowerhead (× 4); B, enclosed fruit (× 2); C, leaves (actual size).

*F. bellidifolia.* White. Alpine, subalpine. Tas.
A small tufted plant of the southern mountains. Leaves in a basal rosette, oblong or spoon-shaped, blunt, rather thick, 8–15 mm long. Flowers solitary on long, slender stalks; the tube of the flower about 2 mm long, the lobes slightly longer, unequal and egg-shaped; the column shorter than the petals and not sensitive. Fruit globose, 5–6 mm long, with fine seeds.

*PHYLLACHNE* (from Gk *phyllon*, a leaf, and *achnous*, without down). Fig. 188.

FIG. 188 *Phyllachne*: flower (× 2).

*P. colensoi.* White. Alpine, montane. Tas.
A densely tufted little plant only a few centimetres high, forming a thick, hard cushion. Leaves stalkless, numerous and closely overlapping, smooth, glossy, rather thick with the tips often brownish-yellow, 3–4 mm long. Flowers solitary, scarcely raised above the plant, about 4 mm in diameter, tubular with 5–6 spreading, rounded petals. Fruit about 2 mm in diameter.

*STYLIDIUM* (from Gk *stylos*, a column, referring to the united stamens and style). Fig. 189.
*S. graminifolium.* Trigger Plant. Pink. Alpine, subalpine, montane. N.S.W., Vic., Tas.

The species of trigger plant which occurs so commonly at these elevations has a wide distribution and is not uncommon in grasslands at lower altitudes. It is, however, so common in the alpine and particularly the subalpine areas, sometimes covering wide stretches of country and sometimes occurring in small, isolated clumps, that it may well be regarded as a feature of these higher parts.

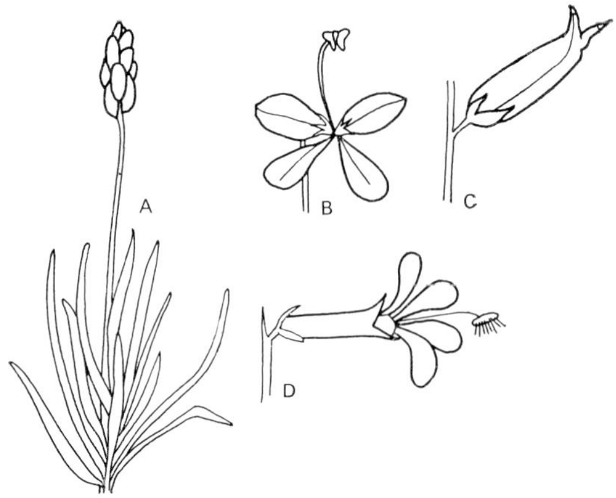


FIG. 189 *Stylidium*: A, bud; B, ripe sticky stigma; C, fruit; D, unripe stigma for holding pollen. (All × $1\frac{1}{2}$.)

Stylidiums form a group of plants very easily distinguished because of their peculiar floral structure. There are 5 joined sepals and 5 joined petals, 4 of these spreading beyond the tube, the fifth smaller, reflexed and isolated from the other 4. There are only 2 stamens joined to the style by the filaments, forming what is called a column. The anthers are sessile on the stigma. The column is bent twice.

This peculiar arrangement is a pollinating device. The flower produces nectar at its base at a stage when the anthers are just ready to burst and the stigma is not yet ripe. Attracted by the nectar, small visiting pollinators alight on the expanded petals and push down into the tube. In doing this they frequently set off the column, which is a kind of trigger. This swings over and hits the visitor lightly on the back, exploding the ready-to-burst anthers and depositing the pollen. A little later the column swings back to its original position and it is once again sensitive. This may happen several times, but eventually all the pollen will be dispersed or blown away. Nectar, however, continues to be produced and so the flower is still attractive to visitors. At this stage, however, the stigma is ripe and sticky. If the flower is now visited by a pollinator which has already received pollen from another trigger flower in an earlier stage of development the hitting column will collect some of this on the sticky stigma. When this happens nectar ceases to be produced and the flower is no longer attractive to pollinators. Seeds may now mature.

The trigger plant is extremely variable. The leaves are in a basal rosette, 5–20 cm long, stiff, narrow, grass-like, with a widened base. The flowering stems are more than twice as long as the leaves, each bearing numerous flowers in elongated heads. The flowers are 0.8–1.5 cm in diameter, the basal petal rounded. The column is longer than the petals. The fruit is oblong, dry, splitting, with numerous seeds.

**Family DONATIACEAE**

*DONATIA* (from Gk *donax, donacus*, possibly referring to the fruit resembling the mussel *Donax*). Fig. 190.

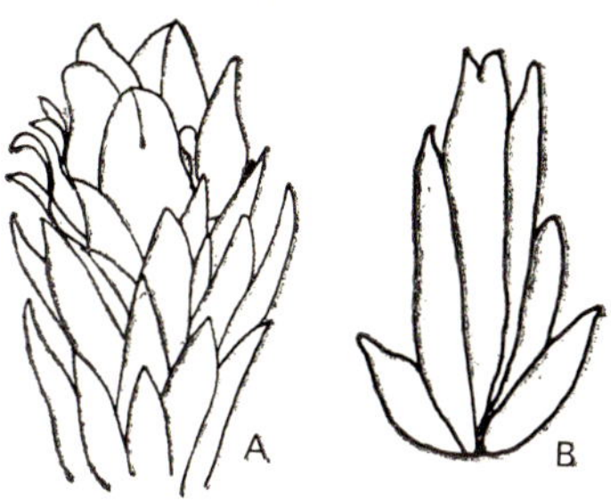


FIG. 190 *Donatia*: A, flowerhead; B, fruit. (Both × 3.)

*D. novae-zelandiae*. White. Alpine. Tas.
This is one of the several cushion bushes of exposed situations on wet scree and other places where snow lies for several months, a bright green plant with tightly packed stems often coalescing with other cushion bushes forming compact mounds 1 metre or more in diameter. Leaves stalkless, persistent, narrow, and pointed but broad at the base, 5–6 mm long. Flowers solitary, 5–6 mm long; sepals 5; petals 5, bluntly pointed, oblong; stamens 2–3 on a fleshy disc covering the top of the seedbox, which is inserted below the other flower parts; seedbox turban-shaped, fleshy.

**Family GOODENIACEAE**

This family, in which the irregular-shaped flowers are easily recognizable, is represented on the highlands by 4 common genera, *Dampiera*, *Goodenia*, *Scaevola* and *Velleia*. The family contains only small plants, for the most part with characteristic flowers of 5 petals and 5 sepals which form a tube split completely or almost completely to the base and often with membranous wings. There are 5 stamens. A peculiarity of these plants is the presence of an indusium or pollen cup surrounding the stigma. The pollen is shed into this cup before the petals open and before the stigma is ripe. At this stage the flower is attractive to pollinators which may accidentally collect pollen. Later, after the pollen has all been collected or has dried up, the style elongates and the stigma enlarges. Pollinators may still call on the attractive flowers, and, if carrying pollen from another flower, may deposit it on the now ripe stigma. The seedbox is inserted above the other parts of the flower in some members, but below them in others.

*DAMPIERA* (after William Dampier, 1651-1715, buccaneer who visited the north-west coast of Australia and collected plants). Fig. 191.

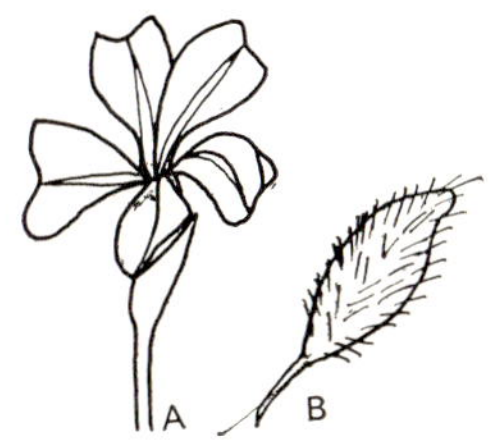


FIG. 191 *Dampiera*: A, flower; B, fruit. (Both × 2.)

*D. stricta*. Blue. Montane. N.S.W. ,Vic., Tas.
A common small plant of coastal habitats also, with angular stems 20–40 cm long. Leaves alternate, stalkless, the upper ones broad, the lower ones narrow, leathery, irregularly broken on the edges, 1–2 cm long. Flowers solitary or

clustered and close together; petals about 15 mm long; anthers joined in a tube round the style; seedbox inserted below the other flower parts. Fruit small, dry, not splitting.

*GOODENIA* (after Samuel Goodenough, 1743-1827, botanist, sometime vice-president of Royal Society of London). Fig. 192. A genus of

FIG. 192 *Goodenia*: A, flowers with leaves; B, fruit. (Both actual size.)

mostly small plants with solitary flowers. The flowers have a small tube, broadly expanded, winged petals and free stamens. The indusium is cup-shaped. The seedbox is inserted below the other parts of the flower with a convex apex which protrudes beyond the base of the sepals. The fruit is a dry, splitting capsule.

*G. elongata.* Yellow. Montane. N.S.W., Vic., Tas.

A herb with a basal rosette of egg-shaped or spoon-shaped leaves 1.5–4.5 cm long with a few leaves also on the erect stems. Flowers 15–22 mm long, the outer surface softly woolly, the upper lobes unequally winged, solitary and terminal on slender stalks 5–30 cm long. Fruit egg-shaped.

*G. hederacea.* Yellow. Alpine, subalpine, montane. N.S.W., Vic.

A dainty little ground cover plant attractive both in leaf and flower, trailing for considerable distances over the ground or hanging over embankments, forming delicate tracery, the dark-green leaves providing a pleasing foil for the brilliant one-sided flowers. Leaves stalked, rounded, rather like ivy in shape (as the specific name indicates), with a wavy or toothed margin, about 1 cm in diameter. Flowers 1–3 together on rather long, thin stalks; tube of the calyx ribbed.

*G. lineata.* Golden yellow. Montane. Vic.

A recently described species, this is a small plant with a rosette of numerous smooth leaves, fairly broad to spoon-shaped, rather thick, 3–8 cm long. Flowering stems erect, slender, 10–30 cm long; flowers in loose terminal heads; petals about 2 cm long, with vertical stripes of deep purple in lower half.

*G. pinnatifida.* Yellow. Subalpine, montane. N.S.W.

A small plant with a basal rosette of finely divided or deeply toothed leaves and a few narrow stem leaves. Flowers large, solitary, on long stalks, the petals covered with stiff hairs, the lobes conspicuously winged and the upper ones deeply cleft. Fruit rounded, about 6–8 mm in diameter.

*SCAEVOLA* (a Roman family name meaning left-handed, alluding to the flowers). Fig. 193.

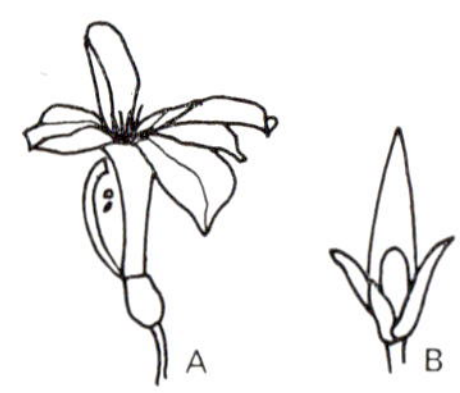


FIG. 193 *Scaevola*: A, flower (× 1½); B, fruit (actual size).

*S. hookeri.* Lilac, white. Alpine, subalpine, montane. N.S.W., Vic., Tas.

A creeping herb with stems 4–30 cm long, covering large areas of wet ground. Leaves variable, broadish or spoon-shaped, blunt and thick, 0.5–5 cm long. Flowers solitary on short stalks scattered over the plant, one-sided; petals 8–10 mm long, with broad membranous margins; anthers free. Fruit succulent, red.

*VELLEIA* (after Thomas Velley, 1748-1806, English botanist specializing in seaweeds). Fig. 194. In this genus the ovary is placed above the

FIG. 194 *Velleia*: A, flower (× 2); B, fruit (actual size).

other parts, whereas in the other members of this family it is below. The petals are not membranous. The fruit is a dry, splitting capsule.

*V. montana.* Yellow, dull reddish-yellow. Montane. N.S.W., Vic., Tas.

A small tufted plant. Leaves in a basal rosette, broad, blunt, thick, 1.5–5 cm long. Flowers in short clusters with 3 sepals; petals 8–9 mm long, split nearly to the base on the upper side.

*V. paradoxa.* Pale yellow. Montane. N.S.W., Vic., Tas.

A small tufted plant. Leaves in a basal rosette, broad, blunt, coarsely toothed, 2–20 cm long. Flowers in clusters, often 2–4 times as long as the leaves; calyx of 5 sepals; petals hairy outside and spurred.

**Family CAMPANULACEAE**

*WAHLENBERGIA* (after George Wahlenberg, 19th-century professor of botany at the University of Uppsala). Fig. 195. A widespread genus with considerable variation and with several forms common in alpine areas. The large-flowering, deeply coloured forms growing in large stands or clumps make a great spectacle and are one of the features of the area. They are dainty little plants with soft foliage and bell-shaped, usually blue, flowers on long, slender stems, with 5 or rarely 3, 4, 6 or 10 petals and usually 5 free stamens with narrow anthers and more or less hairy stems united at the base. The style is divided into 2 or 3 stigmas. The seedbox is inserted blow the other flower parts and the fruit is a capsule with fine seeds and is crowned with the persistent sepals. The genus is still under review, and the species are rather difficult to distinguish.

*W. ceracea.* Pale lilac or blue. Subalpine, montane. N.S.W., Tas.

A slender plant 20–80 cm high. Leaves alternate, nearly opposite, crowded near the base of the stem, almost smooth, narrow, spoon-shaped, 2–3 cm long. Flowers solitary, terminal, nodding in the bud stage, 2.5–4 cm in diameter, the tube 6–9 mm long, shorter than the expanded petals each of which is marked with a conspicuous midvein; style trifid. Fruit egg-shaped, 7–9 mm long, prominently ridged.

*W. communis.* Blue. Montane. N.S.W., Vic.

A slender plant 20–40 cm high. Leaves opposite or scattered, pointed, stalkless, narrow, 10–15

FIG. 195 *Wahlenbergia*: A, flower and bud; B, leaves; C, fruit. (All × ½).

cm long. Flowers sometimes golden-yellow outside, small, 5–8 mm in diameter, in loose heads on long slender stalks, the expanded petals as long as or longer than the tube; style hairy trifid. Fruit 3–10 mm long, smooth, 2–ribbed.

*W. densifolia.* Deep blue inner, light blue outer. Alpine. N.S.W., Vic.
A trailing plant with many erect flowering stems, the lower parts covered with minute hairs, smooth above. Leaves on lower parts of erect stems crowded, stalkless, nearly opposite, with a few scattered hairs, narrow, 3–10 mm long, higher ones broader. Flowers borne singly, 12–18 mm in diameter, with spreading lobes, veined, on long, slender stalks; style bifid. Fruit smooth.

*W. gloriosa.* Royal blue. Alpine, subalpine, montane. N.S.W., Vic., Tas.
A creeping, semi-erect, slender herb, the loveliest of all the bluebells. Leaves alternate, egg-shaped, wavy, often slightly toothed, 1–2 cm long. Flowers solitary, large, with a relatively short tube and broad, spreading lobes, 1–2 cm long; style bifid. Fruit 7–12 mm long, not prominently ribbed.

*W. gymnoclada.* Deep violet-blue. Montane. N.S.W., Tas.
An erect plant 15–30 cm high. Leaves on lower parts of stem only, crowded, opposite, stalkless, broad, upper surface smooth, lower surface hairy, 1.5–4.5 cm long. Flowers usually solitary and terminal, 2–3 cm in diameter, sometimes yellowish in the throat; flower tube about 4 mm long, the lobes about 3 times as long; style trifid. Fruit about 6 mm long, fairly broad.

*W. saxicola.* Pale blue. Alpine, subalpine. Tas.
A creeping, slender plant. Leaves in a basal rosette with a few scattered along the stem, spoon-shaped or egg-shaped, smooth, the upper surface bright green and glossy, the lower surface paler, 1–3 cm long. Flowers solitary on long stalks 18–22 mm in diameter, the tube 4–5 mm long, the lobes partly spreading, 6–8 mm long. Fruit rounded, 6–8 mm long, with prominent veins.

*W. stricta.* Blue. Subalpine, montane. N.S.W., Vic.
A variable plant 6–30 cm high, common in open forest lands. Leaves variable, stalkless, often slightly toothed, usually opposite, 12–25 mm long. Flowers solitary, terminal on long stalks and broadly bell-shaped, the lobes longer than the tube; style trifid. Fruit egg-shaped, usually hairy, prominently veined.

*W. tadgellii.* Blue. Subalpine, montane. N.S.W., Tas.
An erect, fairly stout, branched plant 25–60 cm high. Leaves mainly on the lower part of the stem, stalkless, alternate or nearly opposite, narrow, thickened, midrib prominent on the under-surface, 2–7 cm long. Flowering stalks much branched. Flowers solitary, 12–18 mm in diameter, funnel-shaped, sometimes yellowish at the base; tube about 1–1.5 mm long, lobes 5–6 mm long or longer; style trifid. Fruit broad, 10–ribbed, 9–22 mm long.

## Family LOBELIACEAE

Small plants with alternate leaves and soft tubular flowers with 5 sepals, 5–lipped or irregular petals, and 5 stamens joined at the tip round the style. The ovary is inserted below the other flower parts.

*ISOTOMA* (from Gk *isos*, equal, and *tome*, a cutting, referring to the equally split petals). Fig. 196.

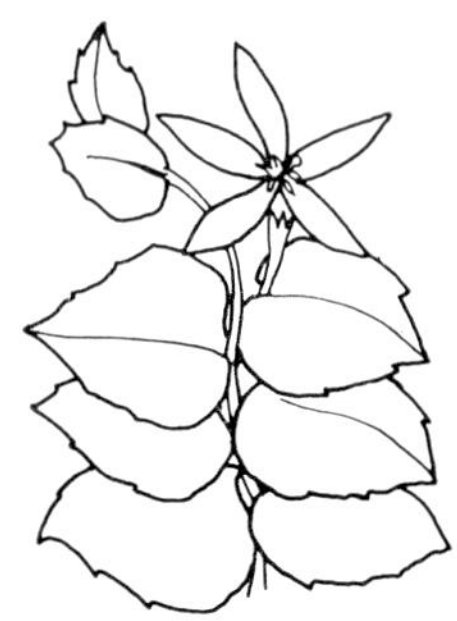

FIG. 196 *Isotoma*: flowerhead (actual size).

*I. fluviatilis* (syn. *Laurentia fluviatilis*). Pale blue, white. Montane. N.S.W., Vic., Tas.
A dainty little mat-forming plant, frequenting moist patches. Leaves variable, egg-shaped, thin, light-green, deep bluish-purple below, 4–10 mm on short stalks. Flowers delicate, stalked, solitary, the tube of the flower split to the base, lobes

nearly equal, petals narrow and obliquely spreading, about 2 mm long; stamens joined to petals. Fruit a turban-shaped splitting capsule 6 mm long.

*LOBELIA* (after Matthias de Lobel, 1528-1616, Flemish physician and plant collector). Fig. 197.

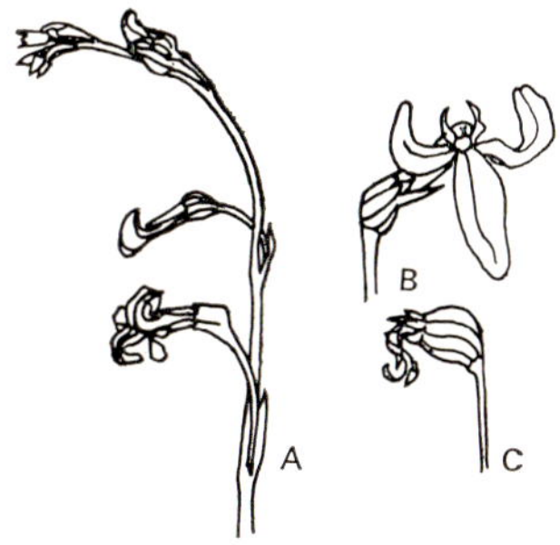


FIG. 197 *Lobelia*: A, flowerhead; B, flower; C, young fruit. (All × ½.)

*L. gibbosa*. Deep blue. Montane. N.S.W., Vic., Tas.
A slender plant 10–40 cm high, of wide distribution at lower elevations also. Branches often reddish and rather succulent. Leaves few, alternate, broad, often withering before the flowers open, 1–4 cm long; petals about 15 mm long, rather similar in form to those of *Pratia*. Fruit a capsule, the upper part swollen.

*PRATIA* (after Charles Prat-Bernon, d. 1817, midshipman on Freycinet's scientific voyage round the world). Fig. 198. A widespread genus of creeping plants with solitary flowers having an oblique, curved, 2–lipped corolla split to the base, the upper lip with 2 deeply serrated lobes, the lower lip with 3 spreading lobes. The fruit is succulent.

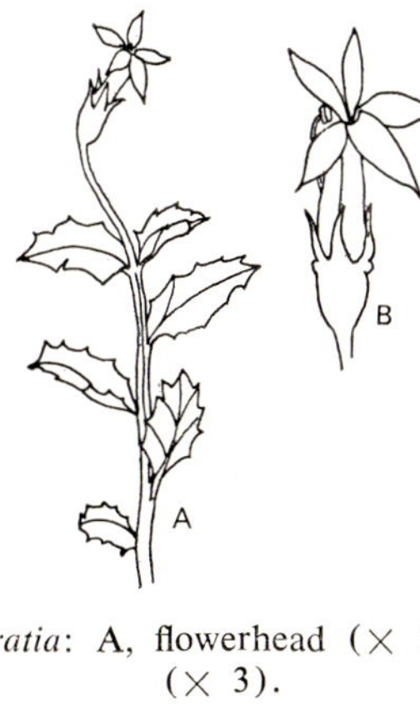


FIG. 198 *Pratia*: A, flowerhead (× 1½); B, flower (× 3).

*P. pedunculata*. Pale blue. Montane. N.S.W., Vic., Tas.
A widespread species at this elevation with short branches sometimes only 2–3 cm long but sometimes much longer, thin and fragile. Leaves alternate, stalkless or shortly stalked, very variable, rounded, 2–20 mm long. Flowers hairy, 4–10 mm long.

*P. puberula*. White. Alpine, subalpine, montane. N.S.W., Vic.
A small trailing plant. Leaves slightly toothed, rounded, 6–8 mm long. Flowers tiny, with nearly equal lobes, on short stalks about 6 mm long.

*P. surrepens*. Pale blue. Subalpine, montane. Vic., Tas.
A smooth, creeping plant with slender, leafy stems. Leaves egg-shaped, blunt, 1–2 cm long. Flowers 8–11 mm long, with lobes narrow at the base, wider in the upper part, spreading on stalks usually shorter than the leaves.

### Family ERICACEAE

This is the true heath family widely represented in northern countries. It has only a few members in Australia and these are mostly Tasmanian. They are shrubs with small tubular flowers, often very showy because the flowers are numerous and close together. The flowers have 4–7 sepals, 4–7 joined petals and the same number of stamens as there are sepals and petals. The petals are persistent on the fruit, which consists of several joined parts but varies in form. The seedbox is inserted above the other parts of the flower. The family differs from the family Epacridaceae, so much better represented in Australia, by the fact that the anthers open by terminal pores whereas in the Epacridaceae they split longitudinally.

*GAULTHERIA* (after Dr. Gaulthier, 18th-century botanist and physician). Fig. 199. A genus of shrubs, often harsh, that are conspicuous by their coloured or white fruits. The flowers

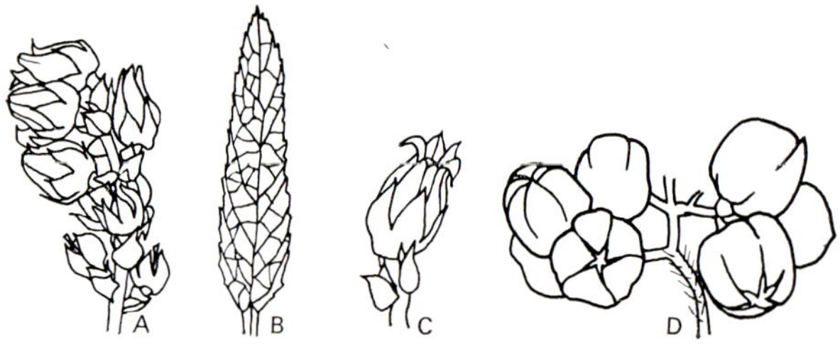


FIG. 199 *Gaultheria*: A, flowerhead (× 2); B, leaf (actual size); C, flower (× 3); D, fruits (actual size).

are tubular or urn- or bell-shaped and have 5 sepals, 5 petals and 10 stamens. The fruit is dry but enclosed by the persistent, swollen, succulent sepals.

*G. depressa*. White; fruit red. Alpine, subalpine. Tas.

A low, spreading shrub ascending to about 20 cm with prostrate, creeping branches with long spreading bristles. Leaves broad, 4–8 cm long, sometimes almost round, smooth, persistent, with toothed margins, teeth tipped by conspicuous reddish bristles. Flowers solitary in the axils of the upper leaves with rounded bracts at the base of each; sepals about 2.5 cm long, enlarging in the fruiting stage; flower tube about as long as the sepals, with short, spreading lobes. Fruit, including the surrounding calyx, about 8–10 mm in diameter.

*G. hispida*. Waxberry. White; fruit white. Subalpine, montane. N.S.W., Vic., Tas.

An erect, spreading shrub 1–2 metres high. Leaves broad, slightly toothed, pointed, leathery with serrate edges, 4–8 cm long, the young leaves a lovely deep-red. Flowers in dense, loose, terminal clusters on short stalks; petals forming an urn-shaped tube, 4–5 mm long. Fruit, including the succulent calyx, 8–10 mm in diameter.

*PERNETTYA* (after Pernetty, historiographer on de Bougainville's voyage to S. America). Fig. 200. A genus of shrubs with alternate leaves and flowers, solitary or in loose heads in the leaf axils. The calyx is 5–lobed and persistent in the fruiting stage, sometimes becoming fleshy. The flowers are tubular with 5 lobes and 10 stamens. The fruit is a succulent berry.

*P. lanceolata*. White; berries rose pink. Alpine. Tas.

A shrub 30 cm high with rough branches. Leaves narrow, leathery, pointed, with serrate margins, 12–22 mm long, the older leaves shining and reticulate. Flowers solitary in leaf axils, 4–6 mm long; flower about 4 mm long. Fruit about 8 mm in diameter, the succulent calyx the same colour as the fruit itself, and half enclosing it.

*P. tasmanica*. White; berries red, yellow or white. Alpine, subalpine. Tas.

A small, smooth shrub with creeping branches forming a dense mat 15–30 cm in diameter. Leaves rather narrow, blunt, 3–6 mm long. Flowers solitary in the axils of the upper leaves; calyx with 5 egg-shaped sepals; corolla urn-shaped with 5 petals, the tube 3–4 mm long, the lobes short, blunt and recurved. Fruit rounded, 6–8 mm in diameter, the succulent calyx the same colour as the fruit itself.

*WITTSTEINIA* (after G. C. Wittstein, 19th-century botanical chemist). Fig. 201.

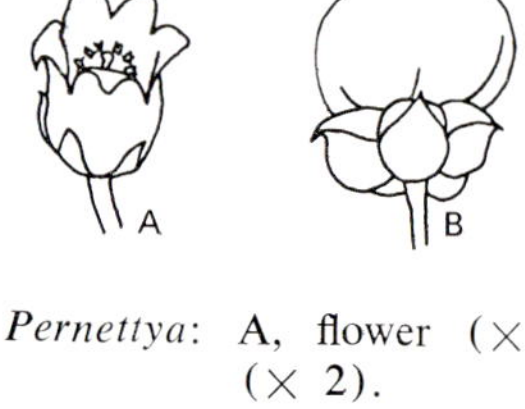


FIG. 200 *Pernettya*: A, flower (× 2½); B, fruit (× 2).

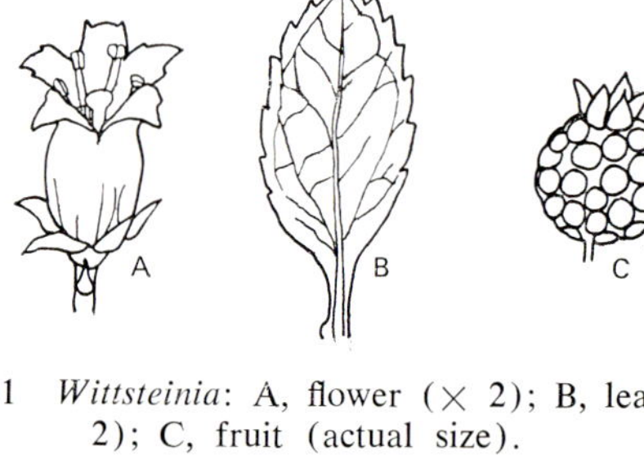


FIG. 201 *Wittsteinia*: A, flower (× 2); B, leaf (× 2); C, fruit (actual size).

*W. vacciniacea*. Baw Baw Berry. Greenish-yellow; fruit reddish. Subalpine. Vic.

A prostrate plant of limited distribution with drooping, solitary flowers. Leaves egg-shaped, smooth, rather large and distantly toothed; sepals 5; petals 5. Fruit nearly globular, berry-like, with a persistent calyx.

## Family EPACRIDACEAE

The epacrids or heaths are well represented in alpine areas, both on the mainland and in Tasmania, many of them sweetly scented. Members of the family are often twiggy or prostrate shrubs with fine, crowded leaves with parallel veins. The flowers are often solitary but so close together in the leaf axils that they give the appearance of spikes. The flowers are usually small, often white, with bracts or bracteoles at the back and with 5 free sepals and 5 petals joined to form a tube expanded at the tip in most species. There are 5 stamens, often joined to the petals. The seedbox is inserted above the other flower parts. The fruit is either a small, inconspicuous, dry capsule or a succulent berry-like one which, especially in several of the Tasmanian species, is richly coloured and most attractive when ripe.

*EPACRIS* (from Gk *epi*, upon, and *akros*, the top, in reference to the high habitat of some species). Fig. 202. A large genus of heath-like, fine-leafed shrubs in which the bracts of the tubular flowers surround the petals closely and are similar to them. The flowers are solitary but abundant and produced in the axils of closely set leaves; the flower parts are in fives, with the stamens joined to the petals. The fruit is a dry, splitting capsule of 5 parts.

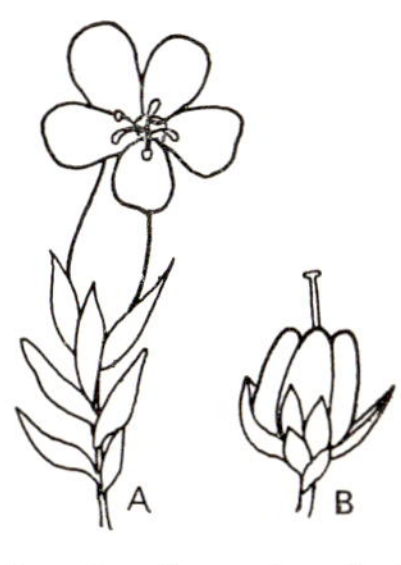


FIG. 202 *Epacris*: A, flowerhead (× 2); B, fruit (actual size).

*E. breviflora*. White. Alpine, subalpine. N.S.W., Vic.

A shrub 50–80 cm high with hairy branches. Leaves egg-shaped, leathery, pointed, almost stalkless, 6–9 mm long, the edges minutely toothed. Bracts and sepals egg-shaped, pointed, minutely fringed on the margins. Flowers clustered at the ends of the branches, small, the tube 1–1.5 cm long, with blunt, egg-shaped lobes about twice as long.

*E. gunnii*. White. Alpine, subalpine, montane. Tas.

Resembling *E. microphylla* and found in wet habitats from the sea-coast to over 4,000 feet. A stiff shrub 20–90 cm high with slender woolly branches. Leaves with a short, wide stalk, spreading, pointed, 3–6 mm long. Flowers almost stalkless in the leaf axils, appearing to form a leafy spike; bracts and sepals blunt, egg-shaped, the sepals about 4 mm long, the flower tube shorter than the sepals and rather wide, the lobes about as long as the tube.

*E. impressa*. Red, pink, cream or white. Subalpine, montane. Vic., Tas.

This species is the floral emblem of Victoria. It occurs at sea-level as well as in subalpine and montane areas and reaches a height of 20 cm–1 metre. It is one of the most attractive and richly coloured species of *Epacris*, the flowers varying in colour even in the same habitat. Leaves fairly narrow, pointed, stalkless, 4–14 mm long. Flowers bell-shaped, 4–6 mm long, solitary in leaf axils, but the leaves so close together and the flowers so abundant that they appear to be in dense heads.

*E. lanuginosa*. Cream. Subalpine, montane. N.S.W., Vic., Tas.

A common plant of wet heaths up to an altitudes of 3,000 feet, 30–90 cm high, with slender stems. Leaves crowded, spreading, rather narrow, pointed, 8–9 mm long. Flowers abundantly produced in the axils of crowded leaves, appearing to be in dense heads; flower tube 8–10 mm long, the flowers sweetly scented.

*E. microphylla*. White. Montane. N.S.W., Vic.

A wiry shrub found in coastal habitats also, 90–100 cm high. Leaves egg-shaped, almost stalkless, tiny, 3–4 mm long. Flowers produced in great abundance in the axils of the crowded leaves; flower tube short and broad, less than 2 mm long.

*E. paludosa.* White. Subalpine, montane. N.S.W., Vic.

A swamp-loving, rather variable stiff shrub, reaching nearly 1 metre at most but commonly only half this height. Leaves rather narrow, pointed, crowded, thick with minutely toothed edges, 6–12 mm long. Flowers solitary but numerous in the axils of crowded leaves at the ends of the branches, giving the appearance of a crowded head and in this way differing from the other alpine *Epacris* species in which the flowers appear as elongated spikes; sepals egg-shaped, blunt, as long as the tube.

*E. petrophila.* White. Alpine. N.S.W., Vic., Tas.

A purely alpine species, often clinging to rocks or sheltering below them, a plant of low growth, 15–40 cm high, usually with trailing branches woolly beneath but also with some erect branches. Leaves numerous, stalkless, overlapping, oval or egg-shaped, blunt and thick, closely pressed against the stem, 1.5–2.5 mm long. Flowers almost stalkless in the upper axils of the crowded leaves, appearing to form small spikes; bracts and sepals turning brown on drying and on the plant as new flowers form; sepals egg-shaped, blunt, hairy on the edges, 2–5 mm long; petals bell-shaped, the tube shorter than the sepals, the lobes about as long as the tube.

*E. robusta.* White. Montane. N.S.W., Vic.

An erect, robust small shrub. Leaves egg-shaped or rounded, very spreading. Flowers crowded near the ends of the branches; flower tube as long as the sepals or shorter.

*E. serpyllifolia.* White. Alpine, subalpine. N.S.W., Vic., Tas.

A variable shrub to 1 metre in height in protected places but the alpine forms almost prostrate. Likes moist situations. Leaves densely crowded, egg-shaped, stalkless, thick, spreading, about 2–4 mm long. Flowers solitary in the axils of the crowded leaves; bracts and sepals sometimes reddish; sepals about 4 mm long; tube of the flower about as long as the sepals, with broad, blunt lobes shorter than the tube.

In the alpine parts of Mount Kosciusko there is a not uncommon little *Epacris* sprawling over the rocks which has not yet been described and named. Branches finely woolly; leaves egg-shaped, very slightly stalked, somewhat blunt, curved at the tip and pressed against the stem. Flowers small, stalkless in the crowded leaf axils but not crowded at the tips of the branches; bracts and sepals blunt, white, the flower tube shorter than the sepals, the expanded tips at least as long, rounded.

*ACROTRICHE* (from Gk *akros*, top, and *thrix*, *trichos*, a hair, referring to the tuft of hairs at the top of the flower tube). Fig. 203.

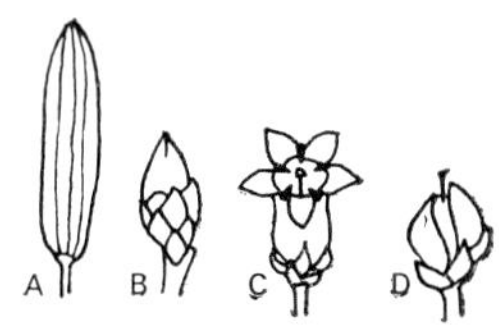


FIG. 203 *Acrotriche*: A, leaf; B, bud; C, flower; D, fruit. (All × 4.)

A genus in which the flower lobes are spreading and have a tuft of long hairs inside at the summit, the lobes themselves being smooth. They are stiff shrubs with small flowers in clusters. The fruit is succulent.

*A. divaricata.* Greenish. Montane. N.S.W.

A small, branched shrub. Leaves rather narrow, sharp-pointed, 10–15 mm long. Flowers about 1 mm long, the base short.

*A. serrulata.* Ants' Delight. Pale green or purplish. Montane. N.S.W., Vic., Tas.

Occurring also in dry habitats at lower elevations. Sometimes prostrate, sometimes attaining a height of 15–30 cm. Leaves narrow, pointed, 5–10 mm long and persistent for 2–3 years. Flowers in short, dense heads; flower tube about 4 mm long, swollen and heavy with nectar, the lobes narrow and blunt.

*ARCHERIA* (after W. Archer, 19th-century Tasmanian botanist). Fig. 204. A Tasmanian genus of much-branched shrubs distinguished by the possession of a seedbox of 5 compartments forming a dry splitting capsule. It is separated from *Epacris*, which also has a capsule, by the fact that the bracts at the back of the sepals are very easily distinguished from the sepals themselves; whereas in *Epacris* they surround

FIG. 204 *Archeria*: A, flowerhead; B, fruit (actual size).

the sepals and are similar to them. The leaves are small and the bell-shaped flowers have short lobes. The stamens are joined to the petals.

*A. comberi.* Pale reddish-pink. Alpine. Tas.
Usually grows in the shelter of other shrubs. Much-branched, 15–100 cm high, covered with scattered hairs. Leaves shortly stalked, spreading, broadly egg-shaped or almost round, blunt with short points and often purplish in colour, 3–4 mm long, leaving raised circular scars on falling. Flowers shortly stalked, solitary or a few together in the leaf axils, usually on last season's wood; sepals usually 5 but occasionally 6–7, elliptical, blunt, finely hairy on the edge with a rounded turgid swelling where they join the stem, about 3 mm long; petals usually 5, occasionally 6–7, smooth and waxy, round, about 6 mm long.

*A. eriocarpa.* White, pink. Subalpine, montane. Tas.
A straggling shrub 1–4 metres high with rough branches. Leaves 6–9 cm long, on very short stalks, flat and spreading, pointed, the margins finely toothed, the veins conspicuous. Flowers solitary, on short stalks in the leaf axils or in short terminal clusters; flower tube 6–7 mm long, the lobes small, spreading. Fruit cylindrical, about 4 mm long.

*A. hirtella.* White, pink. Subalpine, montane. Tas.
A stout, rigid, bushy shrub 15–60 cm high or prostrate in exposed situations. Tips of branches and young leaves often crimson; leaves egg-shaped, thick, shining, flat, shortly pointed, 6–8 mm long, roughly in 2 rows. Flowers on short, hairy stalks, solitary or in short clusters; tube of flower 6–7 mm long, cylindrical, contracted in the throat, with small lobes.

*A. serpyllifolia.* White, pinkish. Alpine. Tas.
A stout, stiff, bushy shrub usually erect and 15–60 cm high but prostrate in exposed situations. Leaves shortly stalked, crowded, usually flat, narrow, elliptical or egg-shaped, blunt, smooth, 4–8 mm long, the margin finely toothed. Flowers on long stalks in short, dense clusters; sepals oblong- elliptical, blunt, about 4 mm long, lightly fringed with hairs; petals scarcely longer than the sepals, forming an almost bell-shaped tube, the lobes shorter than the tube, and spreading, the inner surface of the lobes and the tube woolly.

*ASTROLOMA* (from Gk *aster*, a star, and *loma*, a fringe, referring to the hairs inside the flower). Fig. 205. In this genus there are 5 tufts of hairs in the lower part of the flower tube.

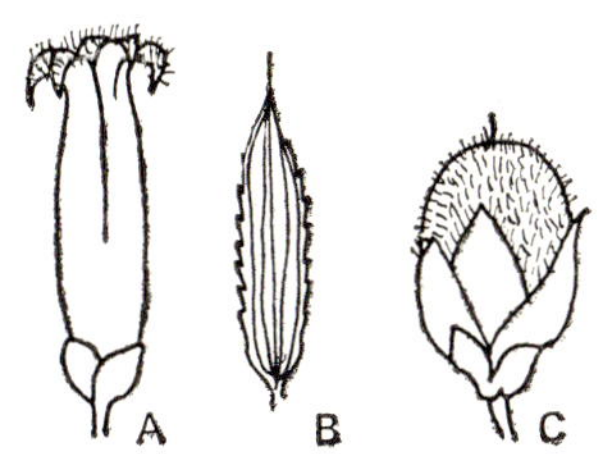


FIG. 205 *Astroloma*: A, flower; B, leaf; C. fruit. (All × 2.)

*A. humifusum.* Red. Montane. N.S.W., Vic.
A small, often prostrate plant with narrow, pointed leaves 5–12 mm long. Flower tube about 9 mm long. Fruit round, succulent, with a single hard stone and about 8–10 mm in diameter with a persistent large stigma.

*BRACHYLOMA* (from Gk *brachys*, short, and *loma*, a fringe, referring to the hairs in the throat of the flower tube). Fig. 206.

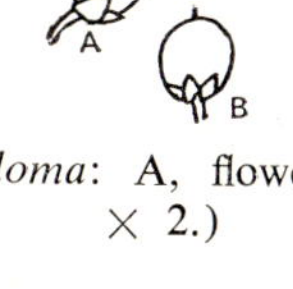


FIG. 206 *Brachyloma*: A, flower; B, fruit. (Both × 2.)

*B. daphnoides*. White. Montane. N.S.W., Vic.
A well-known plant of lower altitudes, this is a bushy shrub 30–100 cm high with broad, blunt, stalkless leaves 4–10 mm long. Flowers tubular with a ring of hairs inside the tube and narrow, short lobes, about 3–4 mm long. Fruit a round, splitting capsule, 2–4 in diameter.

*CYATHODES* (from Gk *kuathodes*, like a cap, referring to the sepals). Fig. 207. A genus of shrubs almost entirely confined to Tasmania,

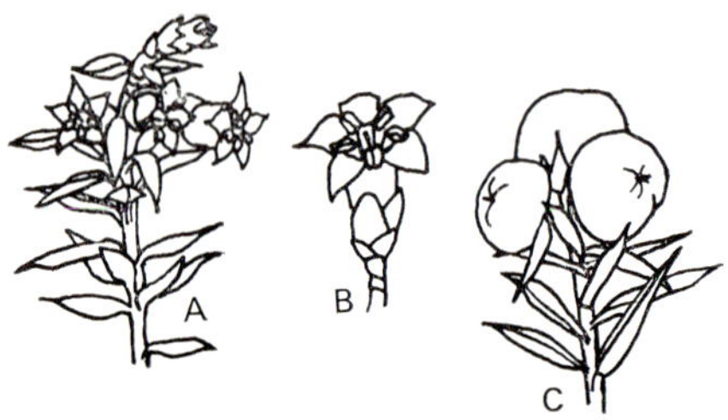


FIG. 207 *Cyathodes*: A, flowerhead (× 4); B, flower (× 5); C, fruits (× 1½).

distinguished from the leucopogons and pentachondras by the fact that the petals are not hairy or are only slightly hairy, and usually red, yellow or cream (not white) and the leaves give the appearance of being whorled. The flowers are small, solitary or in short, leafy clusters, followed by abundant heads of attractive succulent fruit. The leaves are alternate.

*C. glauca*. Cheeseberry. White, red, pink or purple berries, occasionally white. Alpine, subalpine. Tas.
Usually a straggling shrub 60 cm–3 metres high. Leaves 15–30 mm long, clustered at the ends of the branches, narrow-elliptical, sharply pointed and narrowed into a short stalk, the undersurface striate, grey-green. Flowers solitary in the leaf axils, clustered, with the leaves at the ends of the branches, each flower with about 8 overlapping bracts about half as long as the sepals; sepals 3–4 mm long, broad, blunt, striate, lightly fringed with hairs on the edge; corolla tube about 6 mm long, the lobes shorter, scantily hairy. Fruit rounded, somewhat flattened, berry-like, about 8 mm in diameter.

*C. straminea*. White; berries red. Alpine, subalpine. Tas.
A variable shrub 18 cm–1 metre high. Leaves 5–15 mm long, clustered, elliptical with a rounded end and a short, wide stalk, the lower surface striate and often grey-green. Flowers solitary in the leaf axils, clustered, with leaves at the ends of the branches and rather similar to those of *C. glauca*, the tube about 4–6 mm long.

*DRACOPHYLLUM* (from Gk *drakon*, a dragon, and *phyllon*, a leaf, in reference to the sharp-pointed leaves). Fig. 208. A genus of shrubs distinguished from the other members of the family except *Sprengelia* and *Richea* by the broad sheathing bases of the leaves, and by the annular scars left by these. It is distinguished from *Sprengelia* by the tubular flowers and from *Richea* by the absence of the cap formed by the top of the petals which is characteristic of that genus. The fruit is a dry, splitting capsule.

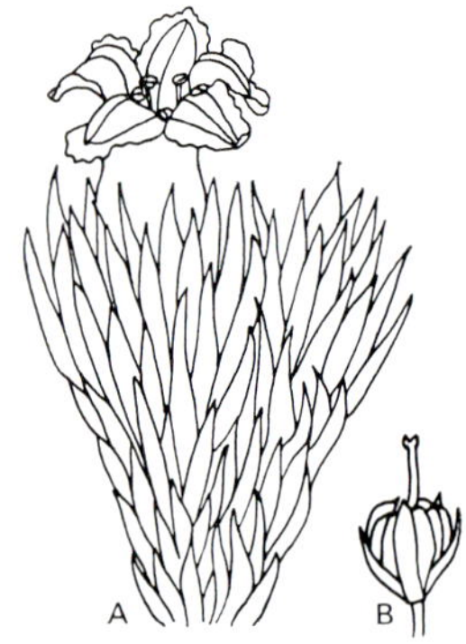


FIG. 208 *Dracophyllum*: A, flowerhead; B, fruit. (Both × 2.)

*D. milliganii*. Pink, pinkish. Alpine, subalpine. Tas.
A stout, woody, unbranched shrub, 15 cm–4 metres high. Leaves narrow, leathery and stiff, tapering to a long point, often coiled or recurved, densely crowded, 15–90 cm long with a broad, sheathing base. Flowerheads 15–40 cm long, on woolly stems with the individual flowers on short stalks along branches; bracts at the base of the flowers very narrow, sepals petal-like, broad, 3–4 mm long; flower tube about as long as the sepals, broad but not contracted at the throat, the lobes shorter than the tube, recurved.

*D. minimum*. White. Alpine. Tas.
A cushion plant forming compact mounds 30–90 cm in diameter at altitudes of about 4,000 feet where snow lies for many months. Erect stems 2–6 cm long, closely packed, much branched. Leaves 3–4 mm long, with a sheathing base, fairly narrow, thick, stiff and sharply pointed. Flowers solitary, terminal, stalkless, surrounded by foliage leaves as long as the sepals; sepals fairly narrow, 4–6 mm long; flower tube round and rather broad, about as long as the sepals, the lobes broad and spreading, slightly shorter than the tube, the base of each lobe with a distinct longitudinal ridge.

*LEUCOPOGON* (from Gk *leukos*, white, and *pogon*, a beard, in reference to the bearded petals). Fig. 209. A genus of variable shrubs

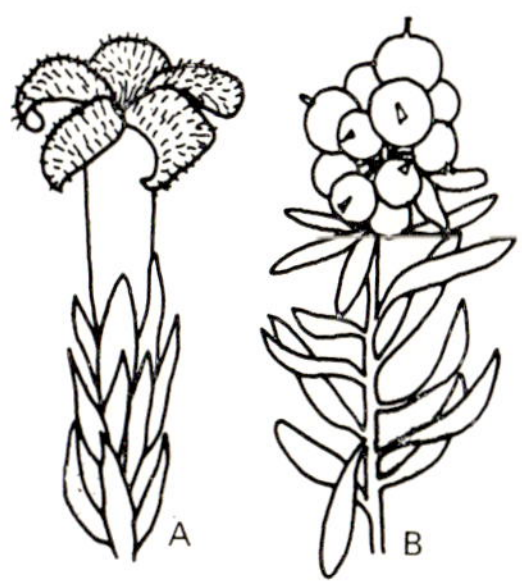

FIG. 209 *Leucopogon*: A, flowerhead; B, fruit. (Both × 2.)

with small tubular flowers densely bearded inside with white hairs. The flowers are in spikes. Each flower has 1 bract and 2 small bracteoles, all deciduous, behind it, distinguishing the genus from *Pentachondra* in which there are 4 or more persistent bracts. The stamens are enclosed and the fruit is succulent, usually with a single seed inside a hard stone.

*L. attenuatus*. White; fruit dark. Montane. N.S.W., Vic.
A spreading shrub. Leaves 1–2 mm long, fairly broad, sharply pointed, closely pressed against the stem. Flowers 1–2 in the leaf axils, almost stalkless, often crowded into a long, leafy spike, with tiny bracts and bracteoles; sepals 2–3 mm long, the flower tube about 2 mm long, the lobes about the same length. Fruit tiny and striate.

*L. collinus*. White; fruit dark. Subalpine, montane. N.S.W., Vic., Tas.
A small, variable, much-branched shrub, prostrate in higher altitudes, up to 60 cm at lower levels, branches often 10–20 cm long, reddish-brown, finely hairy, narrow, oblong or elliptical (in prostrate forms), blunt or with a short point. Flowers in short, dense spikes crowded at the ends of the long branches; sepals blunt, 2–2.5 mm long, with fine hairs on the edge; flower tube scarcely as long as the sepals; lobes longer and narrower. Fruit small, berry-like, almost dry, blackish-green.

*L. ericoides*. White, pinkish; fruit dark. Montane. N.S.W., Vic., Tas.
A slender shrub with spreading branches, often 30–90 cm high but sometimes prostrate. Leaves stalkless or with very short stalks, blunt, with the edges recurved, 5–12 mm long. Flowers 1–5 together in short spikes in the leaf axils, often numerous and appearing like leafy heads; sepals about 2 mm long, the petals longer with short, spreading lobes. Fruit small, ovoid, greenish-black, almost dry.

*L. fraseri*. White; fruit yellowish-orange. Montane. N.S.W., Vic., Tas.
A lovely little compact shrub 7–30 cm high. Leaves 5–8 mm long, closely set, broad, sharp, shining above, striped below. Flowers stalkless, solitary in the leaf axils but appearing to be in masses because the leaves are so crowded, scented; sepals pointed; tube of flower about 6–8 mm long, the lobes shorter, acute and woolly white. Fruit berry-like, 10–12 mm in diameter.

*L. hookeri*. White; fruit red, rarely white. Alpine, subalpine, montane. N.S.W., Vic., Tas.
An attractive little plant, plastering itself over rocks or along the ground at higher elevations and reaching a height of 60 cm as a graceful, erect shrub in selected places at lower elevations. Leaves 5–10 mm long, shortly stalked, crowded, narrow, oblong, with a rounded tip or short point, smooth, the lower surface grey-green. Flowers in short spikes 4–6 together in leaf axils or at the ends of branches, tiny, scarcely opening before falling; petals 2–3 mm long, smooth with a rather wide tube, the lobes about

as long as the tube. Fruit a berrylike drupe, produced in great profusion, most attractive.

*L. juniperinus*. White. Montane. N.S.W.
A widely distributed shrub 1 metre or less in height, erect. Leaves broad, sharp. Flowers few together; corolla tube about 8 mm long, woolly white inside.

*L. lanceolatus*. White. Montane. N.S.W., Vic.
A tall shrub, widely distributed, usually smooth. Leaves 1–4 cm long, with conspicuous longitudinal veins. Flowers in loose interrupted spikes longer than the leaves; petals about 2 mm long.

*L. macraei*. White; fruit red. Subalpine, montane. N.S.W., Vic.
A rather rare tall shrub 2–2.5 metres high, graceful, willowy, with hairy branches, resembling tall forms of *L. hookeri*. Leaves small, stalked, smooth, shining on the top surface, blunt or with a small, thick point. Flowers in spikes in the leaf axils, or at the tips of the branches; the flower tube broad. Fruit nearly globular, succulent, very conspicuous and attractive.

*L. microphyllus*. White. Montane. N.S.W., Vic.
A widely distributed erect or straggling shrub with wiry branches, about 1 metre high. Leaves 2–4 mm long, flat or convex, egg-shaped or narrow. Flowers in leafy spikes, petals about 3 mm long.

*L. milliganii*. White; fruit red. Alpine, subalpine. Tas.
A small species with procumbent or spreading branches 15–40 cm long. Leaves 4–6 mm long, clustered at the ends of the branches, the older ones falling as the new growth appears and leaving the stem between the leaf clusters bare, shortly stalked, narrow, thick and pointed with a covering of white hairs on the undersurface. Flowers 2–3 together in a short spike 3–4 mm long; sepals narrow and pointed, fringed on the edge; flower tube about as long as the sepals, the lobes shorter and spreading. Fruit spherical, berry-like, red, 3–4 mm in diameter.

*L. virgatus*. White. Subalpine, montane. N.S.W., Vic., Tas.
A small shrub with slender, wiry, smooth branches, 15–40 cm high, often spreading. Leaves fairly broad, tapering to a rigid point, about 8–15 cm long. Flowers in very short, dense spikes, usually crowded at the ends of the branches; bracts and bracteoles small; sepals about 2 mm long; flower tube slightly shorter than sepals, lobes longer and spreading. Fruit small, almost dry, greenish-black.

*LISSANTHE* (from Gk, *lissos*, smooth, and *anthos*, a flower). Fig. 210. A genus in which the flower tube is smooth inside.

FIG 210 *Lissanthe*: A, flowerhead (actual size); B, flower (× 2); C, fruit (× 2).

*L. montana*. White. Alpine, subalpine, montane. Tas.
A much-branched smooth shrub 20–60 cm high. Leaves crowded, flat, shortly stalked, narrow, blunt, 5–10 mm long. Flowers 4–6 together in short spikes crowded at the ends of the branches; stamens joined to the edge of the tube, slightly protruding. Fruit succulent, rounded, 3–4 mm in diameter.

*MELICHRUS* (from Gk *melichros*, honey-coloured, referring to the glands). Fig. 211. A genus with smooth flowers, sometimes with a few hairs inside the short, broad tube.

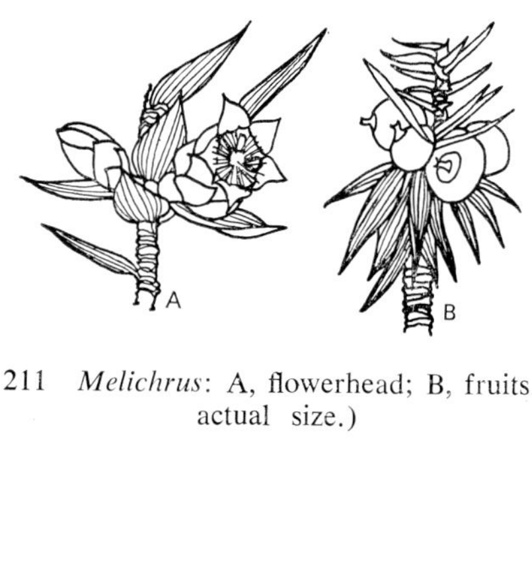


FIG. 211 *Melichrus*: A, flowerhead; B, fruits. (Both actual size.)

*M. urceolatus.* Cream, red. Montane. N.S.W., Vic.

A small stiff shrub 15–90 cm high with rather narrow, sharply pointed leaves, 6–12 mm long without stalks. Flowers solitary. Fruit rounded, succulent.

*MONOTOCA* (from Gk *monos*, one, and *tokos*, a birth, referring to the single-seeded fruit). Fig. 212. A small genus in which the flowers are quite smooth with no hairs in the tube. The fruit is fleshy, with one compartment.

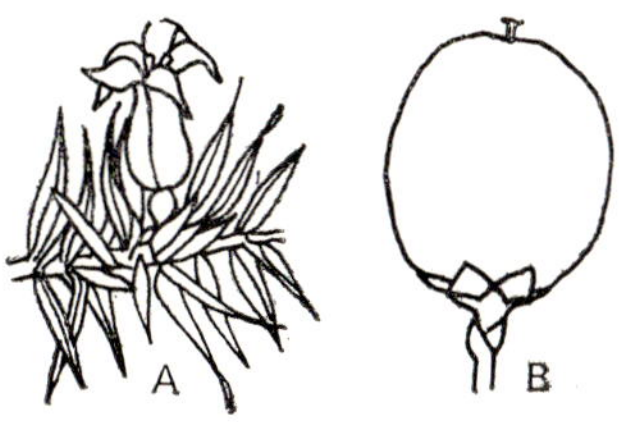


FIG. 212 *Monotoca*: A, flower; B, fruit. (Both × 4.)

*M. rotundifolia.* Creamy yellow. Montane. Vic.

A recently described species, this is a small prostrate shrub with slender, woolly branches. Leaves 3–5 mm long, round but with margins rolled, stalked, veins conspicuous. Flowers solitary in the upper leaf axils, about 2 mm long on stout stalks; anthers yellow.

*M. scoparia.* White. Montane. N.S.W., Vic., Tas.

A smooth, small shrub with narrow, pointed leaves 2–7 mm long. Flowers in clusters in the leaf axils, small, smooth, about 2 mm long, the stamens enclosed. Fruit rounded, about 2 mm in diameter.

*PENTACHONDRA* (from Gk *penta*, five, and *chondros*, a grain, in reference to the 5–parted fruit). Fig. 213. A genus of small shrubs rather like *Leucopogon*, with very hairy flowers but differing in having 4 or more bracts behind the flowers. These often prostrate shrubs have crowded leaves and flowers 2–3 together or solitary. The fruit is fleshy, with a stone separating into 5 or fewer parts.

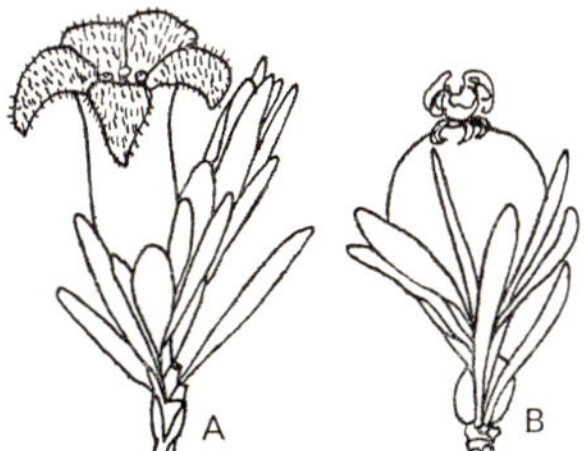


FIG. 213 *Pentachondra*: A, flowerhead; B, fruit. (Both × 4.)

*P. ericifolia.* White. Montane. Tas.

A small heath-like shrub with erect or sprawling branches, 15–60 cm long. Leaves 2–3 mm long, densely crowded and closely pressed against the stem, narrow, usually without stalks, shortly pointed, lower surface with 1–5 prominent veins, margins fringed. Flowers stalkless, solitary in the leaf axils but often several together near the ends of short branches; sepals 2–3 mm long, broad, fringed with hairs, the lobes more than half as long as the tube, spreading.

*P. involucrata.* White. Alpine, subalpine. Tas.

A small shrub with hairy branches, 20–40 cm in height or prostrate, common on Mount Wellington. Leaves crowded, almost stalkless, broad, with marked longitudinal veins and soft, spreading hairs on the margins, 8–18 mm long. Flowers in groups of 1–3 at ends of short branches in the leaf axils; sepals egg-shaped, fringed on the edges, 2–3 mm long; flower tube about as long as the sepals, the lobes spreading widely, 8–10 mm long. Fruit berry-like.

*P. pumila.* White, occasionally red. Alpine, subalpine. N.S.W., Vic., Tas.

A prostrate and matted, rather woody shrub resembling in general appearance the alpine form of both *Leucopogon hookeri* and *Lissanthe montana.* Branches 6–15 cm long; leaves crowded, oblong, smooth, margins fringed with fine hairs, showing 3–7 parallel veins on the underside, usually blunt, 2–4 mm long. Flowers almost stalkless, solitary at the ends of the short branches and scattered all over the plant; sepals about 2 mm long, egg-shaped, blunt, slightly fringed; flower tube cylindrical, about twice as long as the calyx; lobes short, about half as long as the tube, heavily bearded with white wool. Fruit berry-like, spherical, sometimes slightly flattened, 6–8 mm in diameter; petals remaining in position on top of the ripening fruit, a characteristic and very attractive feature.

*PRIONOTES* (from Gk *prionotos*, saw-like or jagged, referring to the leaf edges). Fig. 214.

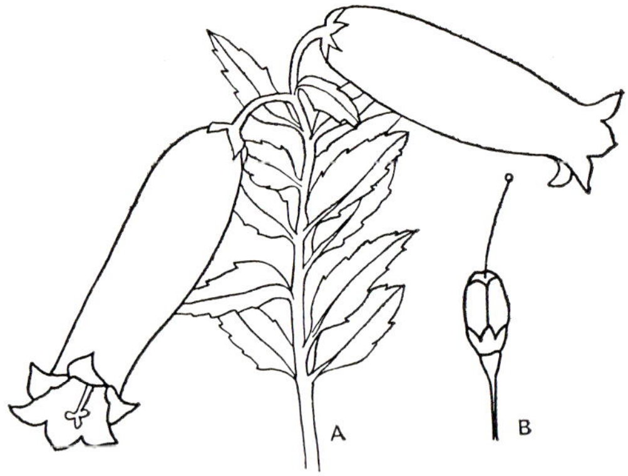


FIG. 214 *Prionotes*: A, flowerhead; B, fruit. (Both × 1½).

*P. cerinthoides*. Crimson. Montane. Tas.
This magnificent plant is a common inhabitant of temperate forest. It is epiphytic, with long, slender branches, climbing over beech and other trees where it displays its flamboyant blossom to best advantage in the late summer months. Leaves oblong, shortly stalked, thick, rounded at the tip with a few blunt teeth, 8–20 mm long. Flowers solitary, profuse, pendulous, tubular, the tube 20–25 mm long, delicately inflated towards the centre and constricted at the throat, the lobes very short and blunt, on slender stalks about 10 mm long. Fruit a round, dry, splitting capsule, 8–10 mm long.

*RICHEA* (after Riche, a naturalist on the voyage of La Pérouse to Australia, 1788). Fig. 215. A very interesting and colourful genus confined to Tasmania with the exception of one species fairly widely distributed on the mainland. It is readily distinguished from other members of the family by the fact that the petals form a cap which separates near the base and is shed to release the stamens. Richeas are shrubs or small trees with crowded leaves and flowers in terminal heads. Fruit a dry, splitting capsule.

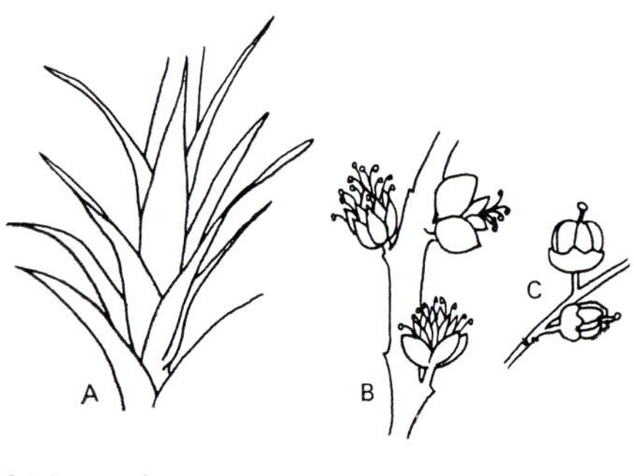


FIG. 215 *Richea*: A, leaves; B, flower; C, fruit. (All actual size.)

*R. acerosa*. White, cream or pink. Alpine, subalpine. Tas.
A stiff, erect shrub 60–120 cm high, found commonly on the central plateau and the west of Tasmania. Leaves usually confined to the current year's growth, overlapping, rather narrow, tapering to a sharp point, 8–14 mm long, with a wide, sheathing base. Flowers in short terminal heads, the individual flowers almost stalkless in the axils of pointed bracts which are brownish at the base, the 2 outer ones about 3 mm long; sepals broadly egg-shaped, pointed, about 4 mm long; cap of the flower broadly conical, about 5 mm long.

*R. angustifolia*. Red. Alpine. Tas.
A rather rare plant similar to *R. scoparia* but with narrow, erect, deeply overlapping leaves, and smaller flowers in shorter heads.

*R. continentis*. Cream. Alpine, subalpine. N.S.W., Vic.
An erect, stiff plant reaching a height of 1 metre and widespread in sphagnum bogs and other wet places, forming dense thickets. Leaves prickly but attractive, stem-clasping, rather narrow, sharp-pointed, 5–10 cm long. Flowering stems a rich red colour. Flowers in little stalkless clumps, each enclosed in a deciduous leafy bract reddish at the base and falling to liberate the flowers; sepals egg-shaped, pointed, reddish-brown, about 3 mm long; petals egg-shaped, conical, about 4 mm long, the deciduous cap separating near the base.

*R. dracophylla*. White. Montane. Tas.
A common species reaching 1.5 metres in height, sparingly branched and marked by annular scars, looking rather like a palm. Leaves crowded, thick, but flexuose, fairly broad, tapering to a slender point, sheathing at the base, 15–30 cm long. Flowerheads 15–25 cm long; the flowers within brown, deciduous bracts with broad, sheathing bases and long points; cap of the flower egg-shaped, 10–12 mm long, the tip bluntly pointed.

*R. gunnii*. Cream. Subalpine, montane. Tas.
Common in the wetter parts. Close to *R. continentis*, differing only in the shape of the bracts, which are membranous, brownish and with a broad base, and the rather smaller flowers. A shrub 30–100 cm high with annular scars. Leaves overlapping, spreading, persisting only on the ends of the branches, stiff and tapering with a broad, sheathing base 2–4 cm long. Flowers 3–5 together on the lateral branches, very shortly stalked, with 2–3 narrow, keeled bracts at the base of each.

*R. procera*. Cream. Montane. Tas.
A stiff shrub 60 cm–3 metres high, sparingly branched. Leaves overlapping, broad, tapering, 10–35 mm long, distantly toothed. Flowers solitary, almost sessile in the axils of persistent bracts, the outer ones prominently keeled; sepals egg-shaped, about 4 mm long, the cap narrow-conical, 10–12 mm long.

*R. scoparia*. White, cream, pink, red. Alpine, subalpine. Tas.
A spectacular plant, widely spread and often occurring in dense stands of varying flower colour. A shrub 1–3 metres or sometimes taller but often only about 40–100 cm in exposed situations. Branches stout with reddish-brown bark peeling in flakes. Leaves remaining on the plant for many years, dying and decaying without falling off, stiff, overlapping, fairly narrow, 3–6 cm long, tapering to a sharp point and with a broad, sheathing base. Flowers stalkless in clusters of 2–5 within bracts shorter than the foliage leaves and very broad, falling before the flowers expand; sepals egg-shaped, 2.5 mm long; cap of the flower 7–8 mm long, somewhat oblong with a bluntly pointed apex. One form at Mount Field has a very pungent odour.

*R. sprengelioides*. Cream. Alpine, subalpine. Tas.
Probably the commonest of the Tasmanian mountain species, forming compact bushes 30–120 cm high, often growing in exposed situations. Leaves overlapping, stiff, spreading, broadly egg-shaped, 8–12 mm long, tapering to a hard, sharp point and with a broad base completely sheathing the stem. Heads of flowers terminal on long, leafy branches; base of each solitary flower surrounded by 2–6 bracts, the outer ones about 5 cm long; sepals egg-shaped, about 4 mm long; petals longer than the sepals; cap of flower narrow-conical, about 10–12 mm long.

*SPRENGELIA* (after Kurt Sprengel, 1766-1833, German botanist and physician). Fig. 216. A genus of smooth plants with overlapping sheathing leaves, persistent for many years. The flowers are solitary with numerous bracts, sometimes crowded into leafy heads. The flower tube is short or the petals free at the base, the lobes narrow and spreading. The fruit is a dry, splitting capsule.

FIG. 216 *Sprengelia*: A, flowerhead (actual size); B, flower (× 1½); C, enclosed fruit (× 1½).

*S. disticophylla*. Pink, white. Alpine, subalpine. Tas.
A small woody plant with branches 2–5 cm long. Leaves small, 3–4 mm broad. Flowers solitary, with broad sepals, 4–5 cm long; petals bell-shaped, the tube about 2 mm long, the lobes slightly longer.

*S. incarnata*. Pink, white. Subalpine, montane. Tas.
On the mainland confined to coastal and coastal mountain habitats. A variable but generally slender shrub. Leaves completely sheathing the stem, very pointed and spreading, 7–14 mm long. Flowers terminal, crowded and surrounded by numerous bracts; sepals and petals similar, spreading widely, about 5–10 mm long.

*TROCHOCARPA* (from Gk *trochos*, wheel, and *karpos*, a fruit). Fig. 217. A genus of usually slender shrubs having terminal spikes of small stalkless flowers, often drooping, the flowers subtended by 4 or more persistent bracts. The fruit is succulent, separating into parts as it ripens.

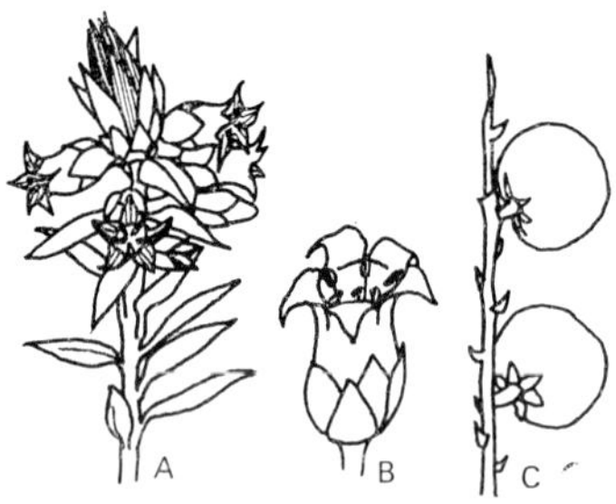


FIG. 217 *Trochocarpa*: A, flowerhead (× 2); B, flower (× 4); C, fruit (actual size).

*T. clarkei.* Greenish; fruit bluish-purple. Alpine, subalpine. Vic.
A small shrub with leaves 6–8 mm long, broad, 3–5 nerved. Flowers bell-shaped with purple lobes and a few tufts of hairs in the tube, in short spikes of a few flowers. Fruit about 6 mm in diameter.

*T. cunninghamii.* Pink, white; fruit bluish-black. Montane. Tas.
A fairly common straggling shrub up to 1 metre high, the young shoots hairy and often brown. Leaves 7–10 mm long, scattered, with twisted stems so that they spread at right angles in opposite rows, egg-shaped, pointed, the young ones often bright-red. Flowers bell-shaped, 3–4 mm long with narrow lobes, crowded in cylindrical spikes about 1.5–2 cm long. Fruit spherical, 10–12 mm in diameter.

*T. thymifolia.* Pink, red or bluish-purple. Subalpine, montane. Tas.
An erect shrub 30–90 cm high, widespread, often clambering over the sheltered side of boulders. Branches woolly. Leaves 2–4 mm long, stalked, alternate, egg-shaped or elliptical, thick, bluntly pointed. Flowers crowded in terminal, recurved, cylindrical spikes about 1.5 cm long; flower tube about 2 mm long, scarcely longer than the sepals, the lobes nearly as long as the tube, bearded on the upper side. Fruit spherical, about 8 mm in diameter.

## Family LOGANIACEAE

*MITRASACME* (from Lat. *mitra*, a mitre, and Gk *akme*, summit, the flower resembling an ecclesiastical mitre). Fig. 218. A genus of

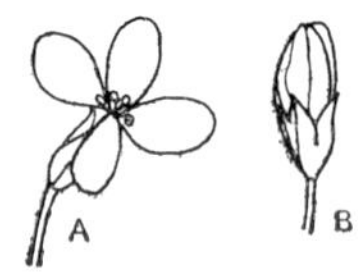


FIG. 218 *Mitrasacme*: A, flower; B, fruit. (Both × 2.)

small, soft plants with opposite leaves. The flowers are regular, usually with 4 sepals forming a tube, 4 joined petals also tubular with spreading lobes and 4 stamens within this tube. The seedbox is inserted above the other parts of the flower with a single style and a bifid stigma at maturity. The fruit is a dry, splitting capsule.

*M. archeri.* White. Alpine, subalpine. Tas.
A small, smooth, tufted plant 2–2.5 cm high. Leaves tough, shining, oblong, blunt, crowded, with transparent margins, 2–3 mm long. Flowers almost stalkless in the axils of the terminal leaves; lobes of the sepals leaf-like; corolla tube broad, 2–3 mm long, the lobes about the same length.

*M. montana.* White. Alpine, subalpine. Vic., Tas.
A dense, slender, prostrate plant. Leaves egg-shaped, thick, blunt, 2–6 mm long; sepals divided to the middle, 2–3 mm long; flower tube short, the lobes spreading, 3–5 mm long.

*M. pilosa.* White. Montane. N.S.W., Vic., Tas.
A rather hairy, prostrate plant. Leaves fairly broad, 4–6 mm long. Flowers almost stalkless; petals slightly longer than the sepals, contracted and bearded in the throat, lobes very short and broad.

*M. serpyllifolia.* White. Subalpine, montane. N.S.W., Vic., Tas.
A slender plant. Leaves egg-shaped to narrow-oblong, blunt, 2–4 mm long. Sepals about 2 mm long, divided to about the middle. Flower tube very short, lobes spreading about 3–4 mm long.

## Family GENTIANACEAE

A small family of smooth herbs with opposite leaves and regular flowers. There are usually 5 sepals forming a very short tube, and 5 petals with 5 stamens inserted in the flower tube. The seedbox is inserted above the other flower parts and the fruits are dry, splitting capsules.

*CENTAURIUM* (possibly derived from one of the Centaurs, Chiron, the legendary father of medicine). Fig. 219.

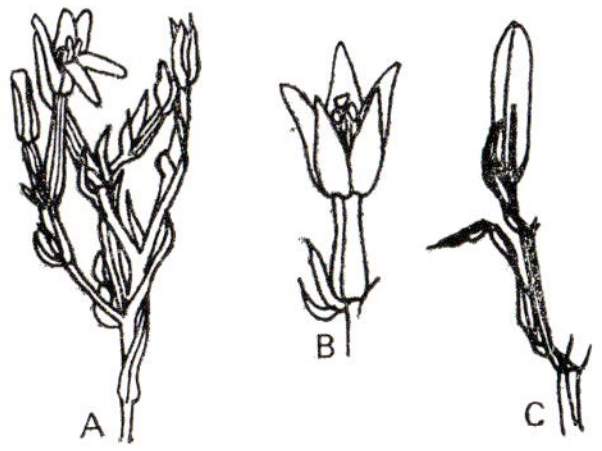


FIG. 219 *Centaurium*: A, flowerhead (actual size); B, flower (× 1½); C, fruit (× 1½).

*C. australe*. Austral Centaury. Red, pink, white. Alpine, subalpine, montane. N.S.W., Vic., Tas.

A widely distributed plant 7–40 cm high, slender, with square stems and broad, stalkless, leaves 1–2 cm long. Flowers in dense heads; sepals deeply divided; petals 4–5, spreading from a narrow tube; stamens 4–5, spirally twisted after flowering. Fruit narrow.

*GENTIANELLA* (Greco-Lat. diminutive from name of yellow gentian, the medicinal properties of which are alleged to have been discovered by Gentius, King of Illyria). Fig. 220.

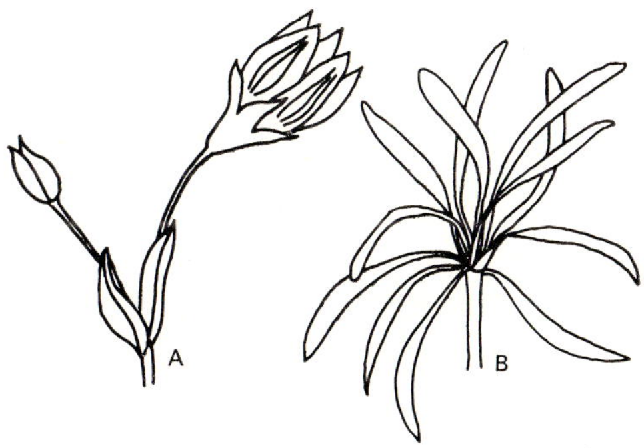


FIG. 220 *Gentianella*: A, fruit; B, basal leaves. (Both actual size.)

*G. diemensis*. White with blue veins. Alpine, subalpine, montane. N.S.W., Vic., Tas.

A dainty little tufted plant closely related to the renowned blue gentians of northern alpine parts, often seen in considerable patches, particularly in the alpine herbfields. A smooth plant with slender branches, about 3 cm in height but in selected spots as high as 60 cm. Leaves soft, oblong or egg-shaped, the upper leaves spoon-shaped, without stalks except for a few of the lower ones, 2–4 cm long. Flowers in loose heads on thin stalks, the tube very short, the lobes long and cupped; sepals 3–6 mm long, rather narrow; petals transparent, 1–3 cm long; sometimes yellowish at the base.

*SEBAEA* (after Albert Seba, 1665-1736, Dutch traveller and collector of natural history specimens). *S. ovata* is an annual herb about 50 cm high with broad leaves and yellow flowers in loose clusters. The fruit is long, enclosed in the calyx. It occurs in montane habitats in New South Wales, Victoria and Tasmania.

## Family GESNERIACEAE

*FIELDIA* (after Barron Field, 19th-century British botanist). Fig. 221.

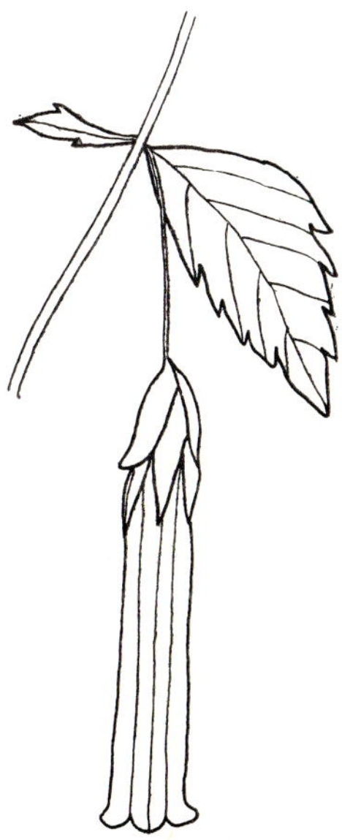

FIG. 221 *Fieldia*: leaf and flower (actual size).

*F. australis*. Greenish-yellow. Montane. N.S.W. A scrambling plant with coarsely toothed leaves in opposite, unequal pairs, the longer leaves shortly stalked, 2.5–7.5 cm long, the smaller ones stalkless, 6–25 mm long. Flowers about 2.5 cm long, irregular, tubular, slightly 2–lipped with 5 sepals, 5 petals and 4 stamens; seedbox inserted above the other flower parts. Fruit a whitish berry.

**Family MENYANTHACEAE**

*NYMPHOIDES* (from Gk Nymphe, goddess of springs (water-nymph), and *oides*, like). *N. geminatum* is a delightful marsh-lover of New South Wales montane waterways. The leaves or blades are broadly kidney-shaped, smooth and floating. The yellow flowers, raised on long stalks above the level of the water, are usually solitary, 5–lobed, with broad, membranous margins. The fruit is dry but not splitting.

**Family BORAGINACEAE**

Small plants with regular flowers of 5 sepals, 5 petals and 5 stamens. The flower is tubular and the fruit dry. The ovary is inserted above the other flower parts.

*CYNOGLOSSUM* (from Gk *kyon*, a dog, and *glossa*, a tongue, referring to the shape of the leaves). A genus differing from *Myosotis* in having rough, barbed fruits and scales in the throat of the flower tube.

*C. australe*. Blue. Montane. N.S.W., Vic., Tas. A rough, hairy plant 15–45 cm high with fairly broad leaves about 10 cm long, and fragrant flowers in small clusters, the tube about 4 mm long. The stalks of the individual flowers are shorter than the calyx.

*C. suaveolens*, with the same distribution, is a rather shorter plant. The stalks of the flower-heads have leafy bracts (none in *C. australe*), and the stalks of the individual flowers are longer than the calyx.

*MYOSOTIS* (from Gk *mus*, a mouse, and *otes*, an ear, the leaves resembling the ears of a mouse). Fig. 222. A genus of soft shrubs with simple, alternate leaves and small flowers with the fruit in the form of 4 small, dry, shining nuts. There are 2 species common in these parts.

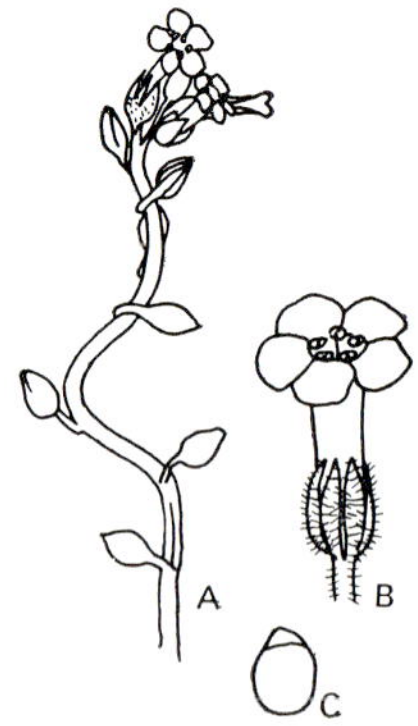


FIG. 222 *Myosotis*: A, flowerhead (actual size); B, flower (× 2); C, fruit (× 2).

*M. australis*. Pale yellow to white. Alpine, subalpine, montane. N.S.W., Vic., Tas.
An often semi-prostrate plant with hairy, broad leaves, the upper ones narrow, 2.5–5 cm long; petals in one-sided heads, 10–15 mm in length.

*M. suaveolens*. White or pale blue. Montane. N.S.W., Vic., Tas.
This species is about 30–60 cm high, with fairly broad leaves 2.5–10 cm long. Flowers sweetly scented, in loose heads, the lobes of the petals about 4 mm long.

**Family CONVOLVULACEAE**

A small family of climbing or creeping herbs with alternate leaves and regular flowers. The flower parts are in fives, the sepals persistent, the petals folded in bud, and the stamens inserted in the tubular flower. The seedbox is inserted above the other parts of the flower. The fruit is usually dry and splitting.

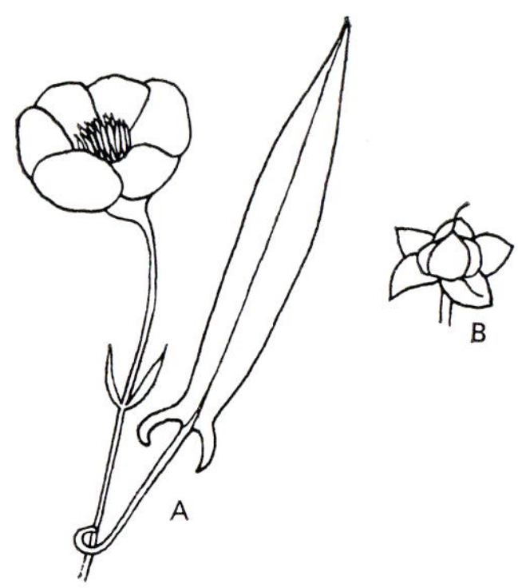


FIG. 223 *Convolvulus*: A, flower; B, fruit. (Both actual size.)

*CONVOLVULUS* (Lat. name of a bindweed, from *convolvere*, to entwine). Fig. 223.

*C. erubescens.* Pink. Montane. N.S.W., Vic., Tas.

A twining plant with bell-shaped flowers and very variable leaves, sometimes broad and heart-shaped, sometimes narrow, wavy or deeply incised. Flowers usually solitary on slender stalks, 10–20 mm in diameter. Fruit usually 2–celled.

*DICHONDRA* (from Gk *di*, two, and *khondros*, a grain, referring to the two fruitlets). Fig. 224.

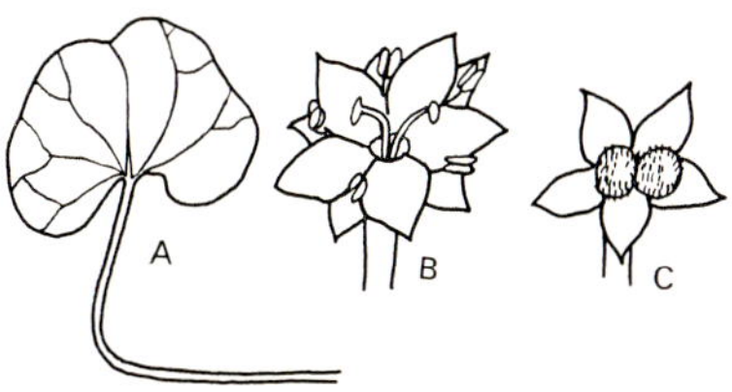


FIG. 224 *Dichondra*: A, leaf (actual size); B, flower (× 2); C, fruit (× 2).

*D. repens.* White. Subalpine, montane. N.S.W., Vic., Tas.

An attractive little creeping plant. Leaves alternate, kidney-shaped, notched at the apex, silky-hairy, 15–30 mm long on long stalks. Flowers tiny, tubular with spreading lobes, 2–3 mm long, often hidden by the leaves. Fruits, 2 separate little fruitlets each with one seed.

**Family SCROPHULARIACEAE**

The Snapdragon family is better represented in northern countries than it is in Australia, but in our alpine areas there are 2 widely distributed genera and a number of small ones. Members of the family are soft herbs with opposite leaves and lipped flowers of 4–5 sepals and petals in loose heads. There are 4 or rarely 5 stamens. The seedbox is inserted above the other parts of the flower. The fruit is a dry splitting capsule with 2 compartments.

*EUPHRASIA* (from Gk, meaning joy—one species was formerly believed to be a remedy for eye troubles). Fig. 225. These are the eyebrights, gems of alpine areas, growing in small, massed clumps and making the herbfields alive with delicate colour in the summer. The flowers are stalkless in terminal spikes, tubular and 2–lipped with 4 sepals, 5 petals and 4 protruding stamens. The seedbox is inserted above the other parts of the flower. The fruit is long and flattened. There are several species, with lilac, purple or white flowers. The genus still requires some revision.

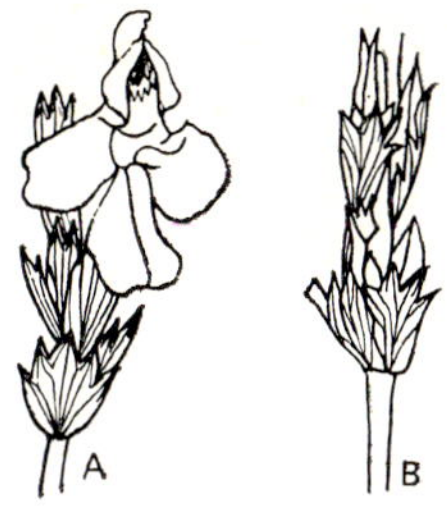


FIG. 225 *Euphrasia*: A, flowerhead; B, fruits. (Both × 2.)

*E. collina.* Lilac, pale purple, white. Montane. N.S.W., Vic., Tas.

A small shrub with hard stems, much branched. Flowering stems 10–50 cm high, slender, tough. Leaves crowded on vegetative branches and the lower part of the flowering stem, more distant above, stalkless, oblong or wedge-shaped with 1–3 pairs of blunt teeth in the upper part, the central lobe larger, thick, smooth, 4–5 mm long, sometimes longer on flowering stems. Flowers in loose terminal heads variable in size; lower lip larger than upper, lobes oblong; anthers reddish-brown. Capsule narrow-oblong, 8–9 mm long.

*E. diemenica.* Lilac, white. Alpine, subalpine. Tas.

A small shrub with a hard base, much branched, with many erect flowering stems 15–30 cm long, often purplish, somewhat fleshy. Leaves stalkless, overlapping in 4 ranks on vegetative branches but distant on the upper parts of flowering stems, 5–12 mm long, oblong or egg-shaped, bluntly 3–lobed at the top, thick, smooth, shining. Flowers showy, in terminal heads, lightly hairy, not striated, the lower lip larger than the upper, the lobes broad and rounded; anthers red-brown. Capsule egg-shaped, 8–9 mm long.

*E. glacialis.* Lilac, white. Alpine, subalpine. N.S.W., Vic.

A small, stout shrub 6–15 cm high forming compact stands on open moors and in the shelter of boulders, very floriferous and spectacular in full flower. Flowering stems slender. Leaves opposite, more crowded towards the base of the branches, without stalks, wedge-shaped, with 3 (sometimes fewer) pairs of blunt-pointed teeth and a rounded apical lobe, hairy on the edges and with glandular hairs on the surface. Flowers in loose terminal heads; calyx hairy; lower lip of corolla larger than upper, deeply and fairly evenly bluntly 3–lobed, 5–8 mm long; anthers brown.

*E. glacialis* var. *eglandulosa* occurs also at high elevations in New South Wales and Victoria. It differs in the complete absence of glandular hairs although it is sometimes densely hairy on the upper parts of the stems, the edges of the leaves and the sepals.

*E. gibbsiae.* White. Subalpine. Vic., Tas.

A small shrub with short, erect branches 15–40 cm long, often growing among cushion bushes and sphagnum in Tasmania. Leaves spreading, stalkless, fairly evenly placed on the stems, those on the flowering stem up to 20 cm long, the others shorter, wedge-shaped, 3–7 lobed, thick soft and densely woolly. Flowers in short terminal heads, showy, hairy, lower lip longer than the upper, with 3 broadly egg-shaped rounded lobes; anthers dark purple-brown. Capsule egg-shaped, about 11 mm long.

One Victorian form, *comberi*, occurring at 5,600 feet, has a markedly glandular covering, egg-shaped to fan-shaped leaves, up to 10 mm long with 5–7 finger-like lobes in the upper part, and large, purplish veined flowers.

*E. hookeri.* White, purple stripes. Alpine, subalpine. Tas.

An annual with erect flowering stems, sparingly branched, 2–10 cm high. Leaves crowded, stalkless, broadly wedge-shaped, 5–9 mm long with 4–8 narrow lobes from about the middle, pointed, smooth or slightly hairy. Flowers in dense terminal heads, the lower lip longer than the upper, rounded. Capsule 6–8 mm long.

*E. scabra.* Creamy yellow. Montane. N.S.W., Vic., Tas.

A stiff, erect annual, sparingly branched, 12–40 cm high, roughly hairy. Leaves stalkless, fairly broad, with 1–3 pairs of narrow, spreading lobes, the whole leaf 8–15 mm long. Flowers in terminal spikes, dense at first, then elongated; lower lip scarcely longer than the upper, with short, broad lobes; anthers brown. Capsule woolly, 5–8 mm long.

*E. scabra* var. *alsa* occurs at high elevations on the mainland. It is a very small form, with conspicuously veined flowers.

*E. scabra* var. *caudata* is a New South Wales subalpine form growing up to 30 cm and with glandular hairs. Upper leaves and floral bracts with long tail-like lobes, about 10 mm long. Flowerheads dense, purple.

*GRATIOLA* (from *gratia Dei*, grace of God, the medieval name for one species used for medicinal purposes). Fig. 226.

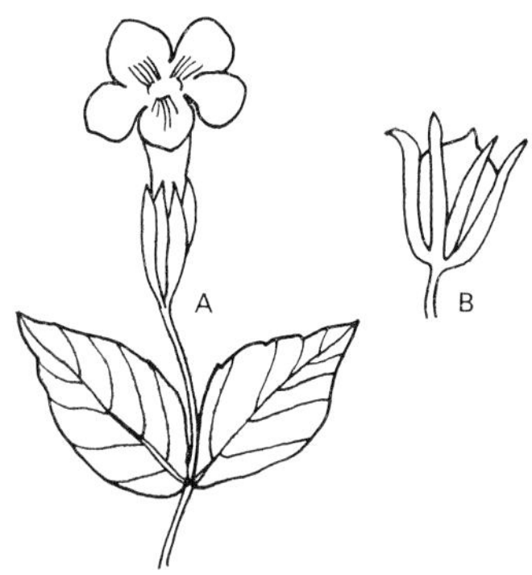


FIG. 226 *Gratiola*: A, flower; B, fruit. (Both × 2.)

*G. nana.* White, pink. Subalpine, montane. N.S.W., Vic., Tas.

A dainty little creeping plant of wet patches with oblong leaves narrowed at the base, toothed, about 6–8 mm long. Flowers solitary, 8–12 mm long, shortly stalked, irregular with 5 sepals, the tube of the flower with 2 petals in the upper lip and 3 in the lower; stamens 2. Fruit egg-shaped, with 4 valves.

*LIMOSELLA* (dim. of Lat. *limosus*, muddy, in reference to the habitat). *L. australis* is a montane water-lover from New South Wales, Victoria and Tasmania. It is a little tufted plant 15–40 mm long, with purple, pink or white flowers 2–4 mm long on short stems.

*PARAHEBE* (resembling *Hebe*, a related plant). A genus closely related to the true Veronicas with species occurring from alpine to montane habitats.

*P. perfoliata.* Deep blue, streaked purple. Alpine, subalpine, montane. N.S.W., Vic.

A widely distributed smooth species 50 cm–1 metre high, somewhat straggling. Leaves very handsome, stem-clasping, grey-green, pointed, rather broad, 2–6 cm long. Flowers on very long, slender spikes at the ends of the branches; flowers 8–10 mm long; sepals 5; petals 5.

*P. derwentiana.* Pale lilac, pink, white. Alpine, subalpine, montane. N.S.W., Vic., Tas.

Perhaps the best known of the species, this is a robust plant ranging from the coast to alpine habitats, about 50 cm–1 metre in height, erect and compact. Leaves broad, pointed, toothed, very handsome, 4–8 cm long. Flowers in dense heads produced abundantly at the ends of the branches; sepals 4; petals 4.

*P. formosa.* Pale lilac, blue. Montane. Tas.

A shrub 50 cm–2 metres high. Leaves numerous, opposite, usually stalkless with the bases fused across the stem. Flowers in short, loose heads; petals 4–5 mm long.

*VERONICA* (origin doubtful, possibly in honour of St Veronica, given by Italian botanist Mattioli to some species). Fig. 227. A genus represented by several species in alpine areas,

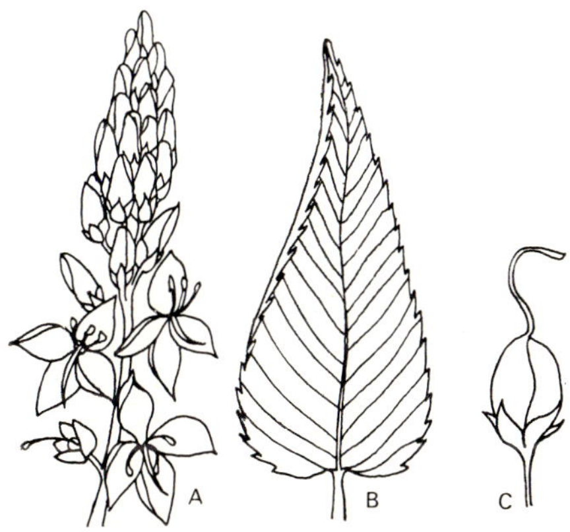


FIG. 227 *Veronica*: A, flowerhead; B, leaf; C, fruit. (All actual size.)

ranging in size from tiny rock-clinging and water-loving forms to erect, bushy shrubs. The sepals are divided into 5 segments, the flower tube is very short and there are 2 protruding stamens. The fruit is furrowed on both sides.

*V. densifolia.* Pink, purple. Alpine. N.S.W.

A dwarf, tufted plant, often rock-clasping. Leaves very small, closely set in 4 rows, smooth, thick, egg-shaped, 2–3 mm long. Flowers solitary, open, with 5 rather rounded petals up to 8 mm in diameter.

*V. nivea.* Pale blue, white streaked with purple. Alpine, subalpine. N.S.W., Vic., Tas.

A slender to robust plant 15–25 cm high with finely divided leaves with narrow leaflets, the whole leaf 0.8–3 cm long. Flowers in long, loose heads at the ends of the branches; sepals 4–5; petals 4–5.

A number of smaller Veronicas, all from New South Wales, Victoria and Tasmania, are also common in these parts: *V. calycina*, a small hairy plant with broad, stalked, coarsely toothed leaves and small pale-blue flowers in short, loose heads, is a montane dweller. *V. gracilis,* from subalpine as well as montane habitats, is a dainty, water-loving, creeping plant 5–15 cm high, with small, open, blue flowers streaked with dark purple, in loose heads of 2–8 flowers, and shortly stalked, narrow leaves. *V. notabilis*, a montane dweller, grows to 30 cm or more, though it is sometimes prostrate. It has stalked, narrow, toothed leaves 2.5–7.5 cm long and pale-blue flowers streaked with purple or black in heads 7–25 cm long. *V. plebeia*, a creeping plant with broad leaves 12–15 mm long, on long stalks with rounded teeth, and violet flowers 6–8 mm long in loose heads of 4–8, is also a montane dweller. *V. serpyllifolia*, reaching alpine habitats, is a shortly creeping plant with rounded leaves 2–15 mm long, and pale blue and white flowers streaked with dark purple, on stalks about 3 mm long.

A common though inconspicuous member of the family from alpine and subalpine Tasmania is *Ourisia integrifolia*, a flat, creeping plant. It has thick, egg-shaped, stalked leaves 6–12 mm long and solitary flowers about 12 mm long, white streaked with purple, with 5 almost equal petals.

## Family LENTIBULARIACEAE

*UTRICULARIA* (from Lat. *utriculus*, a small bladder, referring to the little sacs attached to the submerged leaves). Fig. 228. A genus of slender little water-dwellers which have subterranean bladders formed from the leaves for catching small animals, such as insects, to

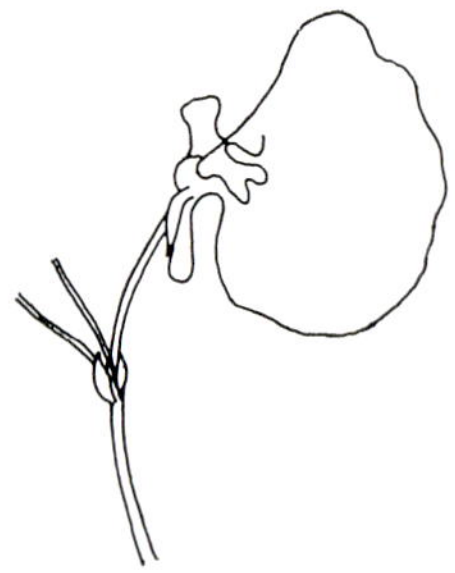

FIG. 228 *Utricularia*: flower (× 1½).

supplement their nitrogen supply. The leaves are submerged or very small at the base of the slender flower stem, usually short, sometimes with a single flower, sometimes with 2–3 flowers. The flowers are irregular, with a short, spurred, 2–lipped flower, the upper lip of which is broad and the lower lip 2–3 lobed, much longer and spreading. There are 2 stamens. The seedbox is inserted above the other flower parts. The fruit is globular, dry and splitting.

*U. dichotoma*. Fairies' Aprons. Purple, blotched yellow. Alpine, subalpine, montane. N.S.W., Vic., Tas.

A widely distributed plant inhabiting marshy lands. Leaves few, spoon-shaped sometimes reaching 20–25 mm in diameter. Flowers delicate, 1–4 on slender stems, 7.5–15 cm in height; lower lip semicircular, 12–20 cm in diameter.

*U. monanthos*. Purple. Subalpine, montane. Tas.

A slender plant about 2.5 cm high. Leaves tiny, narrow, spoon-shaped. Flowers 7–8 mm long on very slender stems, the upper lip rather narrow, the lower lip semicircular with a long spur.

### Family VERBENACEAE

*CHLOANTHES* (Gk, budding or sprouting). Fig. 229.

*C. parviflora*. Bluish with dark spots. Montane. N.S.W.

A bushy undershrub, rather rare, with opposite, rather narrow, rough leaves, often woolly

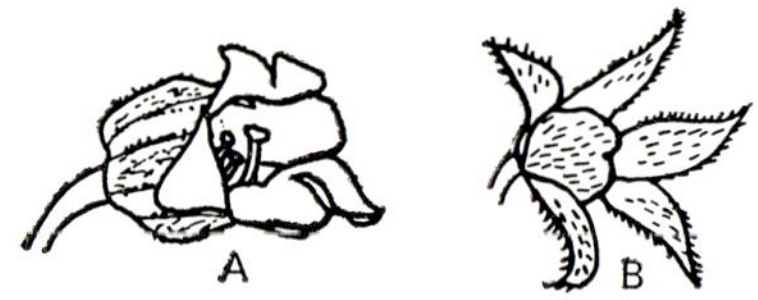


FIG. 229 *Chloanthes*: A, flower; B, fruit. (Both × 2.)

beneath. Flowers solitary in leaf axils, 2–lipped, with 2 lobes in the upper and 3 in the lower lip and about 12 mm long; stamens 4; seedbox inserted above the other parts of the flower. Fruit dry, finally separating into 4 parts.

### Family LABIATAE

This family is readily recognized by the aromatic odour in the leaves, the square stems and opposite leaves. The sepals are persistent and the 2–lipped flowers usually have 4–5 petals. There are 2–4 stamens inserted in the flower tube. The seedbox is inserted above the other flower parts and has 4 lobes ripening into 4 small, single-seeded, non-splitting nuts.

*AJUGA* (probably a corruption of Lat. *abiga*, a plant used in medicine). Fig. 230.

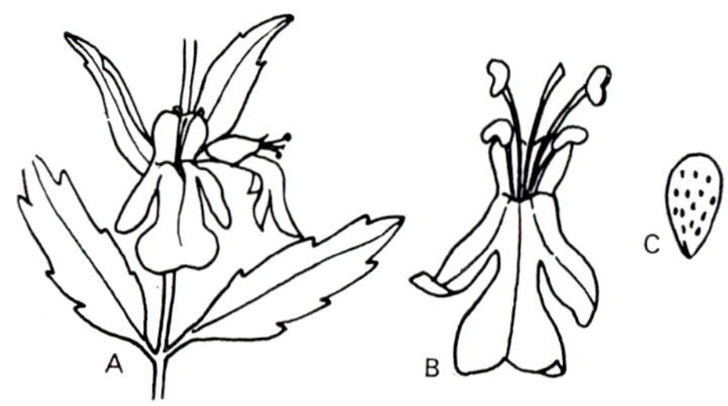


FIG. 230 *Ajuga*: A, flowerhead (actual size); B, flower (× 1½); C, fruit (× 2).

*A. australis*. Australian Bugle. Deep purply blue. Montane. N.S.W., Vic., Tas.

This little plant is found in almost every habitat throughout Australia and is not uncommon in montane areas. It is a variable small plant 4–30 cm high, softly hairy, with numerous branches from the base. Leaves oblong to narrow, mostly on long stalks, the lower ones sometimes 15 cm long, the upper ones not exceeding 2.5 cm. Flowers in loose clusters; sepals 5, very hairy,

5–8 mm long; petals variable, up to 20 cm long, the lower lip long and spreading, the upper lip not conspicuous; stamens 4, in pairs.

*MENTHA* (Lat. name of mint; also Gk name of a nymph). Fig. 231. A small genus of small plants with 5 sepals and 4 nearly equal petals, the upper one often partly divided. There are 4 equal stamens and a bifid style. The nuts are smooth.

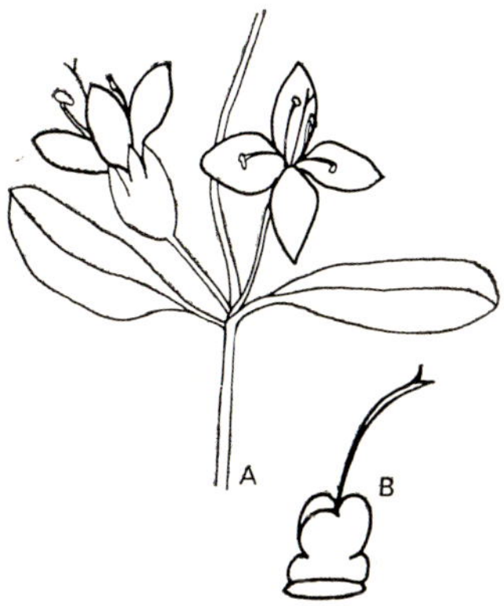


FIG. 231 *Mentha*: A, flowerhead; B, fruit. (Both actual size.)

*M. laxiflora.* Lilac. Montane. N.S.W., Vic., Tas.
Common in damp patches, a coarsely hairy plant 15–30 cm high, often smaller. Leaves fairly broad, the larger ones narrowed into a stalk, the smaller ones stalkless, 6–25 cm long. Flowers numerous, clustered in the leaf axils, about 4–5 mm long, with projecting stamens.

*M. diemenica.* Lilac. Montane. N.S.W., Vic., Tas.
A slender plant with flowering stems 3–30 cm long. Leaves shortly stalked or stalkless, fairly broad, 3–12 mm long. Flowers small.

*M. serpyllifolia.* Lilac. Montane. N.S.W., Vic., Tas.
A small, slender plant, smooth or finely hairy, 10–15 cm high, common in marshy places. Leaves oblong, mostly short-stalked, 4–6 mm long. Flowers few in clusters in the leaf axils, with narrow sepals and short petals.

*PROSTANTHERA* (from Gk *prostheke*, an appendage, and *anthera*, an anther—the connectives of the anthers are spurred). Fig. 232. A genus of mintbushes that may be recognized by the 2–lipped calyx which grows over and encloses the fruit. The upper flower lip of 2 petals is usually shorter than the lower lip of 3 petals. There are 4 stamens and the style is bifid.

FIG. 232 *Prostanthera*: A, flowerhead; B, fruit. (Both × 1½).

*P. cuneata.* White or lilac, blotched yellow. Alpine, subalpine. N.S.W., Vic., Tas.
A lovely little plant occurring also at lower elevations. In higher parts it is prostrate or semi-prostrate, often rock-climbing; at lower elevations it may reach a compact height of 30–40 cm. Leaves soft, light-green, small, rounded, often clustered, 2–8 mm long. Flowers usually solitary in leaf axils, but often so close together as to give the appearance of a head, each flower on a short, slender stalk with narrow bracts behind the sepals; calyx tube conspicuously striped; lips of the petals broad and about as long as the tube, the whole flower about 1 cm long.

*P. decussata.* White, pink or violet with yellow streaks. Subalpine. Vic.
A dense, slightly hairy shrub 30–40 cm high with almost stalkless, crowded, narrow leaves in opposite pairs, about 6 mm long. Flowers shortly stalked, in leaf axils; sepals strongly ribbed, with equal lips.

*P. hirtula.* White or violet. Subalpine. N.S.W., Vic.
A hairy shrub about 80 cm–1 metre high with shortly stalked, fairly broad, blunt, entire leaves,

hairy above, whitish below, 12–25 mm long. Flowers in leafy heads on short stalks; sepals hairy, 5–6 mm long, the lips broad and nearly equal; petals twice as long as the sepals.

*P. incisa* var. *sieberi*. Lilac. Montane. N.S.W., Vic.
A slender, much-branched, slightly hairy shrub 1–2 metres high with small, rounded, smooth, deeply indented, strongly aromatic leaves. Flowers in short heads.

*P. lasianthos*. Pale lilac, white or pink. Subalpine, montane. N.S.W., Vic., Tas.
A fairly tall plant 2–6 metres high, usually growing close to watercourses. Leaves stalked, broad, flat, toothed, pointed, smooth but paler underneath. Flowers fairly large, softly hairy, opposite, in short heads.

*P. phylicifolia*. White or lilac with deep-purple spots. Subalpine, montane. N.S.W., Vic.
Occurring also on the tablelands, a rather dense shrub growing to a height of 60–90 cm with almost stalkless, narrow, light-green leaves, 5–15 mm long. Flowers solitary in leaf axils with 2 narrow bracts behind; sepals striped, about 5–6 mm long; petals forming a broad cup about 8–12 mm long.

*P. rotundifolia*. Purple. Montane. N.S.W., Vic., Tas.
A widely distributed, very variable, tall bushy shrub, 2–6 metres high. Leaves round, small, smooth, soft, wavy, tapering, 6–12 mm long, stalked, very aromatic. Flowers in leafy clusters in the leaf axils, with broad, nearly equal lips, hairy inside.

*P. walteri*. Greenish, streaked violet. Subalpine, montane. N.S.W., Vic.
A robust plant, sometimes straggling, 30–40 cm high though often shorter. Leaves rhomboid, diamond or egg-shaped, smooth, 1–3 cm long. Flowers solitary on slender stalks, the upper lip much longer than the lower and the whole flower much longer and narrower than in most species.

*PRUNELLA* (*prunella* or *brunella*, old Lat. name of a plant, probably from German *braume*, quinsy, which the plant was supposed to cure). Fig. 233.

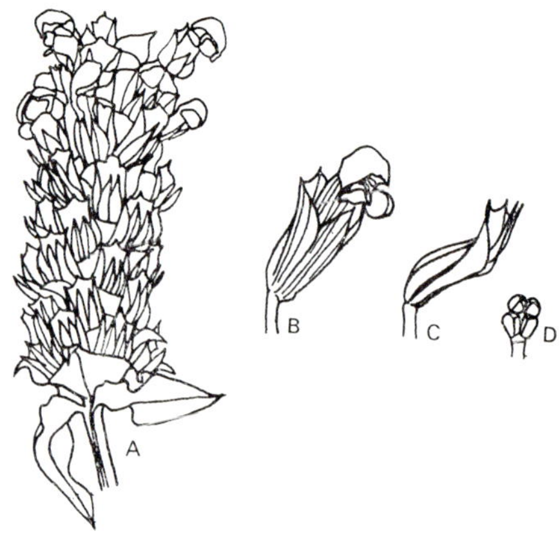


FIG. 233 *Prunella*: A, flowerhead; B, flower; C, enclosed fruit; D, exposed fruit. (All × 2 except A, actual size.)

*P. vulgaris*. Lilac or purple. Montane. N.S.W., Vic., Tas.
A soft, rather straggling, slightly hairy plant 7–30 cm high. Leaves egg-shaped, stalked, entire or with a few teeth. Flowers small in almost cylindrical short spikes with broad, conspicuous bracts at the base of the flowers; upper lip of calyx broad, shortly 3-toothed, lower lip with 2 lobes; upper lip of flower erect, lower lip spreading and 3-lobed.

*SCUTELLARIA* (from Lat. *scutella*, a dish, referring to the form of the calyx). Fig. 234.

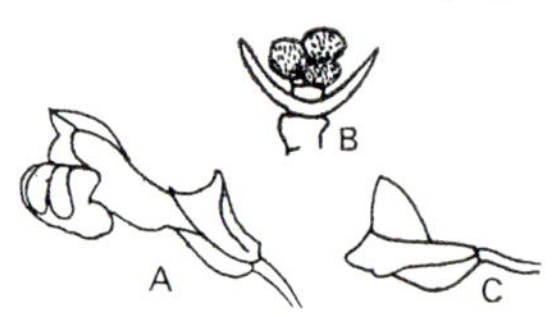


FIG. 234 *Scutellaria*: A, open flower (× 2); B, fruit (× 4); C, bud (× 2).

*S. humilis*. Purple. Montane. N.S.W., Vic., Tas.
A small, smooth, semi-prostrate plant, this has egg-shaped or heart-shaped leaves, stalked, wavy or toothed, 12–15 mm long. Flowers solitary in the leaf axils, shortly stalked; upper lip of calyx with a hollow protuberance on its back; upper lip of flower concave, lower lip spreading with 3 lobes.

*WESTRINGIA* (after John Peter Westring, Swedish writer on lichens). Fig. 235. A closely related genus to *Prostanthera*, but in *Westringia* there are 5 equal sepals (in *Prostanthera* the sepals form 2 lips). The genus comprises bushy shrubs with whorled, usually stiff leaves. The

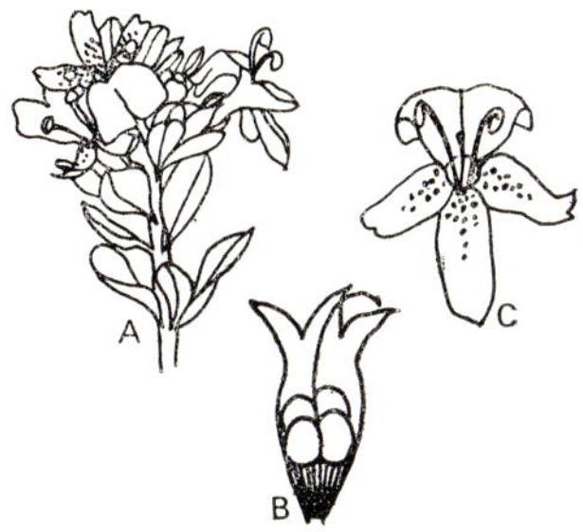


FIG. 235 *Westringia*: A, flowerhead (actual size); B, fruit (× 2); C, flower (× 2).

flower tube is short, dilated upwards, the upper lip flat an broadly 2-lobed, the lower lip spreading and 3-lobed. There are 4 stamens, 2 longer upper ones and 2 shorter lower ones.

*W. angustifolia*. Pale lilac, white. Subalpine, montane. Tas.

A slender, stiff shrub 1–3 metres high with spreading branches. Leaves in whorls of 3, shortly stalked, 1–2 mm long, narrow, the upper surface rough, the lower surface woolly. Flowers solitary in the leaf axils, abundant, about 6–8 mm long.

*W. eremicola* var. *quaterna*. White. Montane. N.S.W.

A slender, variable, slightly hairy shrub. Leaves narrow with rolled margins, pointed, not more than 12 mm long.

*W. lucida*. White, spotted purple. Alpine, subalpine. N.S.W.

A stiff, spreading shrub 1–2 metres high common on rocky outcrops. Leaves in whorls of 4, broadly egg-shaped, shining and rough, very shortly stalked, 5–8 mm long. Flowers solitary in leaf axils, 5–8 mm long; calyx lobes sharply pointed.

*W. rigida*. White. Montane. N.S.W., Vic.

A stiff, rather spreading plant about 80–100 cm high. Leaves in whorls of 3–4, narrow and pointed, stiff, the upper surface usually with tubercles and the undersurface covered with white hairs, the whole leaf 5–15 mm long. Flowers stalkless, solitary in the leaf axils, often very numerous, dotting the plant with white stars; sepals 4–6 mm long, ribbed; petals densely hairy, the lower lip only as long as the upper, about 4–5 mm long, dotted with purple. This species also occurs in Tasmania in coastal situations.

*W. rubiifolia*. White, spotted purple. Subalpine, montane. Tas.

A compact shrub 15–30 cm high with squarish, rough branches. Leaves in opposite pairs, stalkless, rough but not hairy, bluntly egg-shaped, 6–8 mm long. Flowers solitary in the upper leaf axils, close together; sepals rough, 2–3 mm long; petals hairy, 5–6 mm long, the lips nearly equal.

*W. seniifolia*. White, spotted purple. Subalpine, montane. N.S.W., Vic.

Occurring fairly commonly at Mount Buffalo and Mount Hotham in Victoria, not common in New South Wales. A compact shrub 30–60 cm high covered in short, whitish hairs. Leaves stiff, in whorls of 5–6, about 1.5 mm long. Tube of the calyx about 3 mm long; petals slightly hairy, not much longer than the sepals.

## Family PLANTAGINACEAE

Although the flowers are insignificant the genus *Plantago* (Lat. name for some European species) is so widely spread in alpine parts that it cannot be altogether ignored. There are, in addition to the indigenous species, a number of introduced species. They are small plants with parallel veins, basal leaves and tiny stalkless flowers in straight spikes with conspicuous protruding anthers. The flower parts are in fours, the petals are dry and papery. The seedbox is placed above the other flower parts. The fruit is membranous. See Fig. 236.

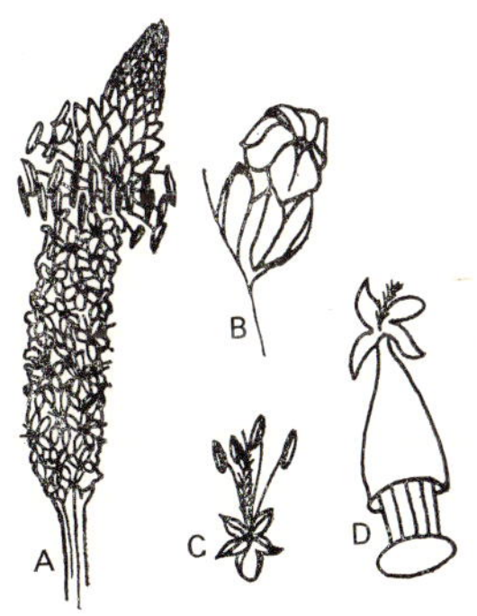


FIG. 236 *Plantago*: A, flowerhead (actual size); B, bud (× 3); C, old flower (× 2); D, fruit (× 3).

Five species, *Plantago antarctica, P. muelleri, P. daltonii, P. gunnii* and *P. tasmanica*, occur at alpine levels and usually not at lower altitudes. *P. daltonii* occurs in Tasmania only, and *P. gunnii* and *P. tasmanica* in Victoria and Tasmania, sometimes descending to lower levels. *P. gunnii* is often associated with the cushion bush *Abrotanella* and others. It is a very distinct and handsome little plant of wet habitats with soft, hairy leaves in a flat, basal rosette, making an attractive pattern on the ground. *P. varia* occurs at subalpine altitudes in New South Wales, Victoria and Tasmania, and *P. glabrata* and *P. paradoxa* occur in montane grasslands in Tasmania.

## Family POLYGONACEAE

Small plants with alternate leaves with sheathing stipules embracing the stem. The flowers are small, usually with 5–6 sepals in 2 rows, no petals and as many stamens as the sepals or an indefinite number. The seedbox is attached above the other flower parts. The fruit is a small dry, non-splitting nut, usually enclosed in the persistent sepals.

*MUEHLENBECKIA* (after Gustav Muehlenbeck, 1798-1845, Alsatian botanist). Fig. 237.

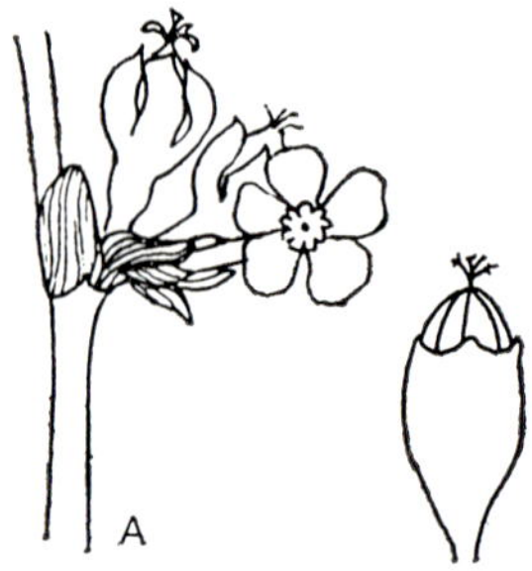


FIG. 237 *Muehlenbeckia*: A, flowerhead; B, fruit. (Both × 3.)

*M. axillaris.* Matted Lignum. Cream or yellow. Alpine, subalpine, montane. N.S.W., Vic., Tas.

A small, smooth, prostrate, matted plant found in other habitats also. Leaves small, rounded, dark-green. Flowers 1–3 together, shortly stalked; sepals 5; stamens 8; styles 3, each with a fringed stigma. Fruit enclosed in the persistent bracts.

*POLYGONUM* (from Gk *polys*, many, and *gony*, a knee, referring to the stem joints). Fig. 238. A genus of small plants with 2 montane species on the mainland and in Tasmania, with very small red or green flowers with white edges. There are 5 sepals and 5–8 stamens. *P. hydropiper* is a slender, knotted, drooping plant of wet habitats with flowers in slender spikes and sepals with prominent glands.

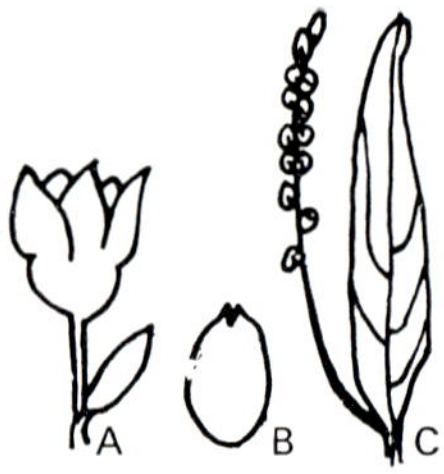


FIG. 238 *Polygonum*: A, flower (× 5); B, fruit (× 5); C, leaf and buds (actual size).

*P. prostratum* is a hairy trailing plant with fairly broad, shortly stalked leaves and flowers in dense spikes in the leaf axils. The nuts have 2 convex sides.

*RUMEX* (old Lat. name for a kind of sorrel, from Lat. *rumo*, to suck, from the habit of Romans sucking sorrel leaves to allay thirst). Fig. 239. A genus that includes a number of introduced species which have caused considerable bother in these parts by encroaching on the habitats of other plants. Nevertheless, with their heads of reddish-brown or bronze fruits they can

FIG. 239 *Rumex*: A, flowerhead (× 2½); B, fruit (× 5).

make a spectacular showing. They are smooth plants with brown, jagged stipules and the flowers in 2 rows, the inner ones larger, in short clusters, with 6 sepals and 6 stamens. There are 3 styles, each with a large fringed stigma.

*R. brownii*. Reddish. Subalpine, montane. N.S.W., Vic., Tas.
A native species, smooth with broad leaves on long stems. Flowers small, often male or female. Nuts with hooked bristles.

### Family MONIMIACEAE

Woody plants with opposite leaves, usually coarsely toothed. The flowers are usually male or female. The calyx is lobed, there are no petals, and the stamens are indefinite in number. There are a number of small, separate seedboxes inserted above the other parts of the flower. The fruits are fleshy.

*ATHEROSPERMA* (from Gk *atheros*, repelling wild animals, and *sperma*, a seed, possibly in reference to the hairy, unpalatable fruits). Fig. 240.

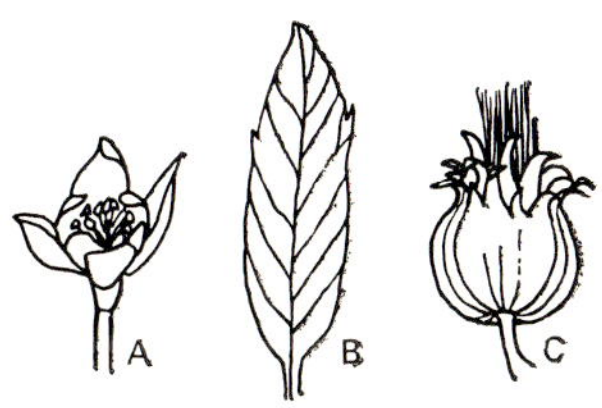


FIG. 240 *Atherosperma*: A, flower (actual size); B, leaf (× 1/3); C, fruit (× 1/3).

*A. moschatum*. Southern Sassafras. Cream. Subalpine, montane. N.S.W., Vic., Tas.
A tree that is especially abundant in the temperate rainforests of Tasmania but also found in the Victorian alps and the brush forests of New South Wales, growing to a height of 6–12 metres with a highly aromatic, pleasing fragrance. Leaves opposite, shining above, greyish-green below, 5–6 cm long. Flowers solitary, male or female, on stalks 6–12 mm long with about 12 stamens in the male flower surrounded by 8–10 petal-like structures; the female flowers with numerous seedboxes tapering into plumose styles.

*HEDYCARYA* (from Gk *hedi*, pleasing; *carya*, a nut or fruit). Fig. 241. *H. angustifolia*, common in brush forests, is found in the denser

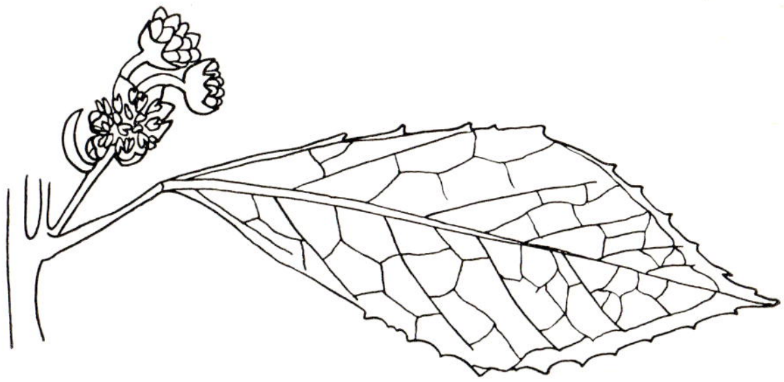

FIG. 241 *Hedycarya*: flowerhead and leaf (actual size).

parts of the mainland montane areas also, and on King Island, Tasmania. It is a smooth, small tree with broad, toothed leaves. The flowers have 6–10 carpels, the male flowers with numerous hairy anthers, the female flowers with many seedboxes. The fruit is succulent, and crowded on the expanded calyx.

### Family PROTEACEAE

There are numerous species of the Waratah family in Australia, but not many representatives in alpine habitats. Members of the family are mostly stiff plants with leathery or thick leaves. The flowers are often small but clustered into showy heads. There are 4 petal-like sepals, no petals, and 4 stamens often held in the tips of the sepals. The seedbox is inserted above the other parts of the flower. The style is often long and conspicuous, sometimes hooked into the tips of the petals and not released until the flower is almost spent. The fruit is dry and splitting or fleshy and berry-like.

*AGASTACHYS* (from Gk *agastachus*, very rich in corn, possibly alluding to heavy seed-bearing). Fig. 242.

*A. odorata*. Cream and white. Subalpine. Tas.
A charming, bushy shrub sometimes reaching the height of a small tree on the button-grass plains. Leaves narrow, shortly stalked, very blunt, thick, shining, 5–8 cm long. Flowers in dense spikes 7–10 cm long, the individual flowers 7–10 mm in length. Fruit a nut 4 mm

long, with 2 well-developed lateral wings and 1 narrow lateral wing.

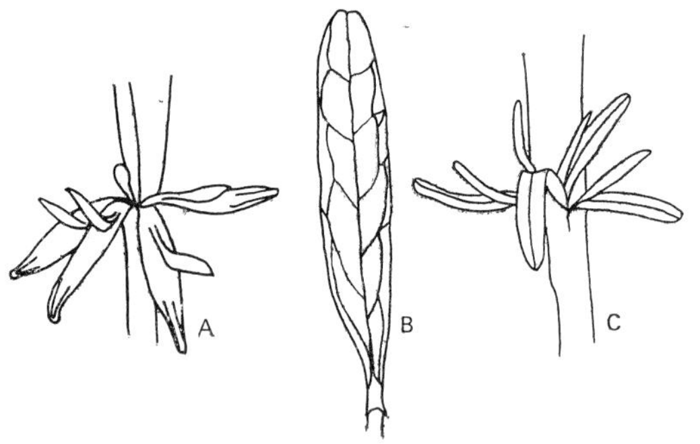


FIG. 242 *Agastachys*: A, flower (× 2); B, leaf (actual size); C, fruit (× 2).

*BANKSIA* (after Sir Joseph Banks, 1743-1820, Father of Australian Botany, who accompanied Cook on the voyage on which New South Wales was discovered). Fig. 243. A genus in which the

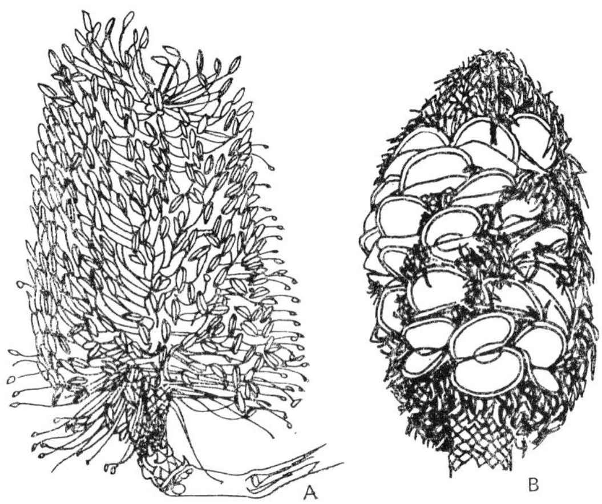


FIG. 243 *Banksia*: A, flowerhead; B, fruiting head. (Both × ½.)

flowers are borne in dense cylindrical heads followed by heads of woody fruits deeply embedded in the axis of the flowerhead. There are 2 species which, while widely distributed elsewhere, occur in montane habitats.

*B. integrifolia*. Yellow. Montane. N.S.W., Vic.
A species with narrow entire leaves 7–10 cm long and spikes of yellow flowers about the same length. It also occurs on King Island, Tasmania.

*B. marginata*. Brownish or yellow. Subalpine, montane. N.S.W., Vic., Tas.
A species with coarsely toothed leaves, woolly-white beneath and 2.5–7.5 cm long. Flowers in heads, 5–10 cm long.

*BELLENDENA* (after John Bellenden-Ker, Scottish botanical writer). Fig. 244.

FIG. 244 *Bellendena*: A, fruiting head with leaves (actual size); B, flowering head (× ½); C, single flower (× 2).

*B. montana*. Mountain Rocket. Cream. Alpine, subalpine. Tas.
A small shrub seldom more than 30 cm high with broadly egg-shaped leaves with 3 or more blunt lobes and about 2.5 cm long. Flowers numerous in a dense terminal head about 4–6 cm long, on a long stalk 5–15 cm in length; sepals about 4–6 mm long, the individual flowers on stalks about the same length. Fruit dry, brilliant red in colour, often accompanied by red foliage leaves.

*CENNARHENES* (from Gk *kenos*, empty, and *arren*, male, referring to the presence of a sterile stamen). Fig. 245.

FIG. 245 *Cenarrhenes*: A, flowering spike (actual size); B, flower (× 2); C, fruit (× 2).

*C. nitida*. Cream. Temp. rainforests. Tas.
This species reaches the height of a small tree, with narrow, shortly stalked, coarsely toothed leaves 10–12 cm long. Flowers in loose spikes 2.5–5 cm long, the individual flowers about 5 mm in length. Fruit attractive, succulent, with a hard stone surrounding a single seed, red or purple, globular, 12–15 cm in diameter.

*GREVILLEA* (after Charles Francis Greville, one-time vice-president of Royal Society of London). Fig. 246. A readily distinguished genus with its leathery fruits, each containing 2 seeds.

*G. alpina*. Red and white. Subalpine, montane. N.S.W., Vic.
A variable, branching plant, sometimes prostrate or up to 60 cm high, woolly all over. Leaves oblong or narrow, almost stalkless, 2–2.5 cm long. Flowers in short heads at the ends of the branches; sepals silky on both sides.

*G. arenaria*. Greenish. Montane. N.S.W.
A bushy shrub with fairly broad, shortly stalked leaves, finely hairy above, the edges recurved. Flowers in loose heads, slightly hairy, pointed.

*G. australis*. White. Alpine, subalpine, montane. N.S.W., Vic., Tas.
At the highest elevations prostrate and rock-clinging, at lower elevations erect and up to 6–8 metres high. Leaves narrow, occasionally oblong, pointed, silky-hairy beneath, 1.2–5 cm long. Flowers in small heads 3–5 mm long, silky on the outside. The only species occurring in Tasmania and the only one reaching true alpine habitats on the mainland. There are several varieties.

*G. juniperina*. Pink and yellowish. Subalpine, montane. N.S.W.
A compact, bushy shrub of wide distribution at lower altitudes also. Leaves narrow, spreading, sharp-pointed, 1.2–5 cm long. Flowers in elegant loose heads borne at the ends of the branches, individual flowers 10–12 mm long.

*G. lanigera*. Pink. Montane. N.S.W., Vic.
A small, compact, densely hairy shrub with narrow, blunt, crowded leaves 1–1.5 cm long.

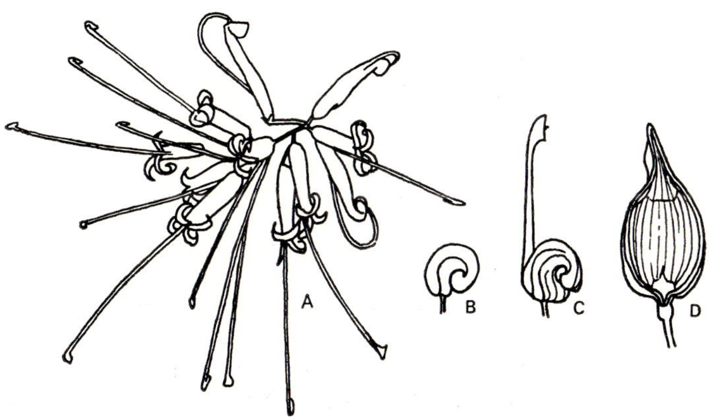


FIG. 246 *Grevillea*: A, flower-head; B, bud; C, mature flower; D, fruit. (All actual size except D × 2.)

Flowers in dense, short heads at the ends of the branches.

*G. miqueliana*. Dull red. Subalpine. Vic.
A very attractive shrub with broad, shortly stalked leaves, hairy beneath, 2.5–6 cm long. Flowers in dense, short heads, hairy outside.

*G. parviflora*. Whitish, grey. Subalpine, montane. N.S.W., Vic.
A shrub about 90–120 cm high, spreading, with pointed leaves 1.8–4 cm long. Flowers in short, dense heads.

*G. rosmarinifolia*. Pink. Montane. N.S.W., Tas.
A tall shrub 1.5–2 metres high with slender, velvety branches and stalkless narrow, pointed leaves 1.2–3.5 cm long. Flowers in short, dense heads at the ends of the branches.

*G. victoriae*. Orange. Subalpine, montane. N.S.W., Vic.
A variable, often spectacular, bushy shrub reaching a height of 1–2 metres in the montane areas, shorter at higher elevations. Branches soft and hairy, leaves stalked, rather narrow, bright-green and conspicuously veined above, silky hairy below, 5–10 cm long, the young foliage often attractively pink or bronze. Flowers in short, slightly elongated, pendant heads; sepals silky-hairy outside.

*HAKEA* (after Baron Christian Ludwig v. Hake, 1745-1818, German patron of botany). Fig. 247. A genus of erect shrubs or small trees with alternate leaves, and hard woody fruits each containing 2 winged seeds. The flowers are often small and clustered.

FIG. 247 *Hakea*: A, fruit; B, single flower. (Both × 1½.)

*H. dactyloides*. Cream. Montane. N.S.W., Vic.
A tall shrub more common at lower elevations. Leaves fairly broad, 5–10 cm long, stiff, flat, prominently veined. Flowers small in clusters in the leaf axils. Fruit smooth or slightly rough, almost globular, 12–25 mm in diameter with a small beak.

*H. epiglottis*. Creamy yellow. Subalpine, montane. Tas.
An erect shrub or small tree 2–4 metres high, smooth. Leaves narrow, stiff, sharply pointed, 2–10 cm long. Flowers in short, stalkless clusters, silky, each 4–5 mm long. Fruit s-shaped, 14–18 mm long, rugose.

*H. eriatha*. Cream. Montane. N.S.W., Vic.
A tall shrub with broad, blunt leaves 7.5–10 cm long. Flowers few, clustered in the leaf axils on silky-hairy stalks. Fruit rather large, nearly blunt, almost smooth.

*H. lissosperma*. Cream. Subalpine, montane. N.S.W., Vic., Tas.
A tall shrub wih stiff, smooth, pointed leaves. The flowers are without stalks in small clusters in the leaf axils.

*H. macraena*. White. Montane. N.S.W.
A smooth shrub with thin, narrow leaves 5.5–15 cm long, pointed. Flowers in very short clusters enclosed in bud in overlapping bracts. Fruit rather large, egg-shaped, the summit compressed, the rest of the fruit deeply rugose.

*H. microcarpa*. White. Subalpine, montane. N.S.W., Vic., Tas.
An erect, stiff shrub up to 2 metres high. Leaves narrow, 2.5–10 cm long, the lower ones sometimes broader. Flowers smooth, clustered in the leaf axils. Fruit small and pointed, 1.2–1.5 cm long.

*ISOPOGON* (from Gk *isos*, equal, and *pogon*, a beard, referring to the hairs on the flowers). Fig. 248.

*I. anemonifolius*. Yellow. Montane. N.S.W., Vic.
A fairly tall, bushy shrub common in lower habitats also. Leaves flat, much divided, 4–10

cm long. Flowers in globular heads without stalks. Fruiting heads globular with numerous small, non-splitting nuts each with a single seed.

FIG. 248 *Isopogon*: A, flowerhead (actual size); B, flower (× 2); C, fruiting head (actual size).

*LOMATIA* (from Gk *loma*, a fringe or border, referring to the winged edge to the seeds). Fig. 249. A genus of shrubs with alternate leaves and flowers in loose, long clusters. The fruit is leathery and contains several seeds (not 2 as in many members of this family), with a yellow powdery substance between them.

*L. fraseri*. Cream. Montane. N.S.W., Vic.
A tall shrub or tree up to 450 cm high. Leaves narrow, slightly toothed, 5–15 cm in length. Flowers in dense terminal heads, the individual flowers very hairy outside.

*L. ilicifolia*. Pale yellow, cream. Montane. N.S.W., Vic.
A tall, shrubby plant up to 450 cm high. Leaves fairly broad, toothed, 10–15 cm in length, covered with soft, rusty hairs when young. Flowers in dense terminal clusters, the individual flowers almost smooth.

*L. myricoides*. Cream. Montane. N.S.W., Vic.
A tall shrub or small tree. Leaves fairly narrow, toothed, 10–20 cm in length, usually smooth. Flowers in long, loose clusters.

*L. polymorpha*. Cream. Subalpine, Montane. Tas.
A small shrub also descending to sea-level. Leaves narrow and tapering, woolly beneath, about 5 cm long. Flowers in loose clusters, individual flowers about 1–2 mm in length.

*L. tinctoria*. Guitar Plant. White. Subalpine, montane. Tas.
A small plant usually 50 cm–1 metre high though sometimes taller. Leaves 4–8 cm long, usually divided into 3–7 pairs of narrow, leathery lobes, occasionally simple or with fewer lobes. Flowers about 1 cm long, in loose clusters, terminal or in the upper leaf axils. Fruit up to 2 cm in length.

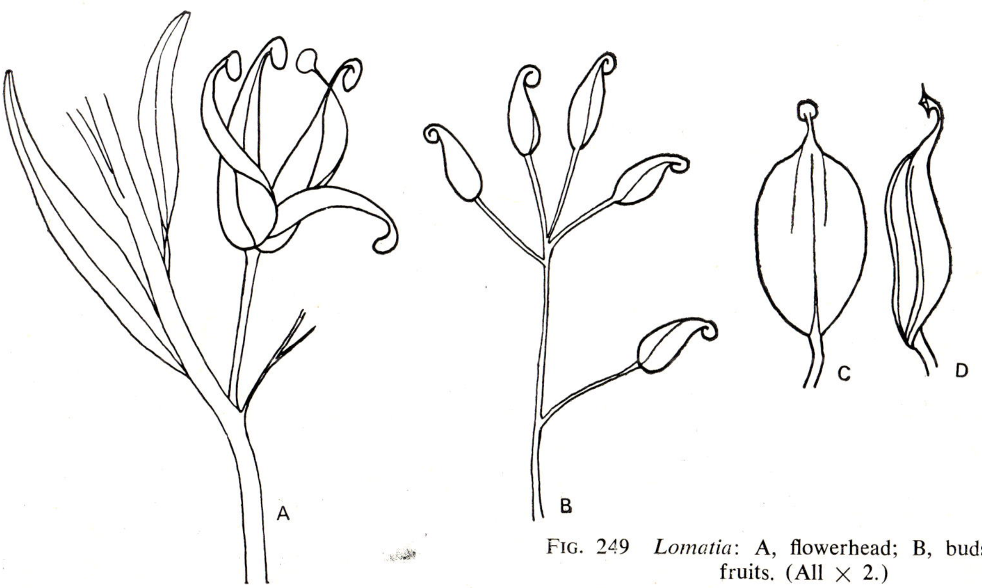


FIG. 249 *Lomatia*: A, flowerhead; B, buds; C, D, fruits. (All × 2.)

*ORITES* (from Gk *oreites*, a mountaineer, referring to the habitat of many species). Fig. 250. A genus with leathery fruits, each with 2 seeds. The leaves are alternate and leathery and the flowers small, in pairs enclosed in a single bract, the petal-like sepals spreading.

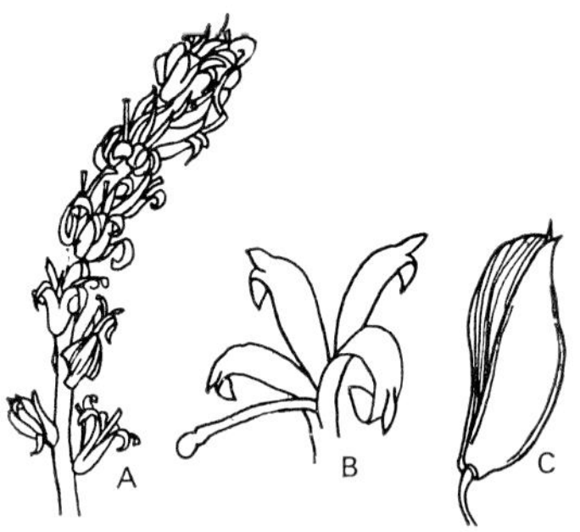


FIG. 250 *Orites*: A, flowerhead (actual size); B, flower (× 2); C, fruit (× 2).

*O. acicularis*. Cream or yellow. Alpine, subalpine. Tas.
A compact, often yellowish shrub 50 cm–5 metres high with narrow, cylindrical, sharply pointed leaves 2.5–5 cm long. Flowers in short, dense spikes about 1–2 cm long; sepals 2–3 mm in length. Fruit narrow-oblong, almost completely smooth with a very oblique point.

*O. diversifolia*. Cream to brown. Alpine, subalpine, montane. Tas.
A much-branched shrub often 16–18 metres high, the branches, leaves and flowers rusty-woolly. Leaves variable from narrow to egg-shaped, thick with a prominent midvein, shining above, pale and slightly hairy beneath, the margins sometimes toothed or lobed, 2.5–8 cm long; sepals about 4–6 mm long. Fruit about 18–20 mm in length.

*O. lancifolia*. Cream. Alpine, subalpine. N.S.W., Vic.
An attractive, bushy small shrub 1–3 metres high, the only species on the mainland. Leaves crowded, tough, rather broad, reticulate beneath, 1.5–2 cm long. Flower spikes 3–5 cm long; sepals smooth, somewhat curly. Fruit smooth, about 18–20 mm in length.

*O. milliganii*. Cream. Alpine. Tas.
A smooth, much-branched shrub 60–80 cm high with oblong, thick, dark, coarsely toothed leaves about 2.5 cm long. Flowers in terminal heads 2.5–5 cm long; sepals 5–7 mm in length. Fruit brown, hairy, about 12–15 mm in length.

*O. revoluta*. White. Alpine, subalpine. Tas.
A shrub 120–150 cm high, the branches covered with rusty hairs. Leaves narrow, smooth or slightly hairy above, densely hairy below, 1.25 cm long. Flowers in terminal spikes about 2.5 cm long; sepals about 4–6 mm in length. Fruit oblong and pointed, very hairy, 14–16 mm in length.

*PERSOONIA* (after Christian Hendrik Persoon, 1755-1837, born in Cape Colony, a writer on fungi). Fig. 251. A genus of shrubs or small trees with entire alternate leaves. The yellow sepals are spreading and recurved at the ends. The fruit is single-seeded and succulent, with a persistent style.

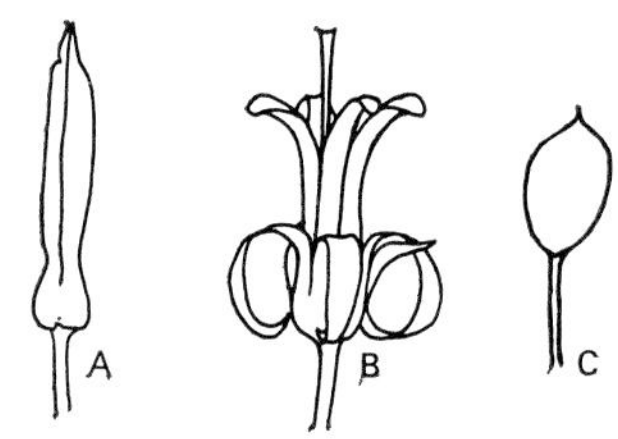


FIG. 251 *Persoonia*: A, bud (× 2); B, flower (× 2); C, fruit (actual size).

*P. arborea*. Cream. Alpine, subalpine. Vic.
A tree over 40 metres in height with shortly stalked, broad leaves 5–10 cm long and hoary young shoots. Flowers solitary in leaf axils, 2–4 mm long.

*P. chamepeuce*. Yellow. Alpine, subalpine. N.S.W. ,Vic.
An attractive, scrambling plant with narrow, pointed, flat leaves 2.5–4 cm long; flowers solitary, rather conspicuous, on short stalks.

*P. gunnii*. Creamy yellow. Subalpine, montane. Tas.
An erect, bushy shrub 2–4 metres high. Leaves crowded, narrow, spoon-shaped, 2–4 cm long on short stems, thick, leathery, usually smooth but hairy in some forms. Flowers fragrant, on short stalks solitary in leaf axils but often clustered towards the ends of the branches, 12–15 mm

long. Fruit egg-shaped, purplish-black, 8–10 mm in length.

*P. juniperina.* Yellow. Montane. N.S.W., Vic., Tas.

A prickly little shrub 60–80 cm high with stiff, stalkless, narrow, pointed leaves 12–25 mm long; flowers solitary, about 1 cm long, shortly stalked.

*P. oxycocceoides.* Cream. Alpine, subalpine. Vic.

A rather diffuse small shrub with small, round, blunt, stiff leaves less than 12 mm long and nodding flowers on short stalks.

*P. rigida.* Yellow. Montane. N.S.W., Vic.

A softly woolly shrub with stiff, egg-shaped leaves 12–36 mm in length and very shortly stalked solitary flowers about 12 mm long.

*P. subvelutina.* Yellow. Subalpine. N.S.W.

A shrub 0.5–5 metres high, sub-erect. Leaves broad, almost opposite, 3–4 cm long, the margins, recurved, softly hairy. Flowers solitary in the leaf axils, on short stalks, velvety. Fruit rounded, 1.2–1.5 cm long.

*TELOPEA* (from Gk, meaning "seen at a distance"). Fig. 252. A genus of tall shrubs with alternate leathery leaves and usually red flowers in large, conspicuous, rounded heads with bracts at the base also red. The fruits are leathery with many seeds.

*T. oreades.* Gippsland Waratah. Red. Montane. N.S.W., Vic.

A tall shrub or small slender tree. Leaves smooth, broad, 10–20 cm long, tapering into a long stalk. Flowers in a dense globular head, 7.5–10 cm in diameter. Fruit 5–7.5 cm long with a persistent style.

*T. truncata.* Tasmanian Waratah. Red. Subalpine, montane. Tas.

An erect or spreading shrub 1.5–3 metres high. Leaves rather thick, egg-shaped, 5–10 cm long. Flowers in flat, terminal heads, 10–12 cm in diameter. Fruit 5–7.5 cm long, curved.

**Family THYMELAEACEAE**

Shrubs with simple leaves and often very tough, stringy bark. There are usually 4 petal-like sepals and no petals. The stamens, 2–8, are inserted on the tube formed by the sepals. The fruit is dry fleshy, usually with one compartment.

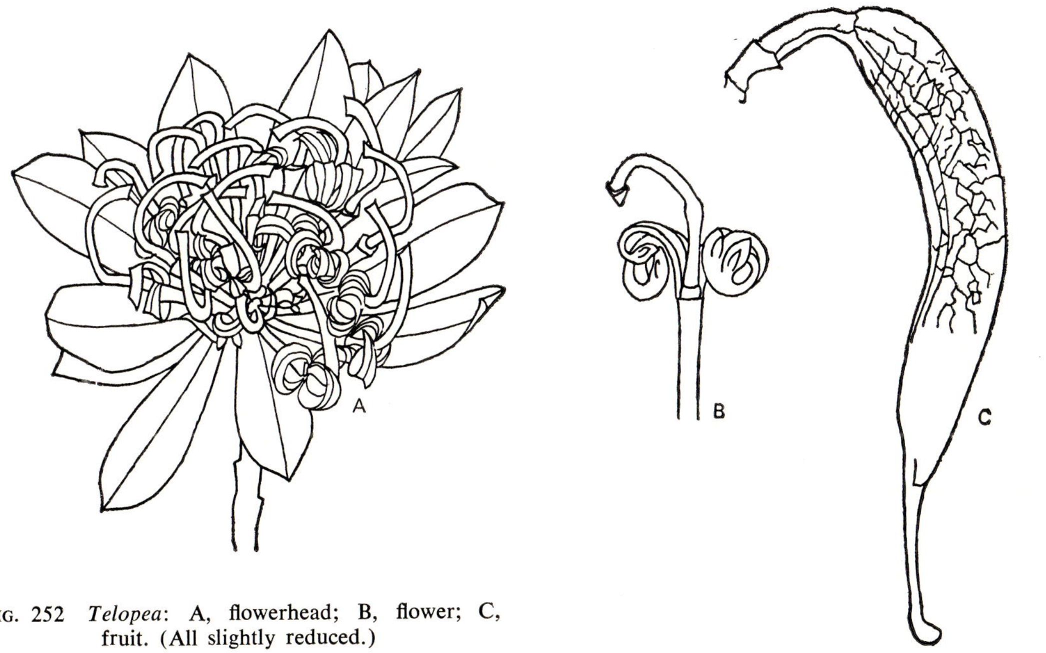


FIG. 252 *Telopea*: A, flowerhead; B, flower; C, fruit. (All slightly reduced.)

*PIMELEA* (from Gk *pimele*, fat, probably referring to the fleshy seed leaves). Fig. 253. A genus represented by a considerable number

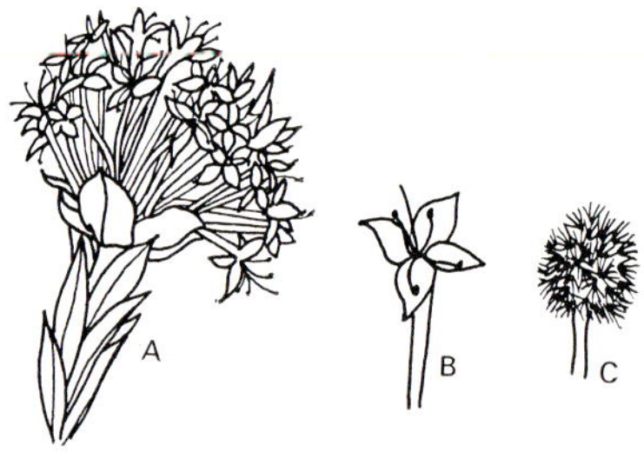


FIG. 253 *Pimelea*: A, flowerhead (actual size); B, flower (× 2); C, fruit (× 2).

of species in these habitats. The small, stalkless flowers are in heads with bracts, often large and showy, at the base. Flowers are male or female. There are usually 2 stamens, but sometimes only one. Fruit fleshy, occasionally dry.

*P. alpina.* Red, pink or white. Alpine, subalpine. N.S.W., Vic.

A prostrate or dwarf plant, often rock-clinging, common in grasslands. Leaves small, opposite, dark-green. Flowers small, open, hairy in rather small heads with 4–6 bracts, rather broader than the leaves at the base. Fruit dry.

*P. axiflora.* Cream. Montane. N.S.W., Vic., Tas.

A slender, small shrub that loves open situations, commonly 3–4 metres in height but often much taller. Leaves narrow, opposite, pointed, 2–4 cm long. Flowers in small heads in the leaf axils with 2–4 small bracts rather like foliage leaves, at the base of the heads. Fruit conspicuous, dry but surrounded by the succulent reddened petals and sepals, about 4 mm long, produced in great abundance and remaining on the plant for some time.

*P. biflora.* Red inside. Alpine, subalpine. N.S.W., Vic.

A prostrate, stout, much-branched shrub. Leaves very dense, forming a mat, opposite, almost stalkless, egg-shaped, smooth, 5–9 mm in length. Flowers hairy, about 5 mm long, in terminal heads, 2 in each head, with 4 bracts at the base resembling the leaves.

*P. collina.* White. Montane. N.S.W., Vic.

A slender, small plant ranging from the coast to the montane area. Leaves prominently 3-veined on the undersurface, opposite, narrow, concave, 8–20 mm long on short stalks. Flowerheads terminal, nodding, with 4 broad, stiff, smooth bracts at the base. Fruit dry.

*P. curviflora.* Yellow. Montane. N.S.W., Vic., Tas.

A shrub 30–40 cm high, extending to the coast, hairy, with broad, oblong leaves, smooth above and hairy beneath, 5–10 mm long. Flowers in small terminal or axillary heads, 4–5 cm long, with 2 or occasionally 3–4 unequal bracts, hairy on the outside, at the base. Fruit dry.

*P. glauca.* White. Montane. N.S.W., Vic., Tas.

A much branched smooth shrub 30–60 cm high, common also in coastal situations, with opposite, thick, oblong, concave leaves 6–20 mm long. Flowers in terminal heads with 4 rather broad bracts at the base. Fruit dry.

*P. humilis.* White. Montane. N.S.W., Vic., Tas.

A small, erect shrub 3–8 cm high with variable, opposite, stalkless leaves 2.5–5 cm long. Flowers many, in fairly large terminal heads, usually nodding, with 4–6 very broad silky-hairy bracts at the base; individual flowers with very slender tubes about 12 mm long. Fruit dry.

*P. ligustrina.* Cream. Subalpine, montane. N.S.W., Vic., Tas.

An erect or scrambling smooth shrub reaching 2 metres at lower elevations. Leaves opposite, without stalks, broadly egg-shaped, variable in size, 2.5–6 cm long, flat, thin, conspicuously veined. Flowers in almost globular heads with 4–6 short, broad bracts at the base, silky-hairy on the inside; flower tube about 12 mm long; stamens protruding and bright-yellow. Fruit dry and rounded.

*P. milliganii.* Cream. Alpine. Tas.

A rather rare, low, much-branched shrub, densely silky-hairy. Leaves opposite, crowded, overlapping, without stalks, fairly broad and blunt, 6–12 mm long, flat. Branches marked by raised scars left by the fallen leaves. Bracts at base of flowerheads similar to leaves; flower tube hairy, 3–5 mm long, persistent round the dry fruit.

*P. nivea.* White. Subalpine, montane. Tas.

A shrub 30–60 cm high covered in white, cottony down. Leaves opposite, almost stalkless, rounded, about 12–14 mm long. Flowers in dense, compact heads surrounded by normal foliage leaves; tube of the flower about 12 mm long, slender, very hairy. Fruit dry.

*P. pauciflora.* Pale yellow. Montane. N.S.W., Vic., Tas.

A much-branched, slender, smooth shrub. Leaves narrow, opposite, 12–25 mm long. Flowers in small heads with 2 bracts at the base; flower tube smooth, about 4 mm long. Fruit slightly succulent, scarlet or orange.

*P. pygmaea.* White. Alpine, subalpine, montane. Tas.

A smooth, prostrate plant forming moss-like hummocks. Leaves mostly opposite and crowded, nearly oblong, egg-shaped, 3–4 mm long. Flowers solitary, about 4 mm long. Fruit dry.

*P. sericea.* White. Alpine, subalpine. Tas.

An erect, branched shrub 3–6 metres high, silky-hairy all over except for the upper surfaces of the leaves. Leaves opposite, almost stalkless, oblong, the undersurface densely covered with silky-velvety hairs and about 1.5 cm long. Flowers in rather large, erect, terminal heads with normal foliage leaves at the base; floral tube slender, about 12 mm long. Fruit dry.

*P. spathulata.* White. Montane. N.S.W., Vic.

A slender, smooth shrub 30–100 cm high with opposite, narrow to spoon-shaped leaves 6–20 cm long. Flowers silky in terminal drooping heads with 4 smooth egg-shaped bracts nearly as long as the head. Fruit dry.

*DRAPETES* (Gk, meaning "a runaway", probably referring to its creeping habit). Fig. 254.

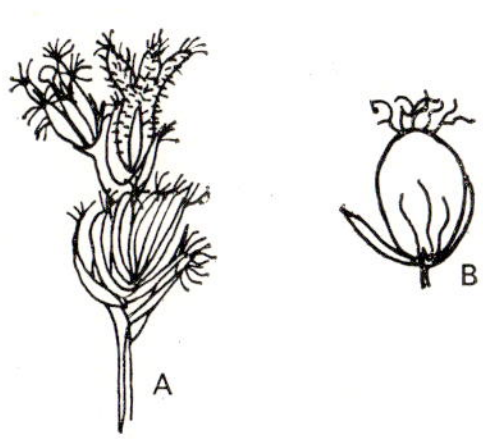


FIG. 254 *Drapetes*: A, flower; B, fruit. (Both × 4.)

This genus differs from *Pimelea* in having 8 scale-like petals (*Pimelea* has none) and 4 stamens (2 in *Pimelea*).

*D. tasmanica* (syn. *Kelleria tasmanica*). White. Alpine. N.S.W., Tas.

A small, compact, prostrate, cushion-like plant with small, alternate, narrow leaves 2–3 mm long. Flowers few in a small dense head on a very hairy base; petals hairy, tubular, with spreading lobes about 4 mm long. Fruit succulent, egg-shaped, about 3 mm in length.

## Family EUPHORBIACEAE

A very variable family with only a few representatives in these habitats, most being small and not very conspicuous plants. The flowers are either male or female, with 4–6 sepals and petals, sometimes one or both missing. The stamens vary from one to many. The ovary usually has 3 parts inserted above the other flower parts.

*ADRIANA* (after Adrien de Jussieu, 1797-1853, botanist). Fig. 255.

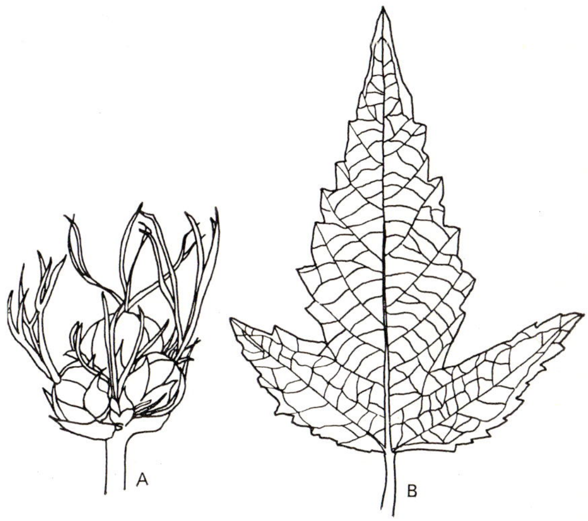


FIG. 255 *Adriana*: A, fruits (× 1½); B, leaf (× ½).

*A. glabrata.* Brownish. Montane. N.S.W., Vic.

A rather scrambling shrub 1–2 metres high. Leaves large, broad, stalked, toothed or trilobed, alternate. Flowers in spikes, male or female; male spike 5–7.5 cm long, female spike shorter; sepals 4–5 in male flowers, 6–8 in female; petals none; stamens numerous and crowded, inserted at base of ovary. Fruit a 3-celled hairy capsule.

*AMPEREA* (after Jean-Jacques-Antoine Ampere, 1800-1864, French historian). Fig. 256.

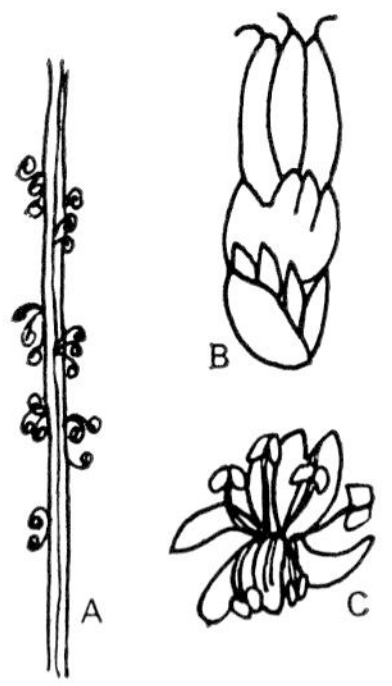


FIG. 256 *Amperea*: A, flowerhead (× ½); B, fruit (× 3); C, flower (× 3).

*A. xiphoclada.* Greenish. Montane. N.S.W., Vic., Tas.

A leafless, angular plant 30–60 cm high with small leaves, from the mainland and Tasmania. Flowers small, almost stalkless in clusters in the leaf axils, surrounded by small brown bracts; petals absent; stamens 10 or fewer. Fruit a rounded capsule.

*BERTYA* (after Count Leonce de Lambertye, 19th-century French botanist).

*B. findlayi.* A tall shrub or small tree with narrow leaves 2.5–7.5 cm long. Flowers solitary or 2–3 together, stalkless, with 3–8 bracts, 5 petals and numerous united stamens; male and female flowers on the same plant; seedbox of 3 parts with 3 styles. Fruit of one compartment.

*MICRANTHEUM* (from Gk *mikros*, small, and *anthos*, a flower).

*M. hexandrum* is a shrub from the mainland and Tasmania, often tall, with narrow or oblong leaves 10–15 mm long. Flowers very small, solitary or few together with 6 petals about 4 mm long, 6–9 stamens and a seedbox with 3 styles; male and female flowers on the same plant. Fruit 3–lobed.

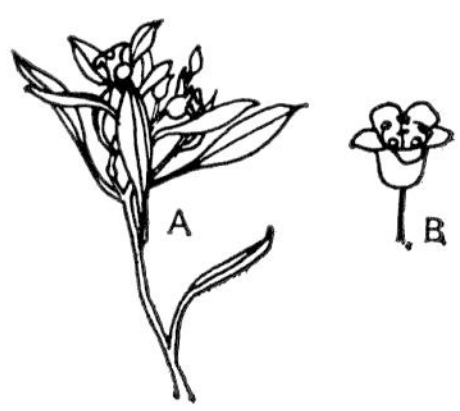


FIG. 257 *Poranthera*: A, flower and fruit (× 2); B, flower (× 4).

*PORANTHERA* (from Gk *poros*, a pore, and *antheros*, an anther). Fig. 257.

*P. microphylla.* White. Alpine, subalpine, montane. N.S.W., Vic., Tas.

A dainty little plant reaching a height of 2.5–10 cm. Leaves small, egg-shaped, 6–8 mm long. Flowers small, in dense flat heads; sepals and petals 5; stamens 5; seedbox with a trifid style; male and female flowers intermingled in the same head. Fruit 6–lobed.

*PSEUDANTHUS* (from Gk *pseudos*, false, and *anthos*, a flower: in *P. pimelioides* many small flowers are clustered together to resemble a single flower). Fig. 258.

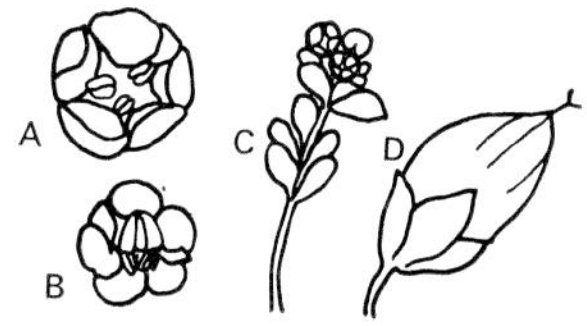


FIG. 258 *Pseudanthus*: A, male flower (× 3); B, female flower (× 3); C, flowerhead (actual size); D, fruit (× 3).

*P. divaricatissimus.* White. Montane. N.S.W., Vic., Tas.

A low, densely branched shrub with egg-shaped or oblong leaves 3–5 mm long. Flowers very small in upper leaf axils; petals 5 or fewer, about 1 mm long; stamens 6 on short stalks. Fruit oblong, with one compartment.

## Family URTICACEAE

These are plants to be wary of, for they often have stinging hairs on the leaves. The flowers are small and usually greenish with 4–5 sepals or petals and 4–5 stamens. The seedbox is inserted above the other parts of the flower. There are 2 genera represented here, *Australina* and *Urtica.* The former does not have stinging hairs and the flowers are tubular. *Urtica* has stinging hairs and irregularly 4–lobed flowers.

*AUSTRALINA* (from Lat. *australis*, southern). Fig. 259. *A. muelleri,* occurring in New South Wales, Victoria and Tasmania, is a succulent

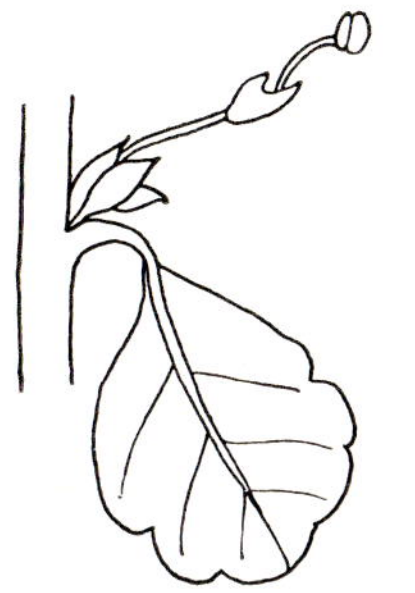

FIG. 259 *Australina*: flower and leaf (actual size).

plant 15–30 cm high with round, wavy, basal leaves 2.5–5 cm long and small greenish male or female flowers in stalkless clusters; calyx 5-toothed; petals absent; one stamen. Fruit hard, with one non-splitting seed.

*URTICA* (Lat. name for nettle). Fig. 260. *U. incisa*, the common nettle, is covered in stinging hairs and is a weak, smooth, trailing plant with

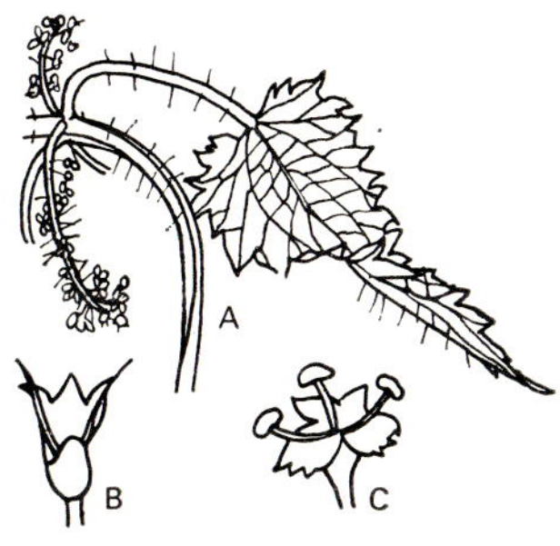


FIG. 260 *Urtica*: A, flowerhead (× ½); B, fruit (× 3); C, flower (× 3).

deeply toothed, fairly narrow, soft leaves 2.5–7.5 cm long. Flowers in small heads; sepals 4; petals none; stamens 4; stigmas tufted. Fruit small, with one non-splitting seed.

## Family CASUARINACEAE

*CASUARINA* (from Lat. *casuarius*, the cassowary, referring to the branches, which resemble birds' feathers). Fig. 261. A characteristic genus of trees or shrubs with long, green articulated branches bearing rows of teeth at each segment which really are reduced leaves. The flowers are either male or female, arranged in spikes. The male flower consists of a single

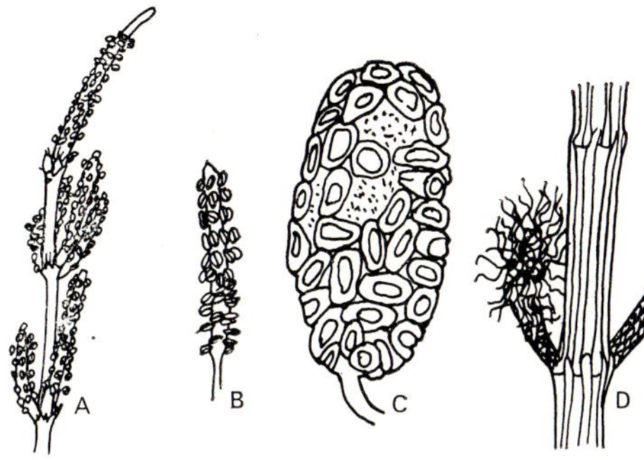


FIG. 261 *Casuarina*: A, flowerhead; B, male flowers; C, fruiting head; D, female flowers. (All × 2 except A, actual size.)

stamen, 1–2 petals or sepals and 2 bracteoles; the female flowers have no sepals or petals, 2 bracteoles and a seedbox of one compartment with 2 conspicuous red styles. The fruit is a winged nut, a group of which are united to form a woody cone. None of the species reaches higher elevations than the montane area.

*C. littoralis* (syn. *C. suberosa*) is a small tree with 6–8 leaves in a ring. Male and female flowers on the same plant. Cone blunt with prominent valves 1–3 cm long, 1–2 cm in diameter. It is widely distributed in New South Wales, Victoria and Tasmania, ascending to montane habitats.

*C. nana* is a dwarf shrub from the mainland, about 1 metre high with 5 or occasionally 6 teeth in each ring. Male and female flowers usually on separate plants. Cone stalkless, 12–15 mm in diameter, 15–20 mm long.

*C. paludosa* is a small shrub 1–2 metres high with more than 6 teeth in a ring. Male and female flowers occur on the same plant. Cone about 2 cm long and about 12 mm in diameter. It is a widely distributed species, ascending to montane habitats in New South Wales and Victoria.

## Family FAGACEAE

Evergreen or deciduous shrubs or trees with flowerheads consisting of either male or female flowers, in the leaf axils of branchlets of the current year's growth, the female flowers above the male, surrounded by bracts. The flower parts are in fours. The leaves are simple, alternate, with stipules, but the branches of the tree are so fine that the small leaves look like leaflets of a divided leaf. The bracts become hard and woody in the fruiting stage and separate into 4 valves. The fruit is a nut, broadly 3–winged.

*NOTHOFAGUS* (from Gk *nothos*, false, and *fagus*, beech). Fig. 262.

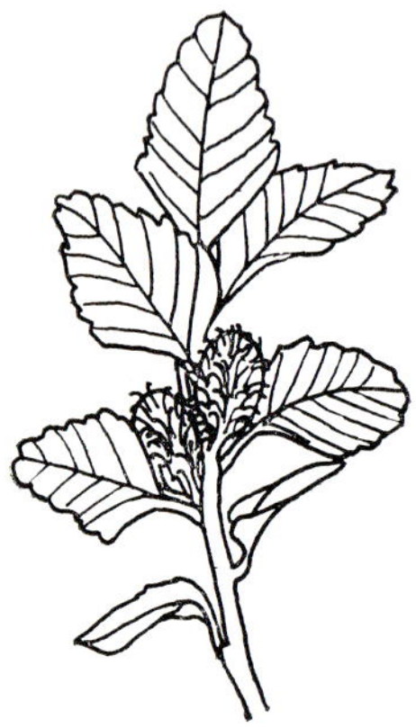

FIG. 262 *Nothofagus*: flowerhead (× 2).

*N. cunninghamii*. Myrtle. Brownish. Alpine, subalpine, montane (temp. rainforests). Tas. An evergreen tree 35–50 metres high in favourable situations but only 6–18 metres high in more exposed situations and reduced to a shrub at the highest elevations. Leaves shortly stalked, egg-shaped or round, thick, leathery, coarsely toothed, 6–18 mm long, veins not conspicuous; young leaves often golden bronze. Male flowers solitary, stalked, unequally 6–lobed, 2.5 mm long, stamens 8–12; female flowers without stalks, 3 together surrounded by bracts about 1.5 mm long, 3–lobed; stigmas 3.

*N. gunnii*. Deciduous Beech. Brownish. Alpine. A small deciduous tree 1.5–2.5 metres high with numerous tangled, wiry branches. Leaves alternate, shortly stalked, egg-shaped or almost round, edge irregularly toothed, deeply furrowed top and bottom, 10–20 mm long, upper surface folded, lower surface with prominent veins and long hairs; they turn golden brown to red before falling. Male flowers solitary or 2–3 together, 3–6 lobed, about 4 mm long; female flowers stalkless, 3 together surrounded by bracts, about 8 mm long.

**Family SANTALACEAE**

A fairly large family of plants with leaves reduced to scales in the alpine species and easily recognized on this account. Flowers small, inconspicuous, often fleshy, usually male or female, with 4–5 sepals, petals and stamens. Fruit dry with a single seed, often on a swollen, succulent base. There are three genera in these habitats, *Exocarpos, Choretrum and Omphacomeria.* In *Exocarpos* the ovary is above the other parts of the flower, with the fruit, which is a small hard nut, resting on an enlarged swollen base. In the other two genera the ovary is below or partly below the other flower parts. *Omphacomeria* has no scale leaves and male and female flowers. *Choretrum* has persistent scale leaves and the flowers are perfect, with both stamens and pistil.

*CHORETRUM* (from Gk *khoris*, separate, and *etron*, abdomen, referring to the rim which separates the top of the flower-stalk from the flower itself). Fig. 263.

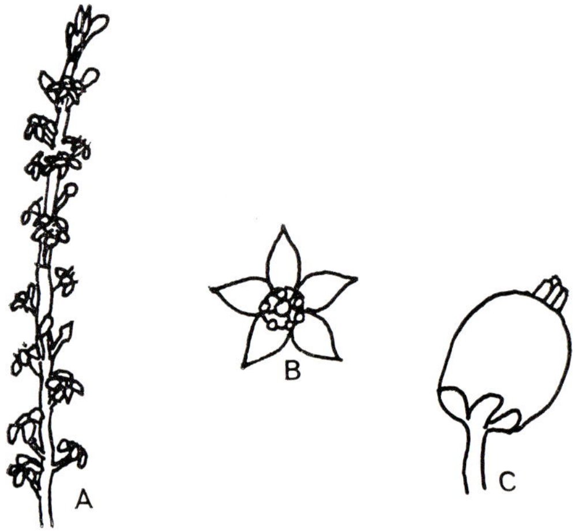


FIG. 263 *Choretrum*: A, flowerhead (actual size); B, flower (× 2); C, fruit (× 2).

*C. pauciflorum*. Cream. Montane. N.S.W. Leaves rudimentary, stems green and drooping. Flowers minute with 2–4 tiny bracts at the base;

sepals 5; petals 5; stamens 5. Fruit small, round, dry.

*EXOCARPOS* (from Gk *exo*, outside, and *karpos*, a fruit, referring to the succulent base of the flower on which the fruit rests). Fig. 264. A genus of shrubs with long, drooping green branches and leaves reduced to small scales. Flowers minute, in clusters; sepals 4–5; petals 4–5; stamens 4–5.

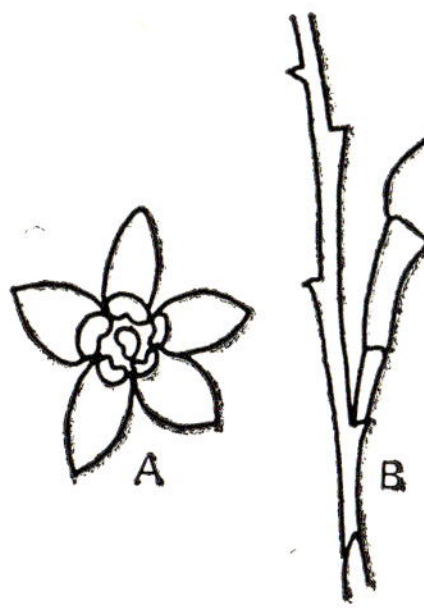


FIG. 264 *Exocarpos*: A, flower (× 3); B, fruit (actual size).

*E. cupressiformis*. Cherry Ballart. Greenish. Montane. N.S.W., Vic., Tas.
A small, drooping, graceful tree with dark-green branches and pale young tips. Fruit about 5–6 mm in diameter, on a bright-red base.

*E. humifusus*. Greenish-yellow. Alpine, sub-alpine. Tas.
A small prostrate plant with coarse, grooved, spreading branches. Leaves tiny, scale-like, persistent. Flowers tiny. Fruit about 3 mm in diameter, on a fleshy white base.

*E. nanus*. Yellowish. Alpine. N.S.W., Vic., Tas.
A prostrate plant with tiny, scale-like, opposite leaves. Flowers minute. Fruit about 5 mm in diameter, on a white, succulent base.

*E. strictus*. Greenish-yellow. Montane. N.S.W., Vic., Tas.
A small shrub with stout, striate branches. Leaves reduced to scales, soon falling. Flowers in short clusters. Base of fruit white, red or pale lilac, 3–4 mm in diameter.

*OMPHACOMERIA* (from Gk *omphax*, unripe, and *homeros*, joined together, in reference to the fruits). Fig. 265.

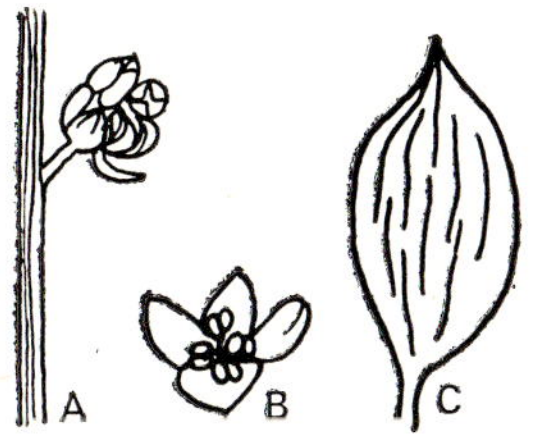


FIG. 265 *Omphacomeria*: A, male flower (× 1½); B, female flower (× 3); C, fruit (× 1½).

*O. acerba*. Sour Bush. Yellow. Montane. N.S.W., Vic.
An erect shrub with wiry, cylindrical, striate branches with prominent nodes, leafless. Flowers male or female, the former in small clusters, the latter solitary, all small with parts in fours or fives. Fruit egg-shaped, greenish, edible, very sour.

# MONOCOTYLEDONS

These are plants in which the embryo within the seed contains only one leaf. Most of them have underground stems (often bulbs), no erect stem, parallel veins in the leaf, and the parts of the flower are in threes or multiples of three.

The monocotyledons include the grasses, rushes and sedges, and the palms, true lilies and orchids, with numerous other plants, usually small. The grasses, rushes and sedges, though represented by a large number of species in alpine areas, have been omitted from this book (except for general references, and these mostly by common names) because they comprise specialist groups beyond its scope.

FIG. 266 *Caladenia*: A, flower; B, leaf. (Both actual size.)

### Family ORCHIDACEAE

Terrestrial orchids are fairly common in the highlands, particularly in the alpine and subalpine areas. They are mostly small plants arising from an underground stem. The flowers are easily recognized. They are irregular with 3 sepals, usually petal-like and free from one another, and 3 petals, one of which, called the labellum, is usually different from the others and often acts as an alighting platform for pollinating agents. Stamens and carpels are fused into a single structure called a column, usually with a single anther of 2 cells at the top. The pollen is waxy or granular, lying in 1–4 masses; the stigma lies in front of these. The ovary is below the other parts of the flower and the fruit is a capsule of 3 parts with very fine seeds.

*CALADENIA* (from Gk *kallos*, beauty, and *aden*, a gland). Fig. 266. A genus of rather small orchids with a solitary leaf at the base and either solitary flowers or a few in a head. The sepals are rather narrow, the petals even narrower and often shorter. The labellum is usually clawed and 3–lobed with thickenings on the central part towards the base.

*C. lyallii*. White, pink. Subalpine. N.S.W., Vic. A slender orchid with a hairy, reddish stem up to 20 cm high. Leaf broadish, ribbed, hairy. Flowers 1–2 on long, slender stalks with a pointed bract at the end of each flower stalk; sepals and petals hairy on the outside; sepals about 12 cm long; labellum egg-shaped, on a short claw, about 9 mm long with 4 rows of yellow or white calli.

*C. angustata*. Dark red and white. Alpine, subalpine, montane. N.S.W., Tas.
A widely distributed species, slender, 15–40 cm high with a long narrow leaf 6–12 cm long. Flowers 1–5 together on a slender stem; sepals and petals hairy; labellum about 7 mm long with entire margins except at the triangular tip which is scantily toothed; calli on labellum in 4 rows, reaching almost to the tip.

*C. carnea*. Pink, white. Montane. N.S.W., Vic., Tas.
A widely distributed slender little orchid 7–24 cm high. Leaf narrow. Flowers 1 or a few on

each thin stalk; sepals and petals pointed; sepals 8–12 mm long. Labellum marked with red bars; it is 3–lobed, 6–8 mm long, with 2 rows of yellow calli along the centre.

*C. cucullata.* Reddish-brown, lighter inside. Montane. N.S.W., Vic.
A slender, slightly hairy little plant. Leaf narrow, slightly hairy, 5–10 cm long. Flowers 1–5 on a slender stalk; dorsal sepal hooded about 7–10 mm long; lateral sepals narrower and longer. Labellum about 7 mm long with 3 lobes on a movable claw; calli in 4 rows.

*C. dilatata.* Yellow-green. Montane. N.S.W., Vic., Tas.
A slender, slightly hairy plant about 30 cm high. Flowers solitary on thin stalks; sepals 4–6 cm long; petals shorter. Labellum 3–lobed, egg-shaped with 4 rows of calli tremulous, on a movable claw. Lateral lobes green. Column incurved, widely winged, with 2 yellow glands at the base.

*C. iridescens.* Bronze, dusky-red. Montane. N.S.W., Vic.
A slender, hairy orchid 10–20 cm high. Leaf oblong, narrow, hairy, 5–8 cm long. Flowers 1–2 in a head, iridescent; dorsal sepal about 10 mm long; lateral sepals about 12 mm long, spreading; petals narrower, about 10 mm long. Labellum egg-shaped, 3–lobed, about 5 cm long, on a short claw 5 mm in length, the calli in 4 rows. Lateral lobes with transverse red stripes, the middle lobe dark purple and glandular. Column with wide wings splotched with red.

*C. patersonii.* Spider Orchid. White, cream. Subalpine, montane. N.S.W., Vic., Tas.
A slender, hairy plant 20–40 cm high. Leaf hairy, oblong, variable in length. Flowers 1–2 in a head; top sepal erect, recurved over the column and bearing a long point; lateral sepals spreading and drooping; petals shorter than the sepals, tapering to fine points. Labellum up to 2 cm long, undivided, with 4–6 rows of calli and a short claw. Column incurved, narrowly winged in the lower part, widely above with 2 stalkless yellow glands at the base.

*C. testacea.* Reddish-brown, white inside. Montane. N.S.W., Vic., Tas.
A slender, hairy plant 15–30 cm high. Leaf narrow, slightly hairy, 7–25 cm long. Flowers 1–6 in a head; top sepal about 10 cm long, lateral sepals longer; petals all spreading, as long as the lateral sepals, narrower. Labellum about 7 mm long, 3–lobed on a movable claw; calli in 4 rows, congested near the tip.

*CHILOGLOTTIS* (from Gk *chillos*, a lip, and *glotta*, a tongue, in reference to the tongue-shaped labellum). Fig. 267.

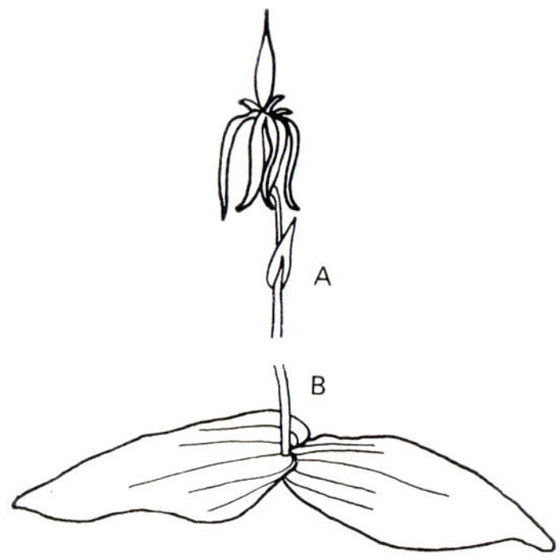


FIG. 267 *Chiloglottis*: A, flower; B, sucker leaves. (Both actual size.)

*C. gunnii.* Brownish-purple or green. Subalpine, montane. N.S.W., Vic., Tas.
A small orchid with a short, stout stem and 2 leaves at the base. Leaves egg-shaped, stalked, 2–7 cm long. Flowers solitary, up to 4 cm in diameter, widely expanding; dorsal sepal egg-shaped, about 8–10 mm long; lateral sepals narrow. Labellum broadly egg-shaped, undivided, on a short claw; calli prominent. Flowering stalk lengthening as fruit ripens.

*CORYBAS* (from Gk *corybant*, one of the dancing priests of Phrygia). Fig. 268.

*C. pruinosus.* Purple. Montane. N.S.W., Tas.
A dwarf herb with a solitary heart-shaped or round leaf, purple underneath, flat on the ground, 18–35 cm in diameter. Flowers solitary, shortly stalked; dorsal sepal large and hooded; lateral sepals and petals small and narrow. Labellum more or less tubular towards the base and then expanded, fringed with a pale-green central boss.

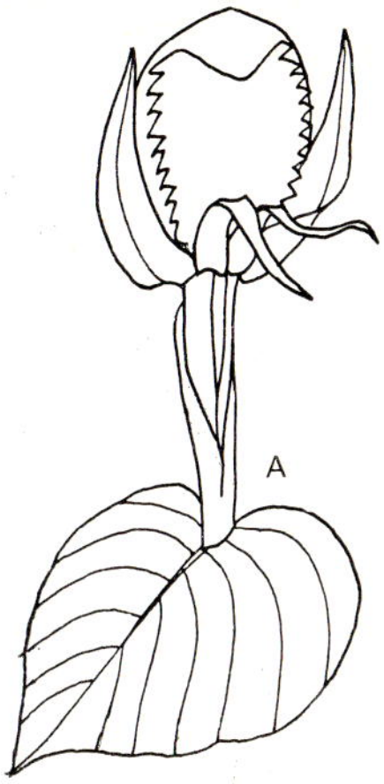


FIG. 268 *Corybas*: A, flowerhead; B, fruit. (Both × 2.)

*DIPODIUM* (from Gk *dis*, twice, and *podion*, a little foot, in reference to the two stalks of the pollen masses). Fig. 269.

*D. punctatum*. Hyacinth Orchid. Pink with dark spots. Montane. N.S.W., Vic., Tas.

One of the loveliest of our ground orchids, this is a leafless saprophytic plant (deriving its food directly from dead organic matter in the soil), with fleshy roots and a dark, reddish stem about 30–90 cm high with loosely sheathing bracts at the base. Flowers numerous on the erect stem, 1–3 cm in diameter; sepals and petals nearly equal, about 17 mm long, the sepals slightly longer than the petals. Labellum 3–lobed, the lateral lobes much shorter than the middle lobe, hairy on the undersurface.

*DIURIS* (from Gk *dis*, twice, and, *oura*, a tail, in reference to the two lateral sepals). Fig. 270. Commonly known as donkey orchid or double-

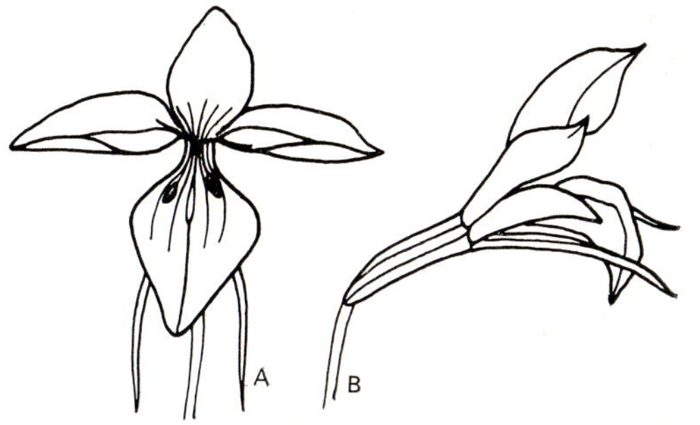


FIG. 270 *Diurus*: A, flower; B, developing fruit. (Both actual size.)

tails, this genus is readily distinguishable by the fact that the 2 lateral sepals are bent downwards and are long and narrow and frequently green. There are 2 or more basal leaves. The flowers are in fairly long heads with bracts on the stem. The labellum is 3–lobed.

*D. maculata*. Yellow, blotched purple. Montane. N.S.W., Vic., Tas.

A widely distributed orchid found at lower elevations also. Stem dark, tall and slender, 20–30 cm high. Leaves slender, sheathing, narrow, 10–15 cm long. Flowers 2–8 in a head; lateral sepals 14–16 mm long, often crossed; petals about as long as the lateral sepals. Labellum 3–lobed with a wavy margin; mid-lobe with 2 ridges.

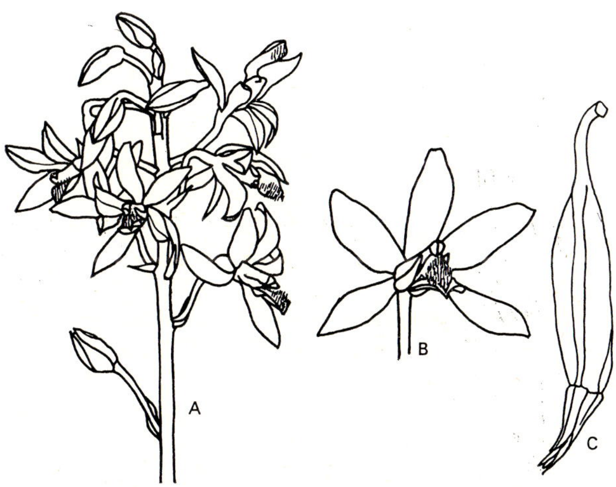


FIG. 269 *Dipodium*: A, flowerhead (× ½); B, flower (actual size); C, fruit (actual size).

*D. penduncunculata* var. *gigantea*. Pale yellow. Montane. N.S.W., Vic., Tas.
A slender, smooth orchid 15–35 cm high, with 5–7 long, narrow leaves. Flowers 1–3 on slender stalks, about 4–6 cm in diameter; dorsal sepal egg-shaped; lateral sepals narrow, green, spreading; petals stalked, elliptical. Labellum 3–lobed, with 2 raised longitudinal lines at the base.

*D. punctata*. Yellow. Subalpine. N.S.W.
A rather robust plant 15–40 cm high with 3 semi-succulent leaves 10–20 cm long. Bracts on stem 2, the lower one leafy, the upper one small. Flowers several on the stem; dorsal sepal 8–14 mm long, broad; lateral sepals green, crossed, as long as the petals, 7–15 mm long. Labellum 3–lobed; mid-lobe with 2 prominent smooth ridges, lateral lobes shorter; column short.

*D. sulphurea*. Yellow, brown spots. Montane. N.S.W., Vic., Tas.
A rather robust species 10–60 cm high with 2–3 narrow leaves. Flowers 1–7; dorsal sepal egg-shaped, about 20 mm long; lateral sepals narrow, pointed and spreading, about 22 mm long; petals egg-shaped, spreading, stalked, slightly shorter than the sepals. Labellum about 12 mm long, with single raised longitudinal line.

*ERIOCHILUS* (from Gk *erion,* wool, and *kheilos*, a lip, referring to the flowers. Fig. 271.

*E. cucullatus*. Parson's Bands. White, pink. Subalpine. N.S.W., Vic.
A small plant 7–25 cm high, with hairy stems and a single, basal, egg-shaped, smooth leaf. Flowers 1–3 on the stem, fragrant; dorsal sepal erect, 5–8 mm long; lateral sepals 11–17 mm long on a slender claw; petals narrow. Labellum equal to or shorter than the petals, on an erect claw with ridges of reddish hairs.

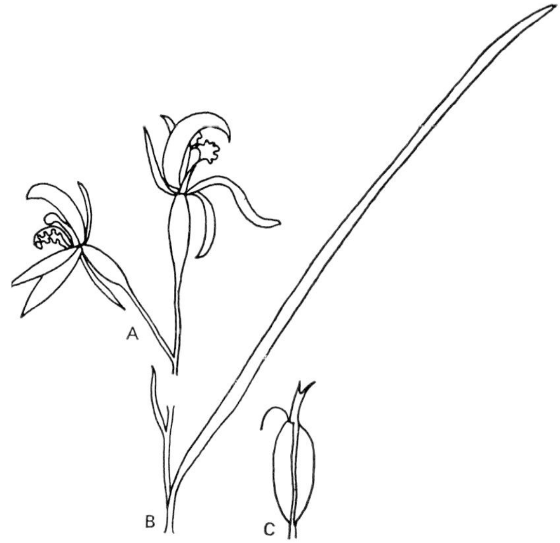

FIG. 271 *Eriochilus*: A, flowers; B, leaf; C, fruit. (All × 1½.)

*GASTRODIA* (from Gk *gastrodes*, pot-bellied, in reference to the shape of the flowers). Fig. 272.

*G. sesamoides*. Dirty-white and brown. Subalpine, montane. N.S.W.
This quaint little orchid occurs in coastal habitats also. Saprophytic, without leaves. Stems rather fleshy, up to 60 cm high with 3–4 bracts.

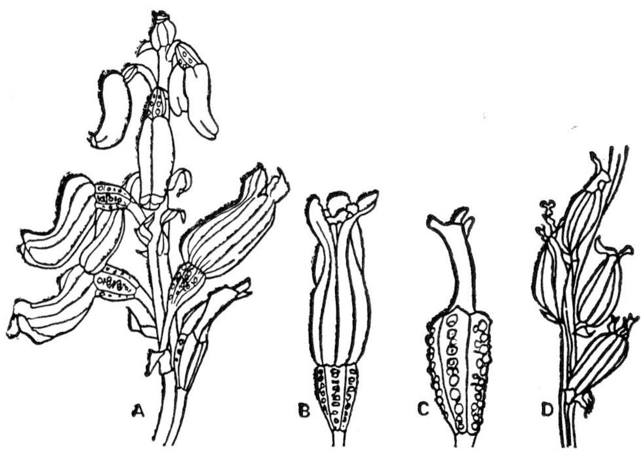

FIG. 272 *Gastrodia*: A, flowerhead (actual size); B, flower (× 1½); C, ripening seedbox (× 1½); D, fruits (actual size).

Flowers in a long head; sepals and petals, except the labellum, joined to form a short, broad tube 10–16 mm long, lobed with a short, blunt spur under the labellum which is short with wavy margins; column elongate, slightly winged.

*GLOSSODIA* (from Gk *glossodes*, tongue-shaped, referring to the appendages at the base of the labellum). Fig. 273.

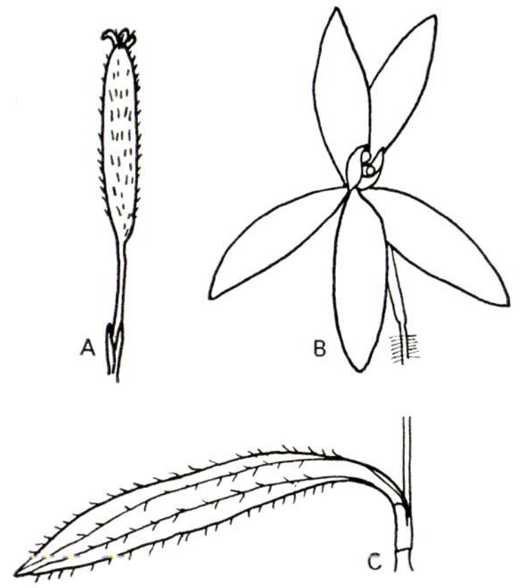


FIG. 273 *Glossodia*: A, fruit; B, flower; C, leaf. (All actual size.)

*G. major*. Blue, purple, white. Montane. N.S.W., Vic., Tas.
A very widely distributed orchid found at lower levels and not ascending higher than montane habitats. Leaves and stems slightly hairy; leaf solitary, basal, oblong, 2.5–5 cm long. Sepals and petals nearly equal, spreading, about 2 cm long. Labellum short, fairly woolly with a solitary, basal, yellow thickening.

*MICROTIS* (from Gk *mikros*, small, and *otes*, an ear, referring to the appendages on either side of the column). Fig. 274. A genus of small, smooth orchids with solitary leaves sheathing from the base for half to two-thirds of the distance to the base of the flowerhead and then forming a long narrow blade. Flowers very small in long terminal spikes.

*M. parviflora*. Green. Montane. N.S.W., Vic.
A quaint little orchid, sometimes growing to 60 cm but usually much shorter, slender and with flowers in a dense spike. Dorsal sepal hooded, 2–3 mm long; lateral sepals fairly broad and not

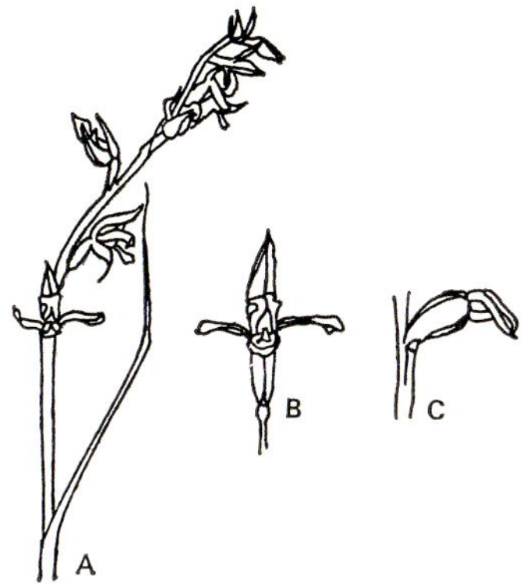


FIG. 274 *Microtis*: A, flowerhead (actual size); B, flower (× 2); C, fruit (× 2).

much longer than the dorsal sepal. Labellum blunt, oblong with 2 calli at the base.

*M. unifolia*. Green. Montane. N.S.W.
A rather robust little plant, 15–40 cm high with fairly dense spikes. Dorsal sepal broad and pointed, about 2 mm long; lateral sepals reflexed, blunt; petals partly hidden by dorsal sepal. Labellum oblong, with crisped margins, blunt; column about 1 mm long.

*PRASOPHYLLUM* (from Gk *prason*, a leek, and *phyllon*, a leaf). Fig. 275. A widely distrib-

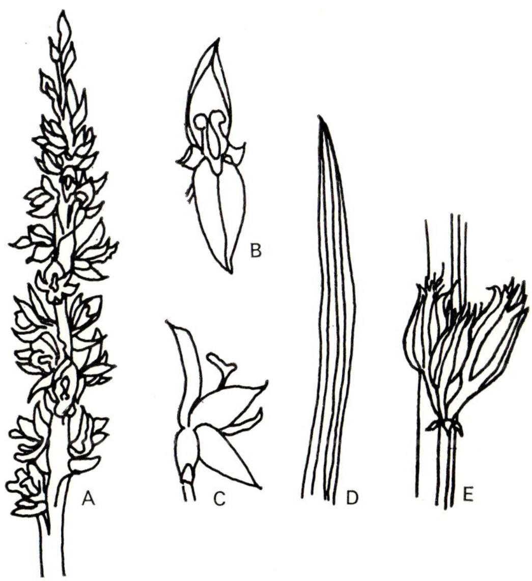


FIG. 275 *Prasophyllum*: A, flowerhead (× $\frac{1}{2}$); B, flower (× 2); C, flower side view (× 2); D, leaf (× 2); E, fruit (actual size).

uted genus. Each plant has a single smooth, narrow leaf and several stalkless flowers in a head. The labellum is undivided with longitudinal thickenings. The dorsal sepal is broad and concave on the inner side; the lateral sepals are narrower. The petals are shorter and narrower than the sepals.

*P. alpinum*. Green, chocolate marks. Alpine, subalpine. N.S.W.

A robust plant 10–23 cm high, the leaf usually longer than the loose, dense flowering stem. Flowering spike 4–8 cm long; dorsal sepal pointed, about 5 mm long; lateral sepals almost as long, united; petals shorter and narrower than the sepals. Labellum conspicuously curved back with the calli almost at the tip and very prominent.

*P. archeri*. Greyish-green, purple marks. Montane. N.S.W., Vic., Tas.

A slender orchid, variable in height and widely distributed. Flowers few in a head; dorsal sepal broad, pointed, about 5 mm long; lateral sepals longer and narrower. Labellum often recurved, divided at the base, with dark calli.

*P. flavum*. Yellowish-green. Montane. N.S.W., Vic.

A small plant with stout brown stems 25–40 cm high with a loosely sheathed leaf. Dorsal sepal broad and pointed, about 4 mm long; lateral sepals joined for the greater part of their length; petals narrower. Labellum erect at base, broadest in middle.

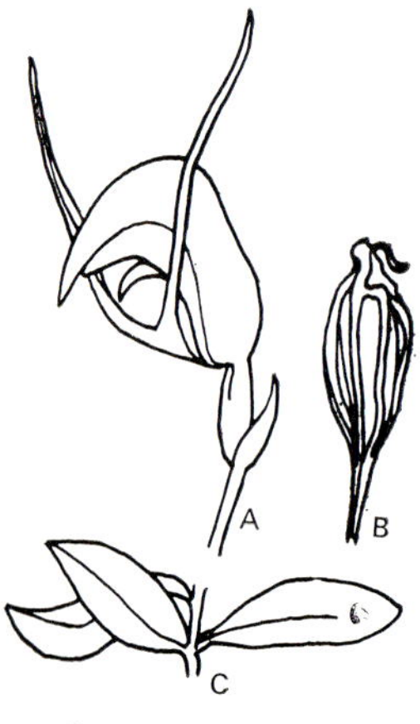


FIG. 276 *Pterostylis*: A, flower (actual size); B, fruit (× 1½); C, leaves (actual size).

*P. fuscum*. Tawny greenish. Subalpine, montane. N.S.W., Vic.

A slender plant 15–30 cm high, fragrant. Dorsal sepal broad, erect, 6–8 mm long, lateral sepals longer; petals shorter than the sepals, narrow, erect. Labellum recurved with wavy margins, the calli sticky and conspicuous.

*P. patens*. Yellow, green, brown. Subalpine, montane. Vic.

A slender plant 30–50 cm high, the leaf well above the base of the plant stem. Flowers few to many in a head; dorsal and lateral sepals about 8 mm long, free; petals shorter and narrower than the sepals, spreading. Labellum flexed, with wavy margins.

*P. suttonii*. Cream. Alpine, subalpine. N.S.W., Vic.

A small orchid 25–30 cm high, often in large stands in grasslands. Leaves narrow. Flowers 6–9 in a head, each shortly stalked and with small bracts at the base; dorsal sepal narrow and hooded; lateral sepals free with a dark stripe down the centre; petals broader and longer than the sepals and also with a dark stripe. Labellum egg-shaped, on a short claw.

*P. tadgellii*. Yellow or greenish. Alpine. Vic.

A short, stout orchid 15–25 cm high, the leaf longer than the flowering head. Flowers 8–12 on short stalks; sepals about 5 mm long; dorsal sepal broad, lateral sepals joined halfway. Labellum with triangular calli.

*PTEROSTYLIS* (from Gk *pteron*, a wing, and *stylos*, a column). Fig. 276. A well-distributed orchid genus with leaves often in a rosette, sometimes absent at flowering time. The orchid flowers are solitary or few in a head, usually green, often with red or red-brown markings. The dorsal sepal is very concave, usually much incurved, the edges often joined with the petals to form a hood over the column. The lateral sepals are joined, forming a lip below the column. The labellum is small and attached to the base of the column by a movable claw, more or less irritable.

*P. alpina*. Green. Alpine, subalpine, montane. N.S.W., Vic., Tas.

A dainty little orchid 13–43 cm high. Leaves along the stem shortly stalked and very similar to the stalkless bracts. Flowers solitary, about 3 cm long; hood curved, blunt; lower sepals erect, united for more than one-third their length. Labellum fairly broad, very straight for three-quarters of its length then sharply pruned down to a blunt point; column not much longer than the labellum, nearly straight.

*P. cucullata.* Green. Subalpine, montane. Vic.
A stout little greenhood 5–20 cm high. Leaves crowded at base of stem, egg-shaped; stem bracts 2–3, large and leafy. Flowers solitary, covered by the leafy bracts; hood about 3 cm long, hairy on the outside, the tip short and pointed; lower lip erect, shortly pointed. Labellum oblong, on a narrow movable claw.

*P. curta.* Green, sometimes tinged brown. Montane. N.S.W., Vic., Tas.
A rather robust little orchid 10–26 cm high. Leaves in a basal rosette, egg-shaped, stalked; stem-bracts 1–2. Flowers usually solitary, up to 3 cm long but often less; hood curving forward for about half its length with a short point; lower sepals erect, united for about half their length. Labellum oblong, 12–14 mm long with a curious twist just above the middle; column very erect, longer than the labellum.

*P. mutica.* Green with rufous markings. Montane. N.S.W., Vic., Tas.
A variable plant with very small flowers in a slightly spiral spike. Leaves 5–14 in a basal rosette, shortly stalked, narrow, egg-shaped; stem bracts 3–7; hood very curved; lower sepals united except at the tips. Labellum oblong, blunt, on a short, very irritable claw; column slightly incurved.

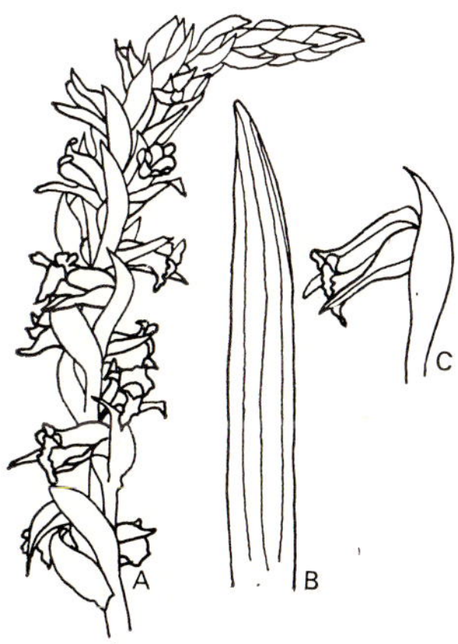


FIG. 277 *Spiranthes*: A, flowerhead; B, leaf (both actual size); C, flower (× 2).

*P. parviflora.* Green, red, brown. Montane. N.S.W., Vic.
A variable plant, usually slender, sometimes robust, 5–35 cm high, extending to other habitats. Basal leaves varying in number; leaves on stem sheathing and scale-like. Flowers 1–10 in a head, very small, 10–13 cm long, very erect; lower sepals united for about two-thirds of their distance, then widely separated, the free portions very short. Labellum blunt, slightly curved in front, 3–5 mm long; column longer than the labellum.

*P. pusilla.* Cream, red, brown. Montane. N.S.W., Vic.
A very variable plant, usually small. Leaves in a basal rosette, numerous, egg-shaped, shortly stalked; stem bracts 1–3. Flowers 1–4 in a head, sometimes more numerous; hood curved for its full length, ending in a short point; lower sepals united for some distance, then turned down and widely expanding. Labellum oblong or egg-shaped, about 5 mm long, with a blunt tip.

*SPIRANTHES* (from Gk *speira*, a wreath, and *anthos*, a flower, in reference to the flower arrangement). Fig. 277.

*S. sinensis.* Ladies' Tresses. Pink. Subalpine, montane. N.S.W., Vic., Tas.
A quaint little orchid that loves boggy places. It grows 15–30 cm high with 3 narrow leaves 4–10 cm long at the base, and 2–3 stem bracts 4–5 mm long. Flowers in a spirally twisted head; sepals and petals almost equal; dorsal sepal egg-shaped, concave; lateral sepals free, spreading. Labellum white; column very short.

*THELYMITRA* (from Gk *thelys*, a woman, and *mitra*, a cap, in reference to the hooded column of some species). Fig. 278. An attractive genus of orchids with solitary leaves and many flowers, often large, in a slender head. The sepals and petals are nearly equal and the labellum is not distinct. The column is broadly winged, the wings united in front at the base, forming an appendage.

FIG. 278 *Thelymitra*: A, fruit (× ½); B, flowerhead (× ½); C, flower (actual size).

*T. aristata.* Mauve. Subalpine, montane. Tas.
A very stout species 60–90 cm high with sword-shaped, channelled sheathing leaves. Flowers in a fairly dense head with bristle-like (aristate) bracts on broad hairy stems.

*T. grandiflora.* Purplish-blue. Montane. N.S.W., Vic.
A robust plant 30–60 cm high with a thick, broad leaf sheathing at the base, and conspicuous stem bracts, the lower ones leafy. Flowers numerous in a head; petals and sepals pointed, broad, about 15 mm long; column erect, hooded; wings wide, inflated.

*T. venosa.* Blue. Alpine, subalpine. N.S.W., Vic., Tas.
One of the daintiest of the alpine orchids, sometimes with delicate pink flowers. A slender plant 15–30 cm high. Leaf narrow, channelled. Flowers 1–5 in a head; sepals and petals thin, veined, about 14 mm long; column erect, about 5 mm long, widely winged, not hooded, the wings produced into 2 blunt appendages.

**Family IRIDACEAE**

In this family the seedbox is placed below the other flower parts and has 3 compartments. There are 3 petal-like sepals, 3 petals, sometimes partly united and often very frail, and 3 stamens. The fruit is a dry, splitting capsule.

*DIPLARRENA* (from Gk *diploos*, double, and *arren*, male, in reference to the two perfect stamens). Fig. 279.

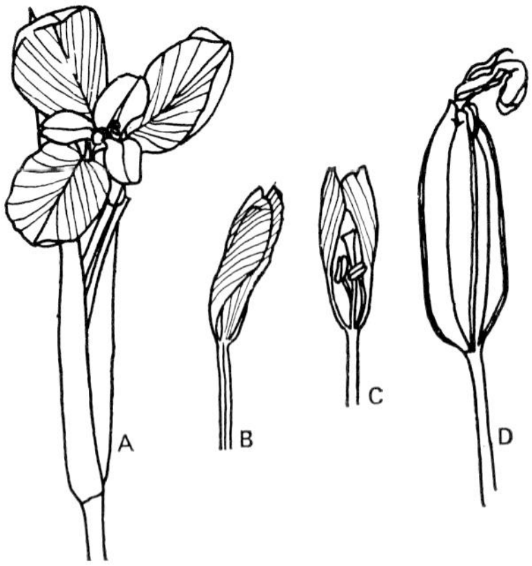


FIG. 279 *Diplarrena*: A, flowerhead; B, bud; C, young flower; D, fruit. (All × ½ except D actual size.)

*D. moroea.* White. Subalpine, montane. N.S.W., Vic., Tas.
A very widely distributed plant found in a variety of other habitats also. Leaves basal, numerous, rather thick, pointed, strap-like, 30–50 cm long. Flowers few in heads, each head enclosed in a pair of narrow bracts; the flowers are fragile, usually opening one at a time, bleaching and fading rapidly; sepals petal-like, oval, spreading, about 5 cm long; petals narrow, about 1–2 cm long; stamens 3, one very short. The alpine form is rather more robust, regarded by some botanists as var. *latifolia*.

*ISOPHYSIS* (from Gk *isos*, equal, and *physis*, nature; reference obscure). Fig. 280.

*I. tasmanica* (syn. *Hewardia tasmanica*). Dark red, purple or yellow. Subalpine. Tas.
This is one of the Tasmanian gems. It is a small, rigid plant with stiff, opposite narrow leaves, usually basal, 10–15 cm long. Flowers solitary on slender stems 11–20 cm high, with a few leafy bracts; sepals and petals similar to each other, very glossy, narrow, spreading, 2.5–4 cm long; ovary only slightly below the other floral parts, a distinction which places it, in the opinion of some botanists, in the family Liliaceae.

*PATERSONIA* (after William Paterson, 1755-1810, early botanical collector in Australia and Lieut.-Governor of New South Wales). Fig. 281.

FIG. 280 *Isophysis*: A, fruiting head (actual size); B, flower (actual size); C, leaf (× ½); D, single fruit (actual size).

FIG. 281 *Patersonia*: A, flower; B, fruits. (Both actual size.)

*P. sericea*. Purple. Montane. N.S.W., Vic.
A small tufted plant with narrow, stiff, grass-like leaves, hairy at the base and 20–30 cm long. The fragile flowers are borne in groups of 1–3 at the ends of stiff, straight stalks, the head enclosed in 2 large brown bracts (spathe); sepals petal-like, large, free and spreading; petals small and inconspicuous. Fruit a triangular capsule.

**Family HYPOXIDACEAE**

*HYPOXIS* (from Gk *hypo*, under, and *oxys*, pointed, referring to the pointed base of the fruit). Fig. 282. A genus of small plants with narrow, grass-like leaves and solitary flowers with a very short tube and then spreading, on slender stalks. There are 3 free sepals, 3 free petals, and 6 stamens. The anthers are lobed at the base and inserted above the other flower parts. The style is trifid and the seedbox has 2–3 compartments. The fruit is a dry, splitting, oblong capsule.

*H. glabella*. Yellow. Montane. N.S.W., Vic., Tas.
A smooth plant of grassy habitats. Leaves narrow, channelled above, 7.5–15 cm high, sheathing at the base. Flowers about 12–15 cm in diameter, solitary, on stalks shorter than the leaves.

*H. hygrometrica*. Yellow. Subalpine, montane. N.S.W., Vic., Tas.

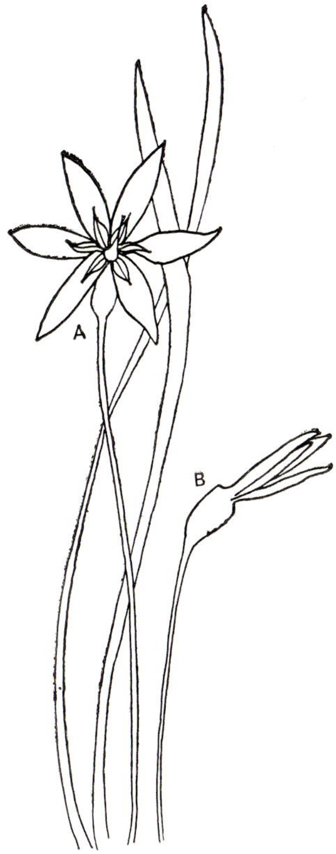


FIG. 282 *Hypoxis*: A, flower; B, young fruit. (Both × 1½.)

A slightly hairy plant up to 25 cm high. Leaves narrow, basal, sprinkled with a few hairs, about 18 cm long. Flowers 8–14 mm in diameter, 1–5 together. Fruit rather elongated.

## Family XANTHORRHOEACEAE

A group in which the flowers are usually arranged in spikes or globular heads. The sepals and petals are 3, often papery. There are 6 stamens and an ovary of 3 parts.

*LOMANDRA* (from Gk *loma*, a fringe, and *andros*, a man, referring to the circular margin of the anthers in some species). Fig. 283.

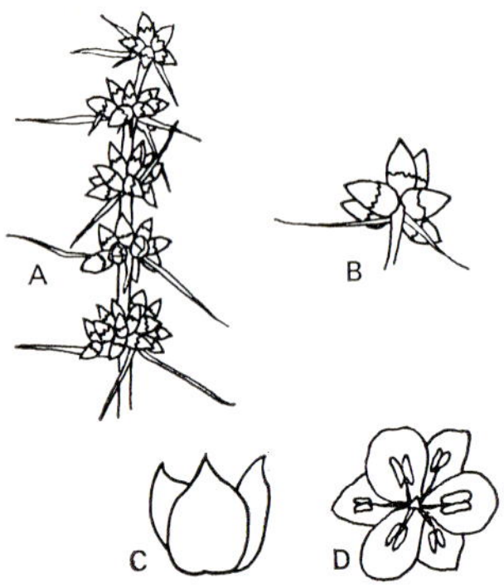


FIG. 283 *Lomandra*: A, flowerhead (× ½); B, fruit (actual size); C, female flower (× 2); D, male flower (× 2).

An interesting genus of little tufted plants, common in a variety of habitats. They have long, radical, harsh leaves and flowers either male or female in long spikes or globular heads or in other arrangements.

*L. filiformis*. Yellow. Montane. N.S.W., Vic.
A small shrub with narrow leaves 15–50 cm long. Flowers solitary or 2–3 together, on short curving stems in the axils of narrow, pointed bracts; flower stems 7.5–20 cm high with short bracts; stamens 6.

*L. longifolia* ssp. *exilis*. Creamish. Montane. N.S.W., Vic.
One of the largest of the lomandras, growing to about 50 cm with stiff, narrow leaves 30–50 cm long, with orange or brown margins at the base; flowers in dense clusters, sweetly scented. It is widely distributed in other habitats also.

*L. multiflora*. Creamish. Montane. N.S.W., Vic.
This is a smaller species, the leaves narrow, usually 40–60 cm long. Male flowers stalked and in clusters; female flowers without stalks. wrinkled. It is widely distributed in other habitats also.

*XANTHORRHOEA* (from Gk *xanthos*, yellow, and *rheo*, flow, in reference to the resin produced). Fig. 284.

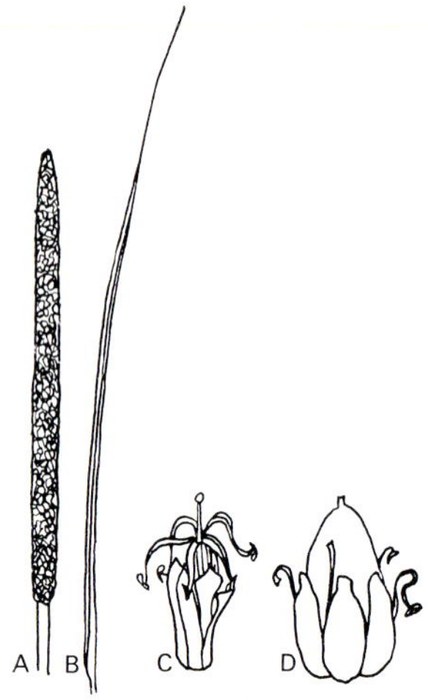


FIG. 284 *Xanthorrhoea*: A, flowering spike (× 1/6); B, leaf (× 1/6); C, flower (× 2); D, fruit (× 2).

*X. australis*. Grasstree Blackboy. Cream. Montane. N.S.W., Vic., Tas.
These majestic plants sometimes occur among the eucalypts of the dry sclerophyll forests and are quaintly attractive against the rocky outcrops. They rise on rugged, stunted and twisted, blackish stems to fan out into a crown of narrow, harsh leaves, long and narrow. From the centre of this crown the tall flower-spikes arise. Each flower is set into the flower spike and surrounded by numerous small bracts with membranous sepals and spreading petals. The fruit is a shiny brown capsule.

## Family AGAVACEAE

This family is represented in the alpine areas by a single genus with 2 species. These are tufted, mat-like plants, the 2 species growing in association with each other, forming an extensive cover.

*ASTELIA* (from Gk *aster*, a star, in reference to the flower shape). Fig. 285.

*A. alpina*. Cream. Alpine, subalpine. N.S.W., Vic., Tas.
Leaves tufted, silvery-grey, about 1 cm wide or wider, with a sheathing base and tapering to a

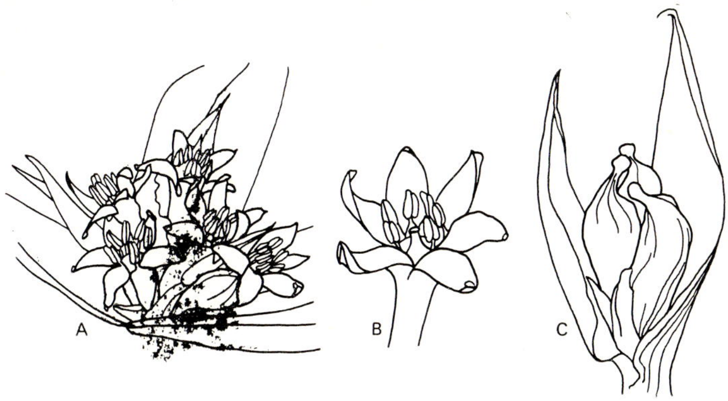


FIG. 285 *Astelia*: A, flowerhead (actual size); B, single flower (× 2); C, fruit (× 2).

fine point, very variable in length, up to 30 cm but usually shorter, covered with fine silver hairs. Flowers small in dense heads, partly concealed in the centre of the overlapping sheathing leaf bases, often male or female; petals and sepals spreading; seedbox of one compartment. Fruit a bright-red, very attractive, oval berry.

*A. psychrocharis*. Cream. Alpine, subalpine. N.S.W.

This species grows in association with *A. alpina*. Leaves narrow, less than 1 cm in width, green, scarcely hairy, the base of the sheathing leaves not silky-hairy but surrounded with the fibrous remains of the old, dead leaves. Seedbox with 3 compartments. Fruit a red oblong berry.

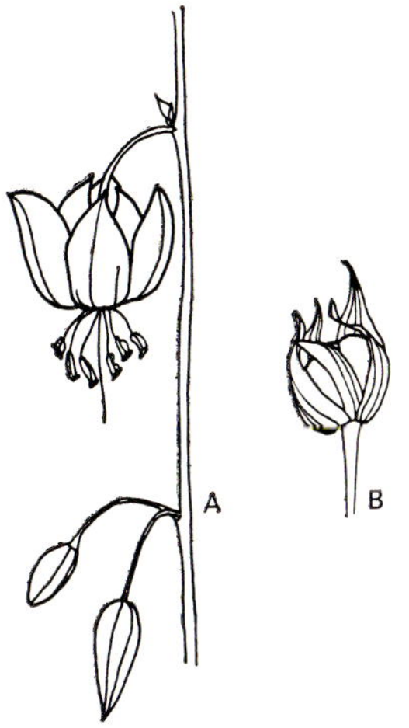


FIG. 286 *Arthropodium*: A, flowerhead; B, fruit. (Both actual size.)

**Family LILIACEAE**

This is a widely distributed group of plants, mainly small, sometimes creeping or scrambling, often with bulbs. There are 3 sepals, commonly coloured or white and looking like petals, 3 petals and 6 stamens, usually all free from one another, and a seedbox of 3 parts inserted above the other flower parts. The fruit is a dry, splitting capsule or a succulent berry. There are a considerable number of members of this family in alpine areas, most of them small plants with attractive flowers.

*ARTHROPODIUM* (from Gk *arthron*, a joint, and *podion*, a little foot, referring to the jointed flower-stalks). Fig. 286.

*A. milleflorum*. Lavender Shower. Lilac. Subalpine, montane. N.S.W.

A delightful little plant widely distributed at lower elevations also, with narrow, basal, grass-like leaves and flowers on long slender stems. It often grows in clumps in grassland, the slender stems moving in the gentlest breeze. Flowers 2–3 together on individual short stalks in long, graceful clusters; sepals and petals turned back to expose dainty, bearded stamens. Fruit a round, dry, splitting capsule.

*BLANDFORDIA* (after George, Marquis of Blandford, 19th-century plant lover). Fig. 287.

*B. punicea*. Christmas Bell. Red, tipped yellow. Subalpine, montane. Tas.

This handsome plant is readily recognized by its large, flamboyant bells, deep-red lined and tipped with bright-yellow, occurring in numerous situations on the south and west of the island. Leaves numerous, basal, 15–40 cm long with

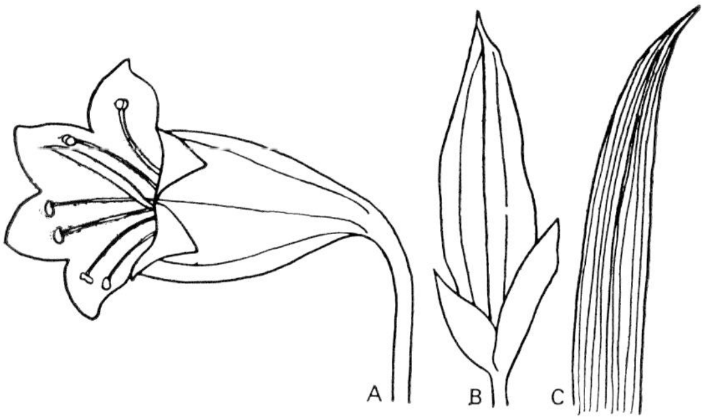


FIG. 287 *Blandfordia*: A, flower; B, fruit; C, leaf. (All actual size.)

finely toothed margins. Flowers numerous in loose heads on stout, smooth, rather fleshy stems; sepals and petals united to form a tube 2.5–4 cm long, bell-shaped, flared at the tip; stamens joined to petals and sepals. Fruit a 3-angled, long, slender, splitting capsule.

*BULBINE* (Lat. name for a kind of bulbous plant). Fig. 288.

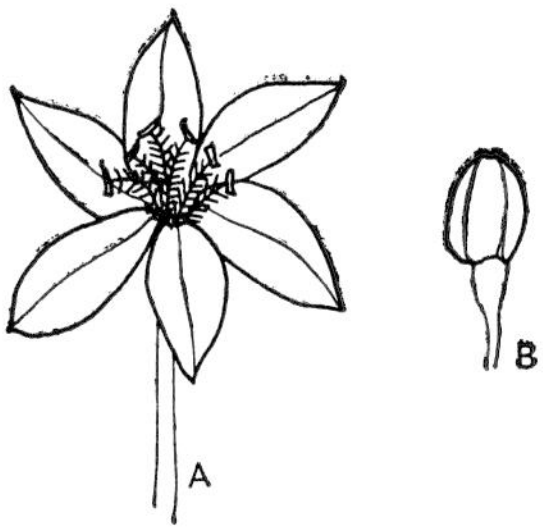


FIG. 288 *Bulbine*: A, flower; B, fruit. (Both × ½.)

*B. bulbosa*. Yellow. Subalpine, montane. N.S.W., Vic., Tas.

Widely distributed in lower elevations also, this is a swamp lover. In a situation to its liking it may cover a large area, and the bright-yellow heads make a spectacular sight. It is a small plant 15–30 cm high. Leaves basal, often fleshy, narrow, grass-like, frequently grooved. Flowers tulip-shaped, 2–3 cm in diameter, several together in rather stiff heads; sepals and petals translucent; stalks of stamens bearded. Fruit a rounded capsule.

*CAESIA* (after Federico Cesi, 1585-1630, Italian naturalist). Fig. 289.

FIG. 289 *Caesia*: A, flowerhead; B, buds. (Both actual size.)

*C. alpina* (syn. *Chlorophyton alpinum*). Cream. Alpine. Tas.

A small plant with grass-like leaves. Flowers solitary or a few together at intervals along the stem on short stalks, the sepals and petals about 5 mm long.

*DIANELLA* (diminutive of Diana, goddess of hunting). Fig. 290.

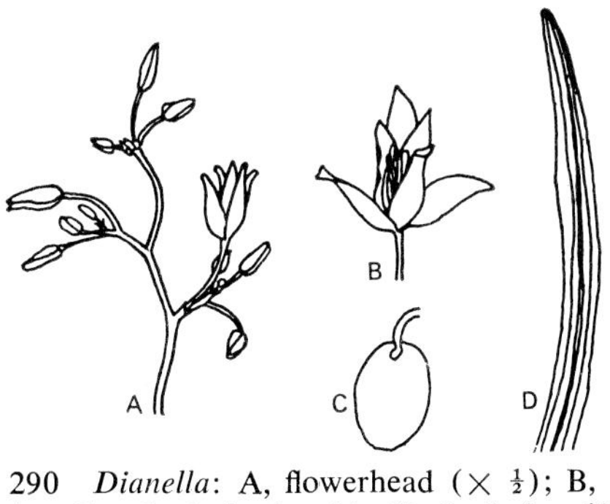


FIG. 290 *Dianella*: A, flowerhead (× ½); B, flower (× 2); C, fruit (× 2); D, leaf (× ½).

*D. tasmanica*. Flax Lily. Blue; berries blue. Subalpine, montane. N.S.W., Vic., Tas.

A straggling but sturdy plant preferring the shelter of taller plants. Leaves basal, stem-clasping and hard, 30–50 cm long. Flowers in heads up to 25 cm long with many flowers to the head; sepals and petals turned back to reveal brilliant yellow stamens. The oval-shaped berries are very attractive.

*DRYMOPHILA* (from Gk *drumos*, a coppice or wood, and *philos*, loving). Fig. 291.

FIG. 291 *Drymophila*: flower and fruit (actual size).

*D. cyanocarpa*. White; berries turquoise. Montane. N.S.W., Vic., Tas.
A small, rather fragile plant about 20–25 cm high with spreading leaves, oblong and almost stalkless, in 2 opposite rows. Flowers solitary or 2 together on recurved stalks in the leaf axils; petals distinctly veined. The berry is about 1 cm in diameter, and round.

*HERPOLIRION* (from Gk *herpo*, to creep or spread, and *leirion*, a lily). Fig. 292.

FIG. 292 *Herpolirion*: A, bud; B. fruit; C, flowerhead. (All actual size.)

*H. novae-zelandiae*. Sky Lily. White, blue. Alpine, subalpine. N.S.W., Vic., Tas.
An interesting little plant growing very close to the ground, sometimes in open country but more often in grasslands and completely hidden from view. It is one of the joys of alpine wandering. Leaves basal, 2.5–4 cm long, grass-like, springing from an underground stem. Flowers starry or funnel-shaped, 5–8 cm in diameter, the sepals and petals free, narrow, and twisted after flowering. Fruit a capsule.

*LAXMANNIA* (from Lat. *laxus*, loose, referring to the heads of flowers). Fig. 293.

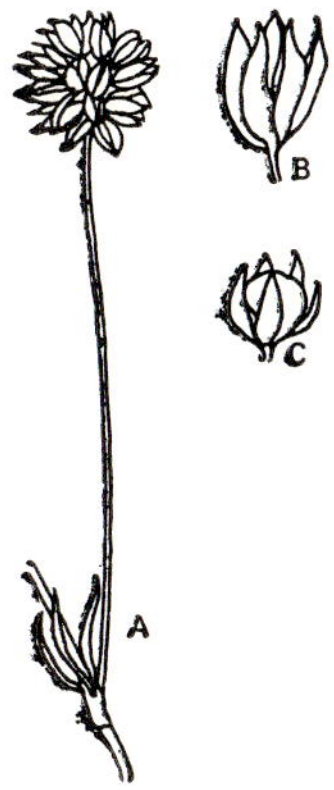


FIG. 293 *Laxmannia*: A, flowerhead (× ½); B, flower (× 2); C, fruit (× 2).

*L. gracilis*. Pink. Montane. N.S.W.
This species is widely distributed in other habitats also. It is a stiff little plant with tufted narrow leaves which are broad at the base and arranged along the stem. Flowers in short heads, almost stalkless, with bracts at the base of the flowers; sepals dry and colourless; petals joined or almost joined; 3 inner stamens joined to petals. Fruit a capsule.

*MILLIGANIA* (after Dr Joseph Milligan, onetime secretary of the Tasmanian Philosophical

FIG. 294 *Milligania*: A, flower (× 1½); B, flowerhead with leaf (actual size); C, fruit (actual size).

Society). Fig. 294. A charming little genus of several species, mostly Tasmanian. The sepals and petals are equal to one another, joined at the base and spreading to display the stamens at the mouth of the tube. Fruit a capsule.

*M. densiflora.* Cream. Alpine, subalpine. Tas.
A compact plant with leaves 15–20 cm long, often broad, with a sheathing base. Flowering stem 10–15 cm long, stout; flowers very numerous in rather narrow dense, branched heads; sepals and petals narrow.

*M. stylosa.* Cream. Alpine, subalpine. Tas.
Very similar to *M. densiflora* but sepals and petals only about half as long and reflexed. Capsule very small.

*SMILAX* (Gk name for a kind of tree). Fig. 295. *S. glyciphylla*, from New South Wales, has broad leaves with 3–4 conspicuous longitudinal veins, grey-green below and without prickles. Flowers either male or female on one plant, arranged in wheel-like heads; sepals and petals free, greenish-white in colour. Fruit succulent, round and black.

*S. australis*, from New South Wales and Victoria, has oblong or rounded leaves with 5 con-

FIG. 295 *Smilax*: A, flowerhead (actual size); B, flower (× 2); C, fruit (× 2).

spicuous longitudinal veins, green on both surfaces, with prickles on the stem. The flowers and fruit are very similar to *S. glyciphylla.*

*STYPANDRA* (from Gk *stype*, tow, and *andros*, a man, referring to the woolly stamens). Fig. 296.

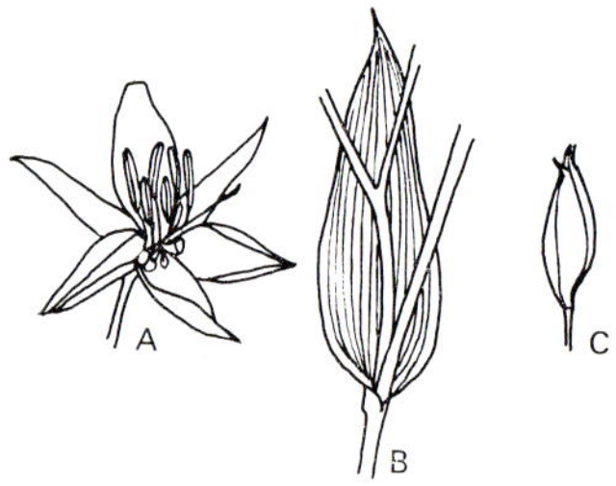


FIG. 296 *Stypandra*: A, flower; B, leaf; C, fruit. (All actual size.)

*S. umbellata.* Yellowish. Montane. N.S.W.
An erect little plant widely distributed in other habitats also. Leaves long and narrow, in 2 opposite rows on the stems, 20–30 cm long. Flowers in loose heads of 2–3 on very slender stalks; sepals and petals free; stalks of stamens covered with yellow hairs; anthers rolled back after flowering. Fruit a capsule.

*THYSANOTUS* (from Gk *thysanotos*, fringed, referring to the petals). Fig. 297.

*T. tuberosus.* Fringed Violet. Purple. Montane. N.S.W., Vic.
A delicate little plant extending far beyond these habitats. It is erect with basal, narrow leaves

10–20 cm long with a few also on the stem. Flowers in loose, branched heads of 1–4 with the petals much larger than the sepals and conspicuously fringed on the margins, free and spirally twisted after flowering. Fruit a capsule.

FIG. 297 *Thysanotus*: A, flowerhead; B, fruit. (Both actual size.)

# GYMNOSPERMS

Gymnosperms are evergreens with unisexual flowers without either calyx or corolla, and with ovules not enclosed in an ovary. The seeds are not enclosed in a fruit but are solitary on the surface of dry scales arranged in cones, the staminal cones being catkin-like. These plants are not very common in alpine areas, with only 2 genera represented on the mainland and 7 in Tasmania.

*ATHROTAXIS* (from Gk *athros*, crowded, and *taxis*, order, referring to the cones). Fig. 298.

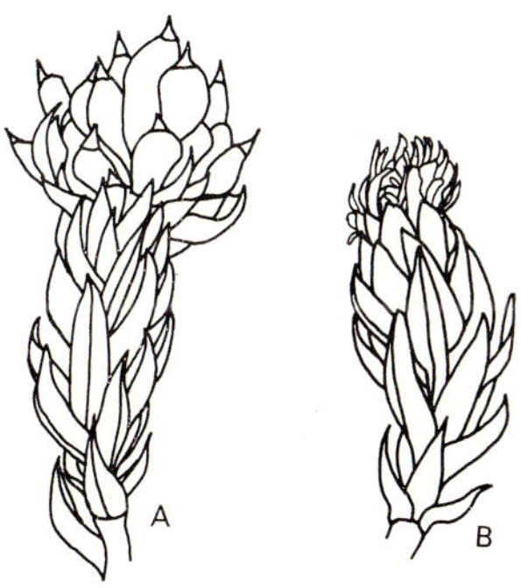


FIG. 298 *Athrotaxis*: A, female cone; B, male cone. (Both actual size.)

*A. cupressoides*, pencil pine, a graceful tree 6–11 metres high and much branched, is a subalpine and montane species from Tasmania. Leaves thick, rhomboid, keeled, blunt and overlapping, pressed against the stem, 2–4 cm long. Male cones small and numerous with overlapping scales; female cones with 8–20 overlapping scales.

*A. laxifolia*, a rare Tasmanian species very like *A. cupressoides*, has slightly spreading leaves with the apex incurved and occurs in the same locality; usually solitary. Probably a hybrid.

*A. selaginoides*. King Billy pine, is a common montane and subalpine Tasmanian species reaching a height of 15 metres or more. Leaves loosely overlapping, thick, slightly keeled, pointed, spreading, 6–12 mm long. Seed cone spherical, 1.8 cm in diameter with thick, broad sides. This species is often associated with *Nothofagus cunninghamii*.

*CALLITRIS* (from Gk *kallos*, beauty, and *treis*, three, in reference to the leaf arrangement). Fig. 299. A genus usually associated with dry

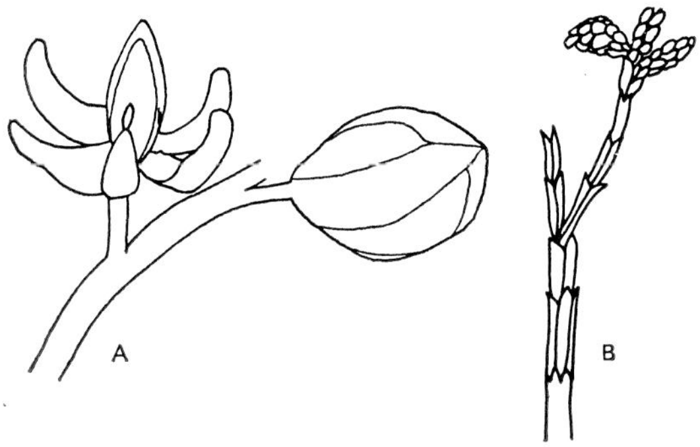


FIG. 299 *Callitris*: A, fruits; B, male cones. (Both actual size.)

inland areas but 2 species are found in montane situations, though they are by no means confined to this habitat. Cypress pines are rather columnar in shape with leaves in rings of 3 or rarely 4, in the mature state adhering to the stem with only the tip free. Male cones usually terminal; female cones borne on short, thick, lateral branches consist of 6 rather woody scales almost globular in shape. Both are mainland species.

*C. endlicheri* (syn. *C. glauca*) is a small, grey-green spreading tree often associated with *Eucalyptus albens*. Leaves 2–4 mm long. Male cone egg-shaped, about 3 mm long; female cone solitary or a few together, up to 2 cm in diameter but often smaller, with 3 alternating scales much shorter than the other 3.

*C. columellaris*, white cypress, is a small grey-green tree with erect or occasionally pendulous branches. Leaves 1–3 mm long. Male cone cylindrical up to 5 mm long; female cone solitary, egg-shaped or globular but flattened at the top, up to 2.5 cm in diameter; cone scales thin, 3 small ones alternating with 3 longer ones, all separating at the base on opening.

*DACRYDIUM* (from Gk *dakros*, a tear, in reference to the weeping habit). Fig. 300.

*D. franklinii*, Huon pine, endemic to Tasmania, grows 20–30 metres, pyramid-shaped with spreading branches and a dense crown. Leaves small, closely set, appearing to be in 4–5 rows, each with a rounded tip. Female cones small, terminal with 8–10 fairly widely separated rather fleshy bracts; male cones also terminal, rounded, about 3 mm long. Occurring mainly in the temperate rainforests, around lakes and on swampy flats up to 2,500 feet. A fine timber, now very rare.

FIG. 300 *Dacrydium*: female cone (× 3).

*DISELMA* (from Gk *dis*, two, and *selma*, a standard or deck, possibly referring to the two sets of scales which occur immediately above one another on the female cones). Fig. 301.

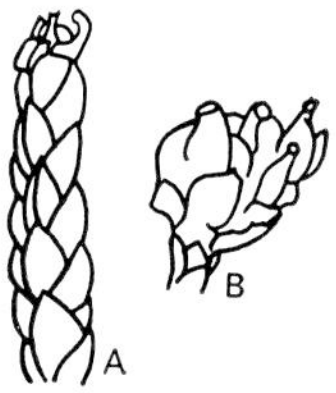


FIG. 301 *Diselma*: A, branch with fruiting head (× 2); B, fruiting head (× 4).

*D. archeri* is an upright shrub 1.5–2.5 metres high, sometimes higher. Leaves very small, blunt, keeled, closely overlapping and pressed against the stem, making the branches seem 4-angled. Male cones small, solitary and terminal, rather like the ends of the normal branches; female cones solitary and terminal. Seeds winged. This species likes wet, subalpine habitats, sometimes occurring as high as 4,000 feet. It is endemic to Tasmania.

*MICROCACHRYS* (from Gk *mikros*, small, and *kachrus*, a cone). Fig. 302.

*M. tetragona* is a creeping pine confined to Tasmania and occurring at alpine and subalpine levels. Leaves very small, blunt, thick, overlapping, pressed against the stem and opposite, making the stems appear 4-angled. Male cones small, ovoid and terminal; female cones terminal, 6–8 cm long. Seeds scarlet on fleshy scales.

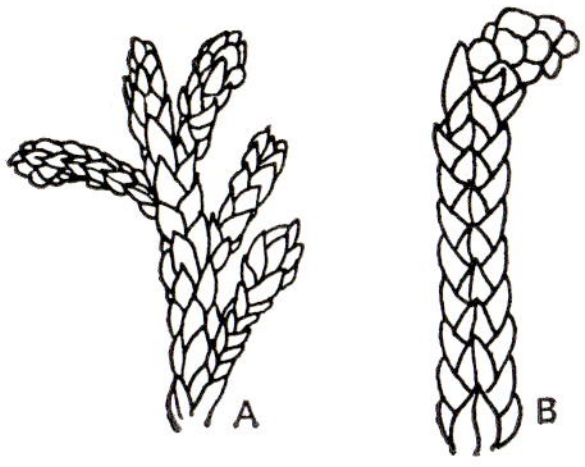


FIG. 302 *Microcachrys*: A, male cone (× $2\frac{1}{2}$); B, female cone (× 3).

*MICROSTROBUS* (syn. *PHEROSPHAERA*) (from Gk *mikros*, small; *strobis*, a cone). Fig. 303. *M. niphophilus* is a small shrubby pine

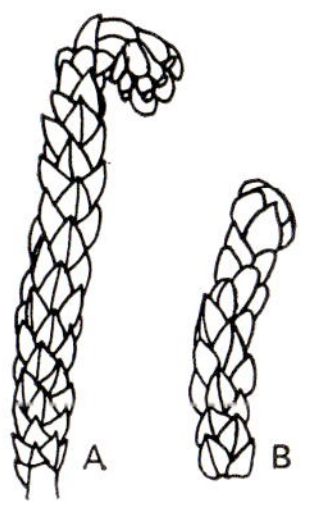


FIG. 303 *Microstrobus*: A, female cone; B, male cone. (Both × 3.)

of Tasmanian subalpine habitats, particularly round tarns and lakes. It grows 1–2 metres high with small, closely overlapping leaves pressed against the stems in 4–5 rows, thick and blunt, 1–5 mm in diameter with purplish-brown scales at the base. Female cones solitary, terminal, pendulous, 2–3 cm long. Seeds hard, brown, produced on upper surface of scales.

*PHYLLOCLADUS* (from Gk *phyllon*, a leaf, and *cladodes,* having many shoots, referring to the leaf-like branches). Fig. 304.

*P. aspleniifolius*, commonly known as celery-top pine, is found in the temperate rainforests and the wet sclerophyll forests of Tasmania. It is an interesting little tree 6–18 metres high with fine pine-like foliage in the very early stages followed by cladodes of a delicate pale-green colour formed by the branches. These are opposite or whorled, thick, leathery, blunt and somewhat indented, 1.5–8 cm long. The male flowers are

FIG. 304 *Phyllocladus*: A, flowerhead (× $1\frac{1}{2}$); B, male cone (× 2); C, female cone (× 2).

in elongated cones 3–5 cm long, 2–3 together at the ends of the branches; the female flowers are in stalked clusters, 3–4 together at the edges of of the cladodes, on pinkish-red bracts. The seeds are black, surrounded by a white fleshy substance.

*PODOCARPUS* (from Gk *podos*, a foot and *carpos*, a fruit, referring to the swollen structure undcr thc fruit). Fig. 305.

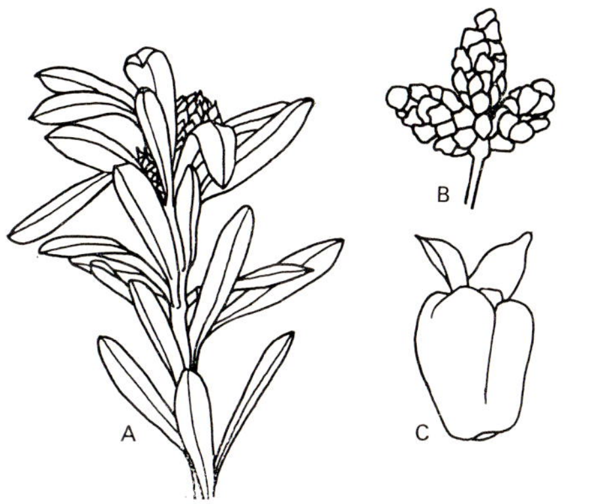


FIG. 305 *Podocarpus*: A, flowerhead (× 2); B, pollen cones (× 3); C, fruit (× 2).

*P. lawrencei* is a much-branched creeping plant from alpine and subalpine parts of the mainland and Tasmania. It often spreads over boulders. Leaves crowded, dark-green, narrow, 6–12 mm long, the new growth a delicate light shade. Male cones small and solitary, or 3–4 together in the axils of the upper leaves with creamy-lilac or crimson scales. The seed is black or greenish. with a succulent red base.

# FERNS AND FERN-LIKE PLANTS
# PTERIDOPHYTA

The ferns and fern-like plants do not bear flowers, even such reduced flowers as the cones of the gymnosperms. They produce spores—in the true ferns usually on the leaves of fronds, and in related plants often in heads or cones. These spores are asexual and develop into small moisture-loving structures which produce egg and sperm cells. These, on fertilization, may develop into individuals similar to the plant of the generation before. Ferns and fern-like plants are not conspicuous among the alpine flora, for only a few types can withstand the subalpine and even few the alpine climate. The protected tree-fern gullies of the mainland montane areas house some smaller species also, and several species occur in temperate rainforests of Tasmania.

**Fern-like Plants**

*AZOLLA* (from Gk *azo*, to dry, and *ollo*, to kill, the plants being unable to live in dry conditions). *A. filiculoides* var. *rubra* is a floating plant of wide distribution in New South Wales, Victoria and Tasmania, spreading a red carpet, sometimes very extensively, over the surface of the water.

*BOTRYCHIUM* (from Gk *botrys*, a cluster of grapes, in reference to the spore masses). *B. australe* lives in many places, including the margins of subalpine streams in Victoria and Tasmania. It has a thick rootstock bearing a few fleshy 3–4 times divided leaves 10–20 cm long; the spore-bearing stem is taller, much branched, 2.5–7.5 cm long.

*ISOETES* (from Gk *isos*, equal, and *etos*, a year, the leaves remaining throughout the year). A genus of small water plants commonly known as quillworts, these have grasslike leaves in tufts. The spore cases are embedded in the upper side of the broadly winged bases of the leaves.

*I. elatior* is a Tasmanian plant growing at 4,000 feet and lower. It has a 3–lobed rootstock and thin grass-like leaves up to 60 cm long.
*I. gunnii* is a Tasmanian alpine plant with a hard 3–lobed rootstock and thick, leathery leaves up to 15 cm long. Spore cases brown and shiny.

*LYCOPODIUM* (from Gk *lykos*, a wolf, and *podos*, a foot, the creeping stems allegedly resembling a wolf's foot). Known as the club mosses, these plants are closely related to ferns. In general appearance they rather resemble giant mosses with straggling or erect branches, carrying very fine, closely set leaves, some of which bear the spores.

*L. deuterodensum* is found in subalpine and montane habitats in New South Wales, Victoria and Tasmania. It grows 60–90 cm high from a creeping base and is bushy at the top with small crowded leaves 2–3 cm long. Spores in stalkless spikes about 10–25 mm long at ends of branches.

*L. fastigiatum* is a Tasmanian and mainland species found along mountain streams in subalpine and montane situations. It has ascending branches and narrow, pointed leaves 2–3 mm long; the spore-bearing leaves form dense spikes 2.5–5 cm long, often 2 or more together.

*L. laterale,* a subalpine and montane dweller in New South Wales, Victoria and Tasmania, is a slender trailing plant with crowded, pointed leaves about 6 mm long. Sporeheads in stalkless, lateral short spikes.

*L. scariosum* is a prostrate, branching plant with broad, spreading leaves 0.3–0.4 cm long in 2 rows on the upper side of the stem, and smaller leaves under the stem. It is an alpine species, rather rare in Victoria but fairly common in Tasmania.

*L. selago* is a mainland subalpine and montane dweller with stems in forked tufts 7.5–15 cm high bearing dark-green crowded leaves about 6–8 mm long. Sporeheads in axils of upper leaves, not forming distinct spikes.

*MARSILEA* (after L. F. Marsigli (Marsileus), d. 1730, an Italian naturalist). *M. drummondii*, commonly known as nardoo, is a mainland plant of wide distribution which occurs in montane situations, sometimes in water, sometimes in mud. Its clover-like leaves are borne on erect stalks, and the spores in little rounded sporeheads arise from the base of the plant.

*OPHIOGLOSSUM* (from Gk *ophis*, a snake, and *glossa*, a tongue, adders' tongues being an old name for these plants). The spores in this group of plants are borne as an offshoot from the stem of the normal foliage, thus giving a forked appearance. The spore cases are embedded in 2 rows along the ends of these fertile branches.

*O. lusitanicum* ssp. *coriaceum* occurs in New South Wales, Victoria and Tasmania in alpine grasslands and at lower elevations. It has a knobby rootstock with leaves 2.5 cm or more long, broad and net-veined, sometimes rather fleshy. The spore-bearing stems are taller, terminating in a short bilateral spike.

## TRUE FERNS

The ferns proper range in size from the giant tree ferns to the tiny filmy ferns, and their general appearance shows a wide range of variation too.

### Tree Ferns

*DICKSONIA* (after Dickson, English cryptogamic botanist). The only representative of the tree ferns here is the soft tree fern, *Dicksonia antarctica*, which occurs in montane gullies in New South Wales, Victoria and Tasmania, sometimes in dense stands. It has a tall, straight stem reaching 6–15 metres in height and covered with matted roots. The divided fronds are 1.5–5.3 metres long.

### Filmy Ferns

*APTEROPTERIS* (from Gk *a*, without, *pteron*, a wing, and *pteris*, a fern) and *HYMENOPHYLLUM* (from Gk *hymen*, a membrane, and *phyllon*, a leaf) are small delicate-looking plants, common in a variety of moist habitats and often growing on the stems of tree ferns and other plants. They are common in the temperate rainforests of Tasmania and occur also in selected moist parts of the montane while 1 or 2 species find their way as high as alpine elevations. They are all delicately beautiful with translucent fronds bearing dark veins, 2–3 times divided with narrow leaflets. The spores are borne in cup-like receptacles on the tips of the leaflets. *Apteropteris malingii* (syn. *Hymenophyllum rarum*) has dainty greyish-green fronds and is a Tasmanian species of fairly wide distribution occurring at subalpine elevations, commonly on the stems of King Billy pine, *Athrotaxis selaginoides*. *Hymenophyllum cupressiformis* (syn. *H. tunbridgense)* is common in montane situations at all elevations, often on tree fern trunks; it has narrow fronds 2.5–5 cm long. *H. peltatum* is similar to *H. cupressiformis* but is smaller and has finer foliage; it occurs as high as alpine elevations on the mainland and in Tasmania.

*MECODIUM* (from Gk *mecodios*, seen by the way). *M. rarum* has narrow, often irregular fronds 5–7.5 cm long and occurs as high as 4,000 feet in Tasmania and at somewhat lower elevations on the mainland.

### Maidenhair Ferns

*ADIANTUM* (from Gk *adiantos*, dry, the fronds when plunged into water coming out dry). *A. aethiopicum*, the common maidenhair fern, is very widely distributed at lower elevations in New South Wales, Victoria and Tasmania and is found in moist situations also. It has dainty light-green fronds 18–40 cm high on reddish-brown stems.

*CHEILANTHES* (from Gk *khelios*, a lip, and *anthos*, a flower, referring to the covering over the spore masses). *C. tenuifolia*, rock fern, is one of the most widely distributed of the maidenhair ferns and extends to montane areas in New

South Wales, Victoria and Tasmania. It is a dainty little plant up to 30 cm high with thrice-divided fronds with glossy brown stems, the leaflets deeply cut. The spore masses are in short, interrupted lines along the edges of the leaflets.

*PELLAEA* (from Gk *pellos*, dusky, referring to the fronds). *P. falcata*, sickle fern, is a New South Wales species reaching to montane habitats. It is a small plant with a scaly rhizome and once-divided fronds on wiry stems, the fronds about 13–40 cm long. The spore masses are in continuous lines on the margins of the leaflet except at the blunt end and often in the pointed tip.

### Polypod Ferns

*GRAMMITES* (from Lat. *gramma*, a line, referring to the arrangement of the spore masses), *MICROSORIUM* (from Gk *mikros*, small, and Lat. *sorium*, a spore case) and *PYRROSIA* (from Gk *pypa, hosos,* very short point). These ferns have creeping above-ground stems, some of them using these to attach themselves to the trunks of trees. The spores are in rounded masses variously disposed on the backs of the fronds.

*Grammites armstrongii* is a fern of wide distribution in New South Wales, Victoria and Tasmania, extending to montane parts. It is epiphytic, common or tree trunks, with simple, slightly lobed short tufted fronds about 2–15 cm long covered with long hairs. The spore masses are rounded and on each side of the mid-vein which is very prominent.

*Microsorium diversifolium* is an abundant species in many habitats in New South Wales, Victoria and Tasmania, sometimes found in crevices at alpine elevations. It has a short stem with leathery fronds, sometimes undivided, sometimes divided, 10–15 cm long.

*Pyrrosia rupestris* is a small epiphyte with a creeping stem and undivided, thick, leathery, hairy, variable fronds 1–2.5 cm long. The spore masses are orb-shaped, irregularly and densely crowded with 1–4 rows in each side of the midrib and becoming confluent with age. It is a mainland montane species.

### Spleenwort Ferns

*ASPLENIUM* (from Gk *asplenon*, name of a fern supposed to cure diseases—from *am*, without, and *splen*, spleen). A genus of ferns with long narrow spore masses running along the veins diverging from the midrib.

*A. flabellifolium,* necklace fern, occurs from alpine habitats to sea-level in New South Wales, Victoria and Tasmania. It is a small, well-known fern with weak, drooping branches bearing fan-shaped bluntly toothed fronds. A rare small species, more common in New Zealand, was recorded once as "On the Upper Hume River at 3,000 to 4,000 feet" and called *A. hookerianum* (N. A. Wakefield, *Ferns of Victoria and Tasmania*).

### Rasp Ferns

*BLECHNUM* (from Gk *blokhon*, old name of a fern). A genus of widely distributed ferns of which a few species are found in high areas. They have divided or simple, often leathery fronds, the spore-bearing fronds often narrower. The spore masses are narrow, almost continuous with one another and are parallel to the midrib. *B. minus* (syn. *B. capense*) confined to the mainland montane and lower habitats, has once-divided fronds up to 90 cm high. The fertile fronds are very narrow.

*B. nudum*, with the same distribution, has very stiff, once-divided leaves, sometimes grey-green or reddish, up to 60 cm high, and narrow fertile fronds.

*B. penna-marina*, sometimes called alpine fern, extends from 6,000 feet to montane habitats but does best at the higher elevations, being strictly confined to these in Victoria. The fronds are once-divided, from a few centimetres to 40 cm high, the fertile fronds usually taller with narrower segments.

*B. wattsii* (syn. *B. procerum*) is a moisture-loving plant from mainland montane habitats. The young fronds are a delicate pink, the once-divided, mature fronds are drooping, leathery and may reach 1–1.5 metres in height. The fertile fronds are more upright, usually shorter and with narrower leaflets.

### Bracken and Bat's-wing Fern

In this group the spore masses are in continuous lines on the leaf margins.

*PTERIDIUM* (from Gk *pteris*, a fern). *P. esculentum* (syn. *P. aquilinum*), common bracken fern, is found in New South Wales, Victoria and Tasmania, extending to montane areas. The fronds are several times divided, stout, 0.6–3 metres high. The spore masses are narrow and continuous near the ends of the leaflets.

*HISTIOPTERIS* (from Gk *histion*, tissue, and *pteris*, a fern). *H. incisa* is a mainland montane species with fronds 2–3 times divided and up to 180 cm long with blunt leaflets. The spore masses are in interrupted groups round the margins of the leaflets.

### Shield Ferns

*POLYSTICHUM* (from Gk *polys*, many, and *stichos*, a row, referring to the many rows of spore masses). Ferns usually with wedge-shaped or triangular fronds with small round spore masses situated away from the leaf margin.

*P. proliferum* extends from alpine to montane and lower habitats in New South Wales, Victoria and Tasmania. It is a handsome fern with twice-divided, sharply toothed, rather soft fronds. Before uncoiling, the fronds are a rich nut-brown. The spore masses are rounded, usually in 2 rows between the midrib and the leaf margins.

### Fan Ferns and Coral Ferns

*GLEICHENIA* (after F. W. Gleichen, 1717-1783, German botanist). A genus of ferns with spreading and dichotomously branched fronds, each divided equally into 2 branches. The spore masses are large and clustered.

*G. abscida*, dwarf coral fern, has fronds up to 30 cm long on stout, stiff stems each divided into 2 branches bearing fronds with numerous tiny lobes rolled under towards the midrib, each with 2 spore-cases. It is a rare species found only in the mountains in the south of Tasmania.

*G. alpina,* alpine coral fern, has fronds varying in length from a few centimetres to over 60 cm. The stems are covered with dark, fringed, overlapping scales and are shortly branched. The leaflets are convex on the upper surface and pouched beneath, bearing 2 spore-cases. It is a Tasmanian species fairly common in alpine and subalpine situations.

*G. microphylla* (syn. *G. circinata*) is widely distributed on the mainland and reaches montane heights. The fronds are 30 cm–3 metres long, repeatedly forked, the leaflets often whitish below. The spores form 3–4 rosette-like masses towards the centre of the leaflets.

*G. dicarpa* is a smaller plant than *G. microphylla* and is found in New South Wales, Victoria and Tasmania, reaching subalpine habitats and in Tasmania sometimes even higher. In the latter State it is very abundant in wet habitats such as fens and its brownish-pink folded young fronds are very attractive. The fronds are 15–90 cm long. The margins of the leaflets roll under to form little pouches and the spore masses are grouped 2–3 together towards the centre.

### Comb Ferns

*SCHIZAEA* (from Gk *schizo*, to cut, in reference to the split fronds). There is no mistaking these quaint little ferns which have their leaflets so closely pressed together at the end of a thin stem as to give a comb-like appearance. The spore masses are readily seen covering the exposed backs of the fronds.

*S. bifida* has an erect thin stem up to 30 cm high mounted by the comb-like segments of crowded leaflets in 2 sterile rows with the spore cases between them. It is a montane species from New South Wales, Victoria and Tasmania.

### A Miscellany of Ferns

A few species which do not fit into any of the above categories appear below.

*CYSTOPTERIS* (from Gk *kostion*, bearing fruit in a bladder, and *pteris*, a fern). *C. fragilis,* a tender little fern from montane to alpine habitats in New South Wales, Victoria and Tasmania, has twice-divided fronds up to 30 cm high with toothed leaflets. The spore masses are borne on the backs of the veins near the margins.

*LINDSAEA* (after J. Lindsay, a Jamaican surgeon). *L. linearis,* distributed throughout most of New South Wales, Victoria and Tasmania, extends to montane habitats. It is a very dainty little fern with short drooping fronds 5–30 cm long on wiry stems, 2–3-pinnate with fan-shaped leaflets. The spore masses form a continuous line on the margins of the leaflets.

*PLEUROSORUS* (from Gk *pleuros*, once, and *soros*, a heap, referring to the spore masses). *P. rutifolius* is a small tufted fern from montane habitats in New South Wales, Victoria and Tasmania. The fronds are divided, 10–15 cm long, very hairy, with finely toothed fan-shaped leaflets. The spore masses are narrow and divergent from the base of the fan-shaped leaflets.

# GLOSSARY

*Alternate*, scattered, not opposite, used here in reference to leaves.
*Anther*, the top of the stamen, holding the pollen.
*Aril*, an extension of the stalk of the seed on some plants, such as *Acacia* (wattle).
*Axil*, the angle formed by a leaf with a branch.

*Berry*, a succulent, usually non-splitting fruit of any number of carpels, with seeds scattered in a fleshy pulp.
*Bract*, a small leaf used for a special purpose and usually differing from a foliage leaf.
*Bracteole*, a small bract usually inserted below the calyx.

*Calyx*, the outer whorl of a flower, consisting of sepals.
*Capsule*, a dry splitting fruit of one or more joined carpels.
*Carpel*, a modified leaf forming the whole or part of an ovary and extended to form the style and stigma. The leaf is folded or united with others to form an enclosure for the seeds.
*Corolla*, the whorl of the flower outside the calyx, consisting of petals.

*Deciduous*, falling off, opposite to persistent, remaining throughout the period of maturity.
*Dicotyledon*, a plant whose embryo possesses two seed-leaves.

*Endemic*, peculiar to a particular region.
*Epicalyx*, a calyx-like structure immediately behind the true calyx.
*Epiphyte*, a plant which grows upon another plant but does not derive nourishment from it.

*Ericoid*, narrow, needle-like, rolled back.
*Exserted*, projecting beyond the surrounding parts.

*Falcate*, scythe-shaped.
*Family*, a group of plants related botanically.
*Filament*, stalk of a stamen.
*Floret*, one of the small flowers forming a compact head in family Compositae.
*Follicle*, a dry, non-splitting fruit of one carpel with any number of seeds and splitting down one side.
*Fruit*, a matured ovary with fertilized seeds and any other parts of the flower which adhere to it.

*Genus*, the smallest natural group of species having certain essential characters in common.

*Habitat*, the natural home of a plant.
*Herb*, a plant which does not develop a woody stem.

*Keel*, applied to the lowest petal of pea-flowers.

*Leaflet*, a division of a divided leaf.
*Legume*, a dry, splitting fruit of one carpel and any number of seeds, splitting down one side, characteristic of families Papilionaceae, Caesalpiniaceae and Mimosaceae.

*Monocotyledon*, a plant whose embryo possesses only one seed leaf.

*Nectary*, a glandular (secreting) swelling producing nectar.
*Node*, a point where buds arise.

*Ovary*, the lowest part of a carpel containing the potential seeds and, on fertilization, becoming the fruit.

*Pappus*, the persistent limb of the calyx of some members of family Compositae, often used in fruit dispersal.
*Petal*, a segment of the corolla.
*Phyllode*, a flattened leaf-stalk resembling and functioning as a leaf, found particularly in family Mimosaceae.
*Pistillate*, applied to a flower which contains pistils (carpels) but no stamens.
*Pollen*, a substance produced by the anthers by means of which the potential seeds are fertilized.

*Regular*, a term applied to flowers in which the segments of each whorl are all similar and are arranged regularly round the axis.

*Saprophyte*, an organism living on dead organic matter.
*Sepal*, a segment of the calyx.
*Shrub*, a woody plant not attaining the size of a tree and bearing several branches from near the roots.
*Simple*, a term applied to an undivided leaf.
*Species*, a division of a genus, each species possessing characteristics distinguishing it from all other species of the same genus.
*Stamen*, the male organ of a flower consisting of a stalk (sometimes absent) supporting an anther (pollen box).
*Staminate*, a term applied to flowers which have stamens but no pistil (carpel).
*Staminode*, an aborted or sterile stamen.
*Standard*, the large posterior petal of flowers of family Papilionaceae.
*Stipule*, an appendage at the base of the leaf stalk.
*Stigma*, the top of the carpel or carpels, which acts as a receiving surface for pollen.
*Style*, the stalk joining the ovary to the stigma.

*Terete*, nearly cylindrical, not angular.
*Trifoliate*, a leaf of three leaflets.

*Valve*, a distinct portion of certain parts of a fruit or an anther which becomes detached on splitting.

# BIBLIOGRAPHY


COSTIN, A. B., *A study of the Ecosystems of the Monaro Region of New South Wales.* Government Printer, N.S.W., 1954.

KING, H. J., and BURNS, T. E., *Wildflowers of Tasmania.* Jacaranda Press, Brisbane, 1968.

STONES, Margaret, and CURTIS, Winifred, *The Endemic Flora of Tasmania.* Ariel Press, London, 1967.

CHILD, John, *Australian Alpine Life.* Periwinkle Books, Melbourne, 1969.

# INDEX